Explorations in Australia
The Journals of John McDouall Stuart

By

John McDouall Stuart

The Echo Library 2006

Published by

The Echo Library

Echo Library
131 High St.
Teddington
Middlesex TW11 8HH

www.echo-library.com

Please report serious faults in the text to complaints@echo-library.com

ISBN 1-40680-794-X

PREFACE BY THE EDITOR.

THE explorations of Mr. John McDouall Stuart may truly be said, without disparaging his brother explorers, to be amongst the most important in the history of Australian discovery. In 1844 he gained his first experiences under the guidance of that distinguished explorer, Captain Sturt, whose expedition he accompanied in the capacity of draughtsman. Leaving Lake Torrens on the left, Captain Sturt and his party passed up the Murray and the Darling, until finding that the latter would carry him too far from the northern course, which was the one he had marked out for himself, he turned up a small tributary known to the natives as the Williorara. The water of this stream failing him, he pushed on over a barren tract, until he suddenly came upon a fruitful and well-watered spot, which he named the Rocky Glen. In this picturesque glen they were detained for six months, during which time no rain fell. The heat of the sun was so intense that every screw in their boxes was drawn, and all horn handles and combs split into fine laminae. The lead dropped from their pencils, their finger-nails became as brittle as glass, and their hair, and the wool on their sheep, ceased to grow. Scurvy attacked them all, and Mr. Poole, the second in command, died. In order to avoid the scorching rays of the sun, they had excavated an underground chamber, to which they retired during the heat of the day.

When the long-expected rain fell, they pushed on for fifty miles to another suitable halting-place, which was called Park Depot. From this depot Captain Sturt made two attempts to reach the Centre of the continent. He started, accompanied by four of his party, advancing over a country which resembled an ocean whose mighty billows, fifty or sixty feet high, had become suddenly hardened into long parallel ridges of solid sand. The abrupt termination of this was succeeded at two hundred miles by what is now so well known as Sturt's Stony Desert, to which frequent allusion is made by Mr. Stuart in his journals. After thirty miles more, this stony desert ceased with equal abruptness, and was followed by a vast plain of dried mud, which Captain Sturt describes as "a boundless ploughed field, on which floods had settled and subsided." After advancing two hundred miles beyond the Stony Desert, and to within one hundred and fifty miles of the Centre of the continent, they were compelled to return to Park Depot, where they arrived in a most exhausted condition.

A short rest at the Depot was followed by another expedition, Captain Sturt being on this occasion accompanied by Mr. Stuart and two men. The seventh day of their journey brought them to the banks of a fine creek, now so well known as Cooper Creek in connection with the fate of those unfortunate explorers, Burke and Wills. At two hundred miles from Cooper Creek Captain Sturt and his party were again met by the Stony Desert, but slightly varied in its aspect. Before abandoning his attempt to proceed, the leader of the expedition laid the matter before his companions, and he writes as follows: "I should be doing an injustice to Mr. Stuart and my men, if I did not here mention that I told them the position we were placed in, and the chance on which our safety would depend if we went on. They might well have been excused if they expressed an opinion contrary to such a course; but the only reply they made me was to assure me that they were ready and willing to follow me to the last."

With much reluctance, however, Captain Sturt determined to return to Cooper Creek without delay. They travelled night and day without interruption, and on the morning of their arrival at the creek, one of those terrible hot north winds, so much dreaded by the colonists, began to blow with unusual violence. Lucky was it for them that it had not overtaken them in the Desert, for they could scarcely have survived it. The heat was awful; a thermometer, graduated to 127 degrees, burst, though sheltered in the fork of a large tree, and their skin was blistered by a torrent of fine sand, which was driven along by the fury of the hurricane. They still had fearful difficulties to encounter, but after an absence of nineteen months they returned safely to Adelaide.

The discouraging account of the interior which was brought by Captain Sturt did not prevent other explorers from making further attempts; but the terrible fate of Kennedy and his party on York Peninsula, and the utter disappearance of Leichardt's expedition, both in the same year (1848), had a very decided influence in checking the progress of Australian exploration. Seven years later, in 1855, Mr. Gregory landed on the north-west coast for the purpose of exploring the Victoria River, and after penetrating as far south as latitude 20 degrees 16 minutes, longitude 131 degrees 44 minutes, he was compelled to proceed to the head of the Gulf of Carpentaria, and thence to Sydney along the route taken by Dr. Leichardt in 1844. Shortly after his return Mr. Gregory was despatched by the Government of New South Wales in 1857, to find, if possible, some trace of the lost expedition of the lamented Leichardt; his efforts, however, did nothing to clear up the mystery that enshrouds the fate of that celebrated explorer.[1]

The colonists of South Australia have always been distinguished for promoting by private aid and public grant the cause of exploration. They usually kept somebody in the field, whose discoveries were intended to throw light on the caprices of Lake Torrens, at one time a vast inland sea, at another a dry desert of stones and baked mud. Hack, Warburton, Freeling, Babbage, and other well-known names, are associated with this particular district, and, in 1858, Stuart started to the north-west of the same country, accompanied by one white man (Forster) and a native. In this, the first expedition which he had the honour to command, he was aided solely by his friend Mr. William Finke, but in his later journeys Mr. James Chambers also bore a share of the expense.[2] Mr. Stuart now

---

[1] It is possible that Mr. McKinlay has been hasty in the opinion he formed from the graves and remains of white men shown to him by Keri Keri, and the story related of their massacre. May they not belong to Leichardt's party?

[2] It is greatly to be regretted that both these gentlemen are since dead. Mr. Chambers did not survive to witness the success of his friend's later expeditions, and the news of Mr. Finke's death reached us while these sheets were going through the press.) This journey was commenced in May, 1858, from Mount Eyre in the north to Denial and Streaky Bays on the west coast of the Port Lincoln country. On this journey Mr. Stuart accomplished one of the most arduous feats in all his travels, having, with one man only

turned his attention to crossing the interior, and, with the assistance of his friends Messrs. Chambers and Finke, he was enabled to make two preparatory expeditions in the vicinity of Lake Torrens—from April 2nd to July 3rd, 1859, and from November 4th, 1859, to January 21st, 1860. The fourth expedition started from Chambers Creek (discovered by Mr. Stuart in 1858, and since treated as his head-quarters for exploring purposes), on March 2nd, 1860, and consisted of Mr. Stuart and two men, with thirteen horses. Proceeding steadily northwards, until the country which his previous explorations had rendered familiar was left far behind, on April 23rd the great explorer calmly records in his Journal the following important announcement: "To-day I find from my observations of the sun that I am now camped in the CENTRE OF AUSTRALIA." One of the greatest problems of Australian discovery was solved! The Centre of the continent was reached, and, instead of being an inhospitable desert or an inland sea, it was a splendid grass country through which ran numerous watercourses.

Leaving the Centre, a north-westerly course was followed, but, after various repulses, a north-easterly course eventually carried the party as far as latitude 18 degrees 47 minutes south, longitude 134 degrees, when they were driven back by the hostility of the natives. As has already been stated, Mr. Gregory in 1855, starting from the north-west coast, had penetrated to the south as low as latitude 20 degrees 16 minutes, longitude 127 degrees 35 minutes. Mr. Stuart had now reached a position about half-way between Gregory's lowest southward point and the head of the Gulf of Carpentaria. Without actually reaching the country explored by Gregory, he had overlapped his brother explorer's position by one degree and a half, or more than one hundred miles, and was about two hundred and fifty miles in actual distance from the nearest part of the shores of the Gulf. It is important to remark that the attack of the savages which forced Mr. Stuart to return occurred on June 26th, 1860, so that he had virtually crossed the continent two months before Messrs. Burke and Wills had left Melbourne.[3]

On New Year's day 1861, Mr. Stuart again left Adelaide, aided this time by a grant from the Colonial Government of 2500 pounds, in addition to the assistance of his well-tried friends Messrs. Chambers and Finke. He made his former position with ease, and advanced about one hundred miles beyond it, to latitude 17 degrees, longitude 133 degrees; but an impenetrable scrub barred all further progress, and failing provisions, etc., compelled him, after such prolonged and strenuous efforts that his horses on one occasion were one hundred and six hours without water, most

---

(the black having basely deserted them), pushed through a long tract of dense scrub and sand with unusual rapidity, thus saving his own life and that of his companion. During this part of the journey they were without food or water, and his companion was thoroughly dispirited and despairing of success. This expedition occupied him till September, 1858, and was undertaken with the object of examining the country for runs. On his return the South Australian Government presented him with a large grant of land in the district which he had explored.

[3] They did not leave Cooper Creek until December 14th, rather more than a fortnight before Mr. Stuart started on his fifth expedition.

reluctantly to return. The expedition arrived safely in the settled districts in September, and the determined explorer, after a delay of less than a month, was again despatched by the South Australian Government along what had now become to him a familiar road. This time success crowned his efforts; a passage was found northwards through the opposing scrub, and leaving the Gulf of Carpentaria far to the right, the Indian Ocean itself was reached. Other explorers had merely seen the rise and fall of the tide in rivers, boggy ground and swamps intervening and cutting off all chance of ever seeing the sea. But Stuart actually stood on its shore and washed his hands in its waters! What a pleasure it must have been to the leader when, knowing well from his reckoning that the sea must be close at hand, but keeping it a secret from all except Thring and Auld, he witnessed the joyful surprise of the rest of the party!

The expedition reached Adelaide safely, although for a long time the leader's life was despaired of, the constant hardships of so many journeys with scarcely any intermission having brought on a terrible attack of scurvy. The South Australian Government in 1859 liberally rewarded Mr. Stuart and his party for their successful enterprise.[4]

Mr. Stuart's experiences have led him to form a very decided opinion as to the cause of the well-known hot winds of Australia, so long the subject of scientific speculation. North and north-west of Flinders Range are large plains covered with stones, extending as far as latitude 25 degrees. To the north of that, although the sun was intensely hot, there were no hot winds; in fact from that parallel of latitude to the Indian Ocean, either going or returning, they were not met with. "On reaching latitude 27 degrees on my return," writes Mr. Stuart, "I found the hot winds prevailing again as on my outward journey. I saw no sandy desert to which these hot winds have been attributed, but, on lifting some of the stones that were lying on the surface,[5] I found them so hot that I was obliged to drop them immediately. It is my opinion that when a north wind blows across those stone-covered plains, it collects the heat from them, and the air, becoming rarified, is driven on southwards with increased vehemence. To the north of latitude 25 degrees, although exposure to the sun in the middle of the day was very oppressive, yet the moment we got under the shade of a tree we felt quite alive again; there was none of that languid feeling which is experienced in the

[4] Mr. Stuart's qualities as a practised Bushman are unrivalled, and he has always succeeded in bringing his party back without loss of life.) On the 10th of March a resolution was passed to the effect that a sum of 3500 pounds should be paid as a reward to John McDouall Stuart, Esquire, and the members of his party, in the following proportions: Mr. Stuart 2000 pounds; Mr. Keckwick 500 pounds; Messrs. Thring and Auld 200 pounds each; and Messrs. King, Billiatt, Frew, Nash, McGorrerey, and Waterhouse, 100 pounds each. Perhaps this is the most fitting place to express Mr. Stuart's appreciation of the honour done him by the Royal Geographical Society of London, in awarding him their gold medal and presenting him with a gold watch. He wishes particularly to express his hearty thanks to Sir Roderick Murchison, and the other distinguished members of the society, for the lively interest they have evinced in his welfare.

[5] On the surface, as I suppose, of the large plains North of Flinders Range. ED.

south during a hot wind, as for example that which blew on the morning after reaching the Hamilton,[6] in latitude 26 degrees 40 minutes.

Mr. Stuart is anxious to direct attention to the establishment of a Telegraph line along his route. On this subject he writes as follows:—

"On my arrival in Adelaide from my last journey I found a great deal of anxiety felt as to whether a line could be carried across to the mouth of the Adelaide river. There would be a few difficulties in the way, but none which could not be overcome and made to repay the cost of such an undertaking. The first would be in crossing from Mr. Glen's station to Chambers Creek, in finding timber sufficiently long for poles, supposing that no more favourable line than I travelled over could be adopted, but I have good reason for supposing that there is plenty of suitable timber in the range and creek, not more than ten miles off my track: the distance between the two places is one hundred miles. From Chambers Creek through the spring country to the Gap in Hanson Range the cartage would be a little farther, in consequence of the timber being scarce in some places. There are many creeks in which it would be found, but I had not time to examine them in detail. Another difficulty would be in crossing the McDonnell Range, which is rough and ragged, but there is a great quantity of timber in the Hugh; the distance to this in a straight line is not more than seven miles; from thence to the Roper River there are a few places where the cartage might be from ten to twenty miles, that is in crossing the plains where only stunted gum-trees grow, but tall timber can be obtained from the rising ground around them. From latitude 16 degrees 30 minutes south to the north coast, there would be no difficulty whatever, as there is an abundance of timber everywhere. I am promised information, through the kindness of Mr. Todd, of the Telegraph department, as to the average cost of establishing the lines through the outer districts of this colony, and it is my intention to make a calculation of the cost of a line on my route, by which the comparative merits and expense will be tested, and I am of opinion I shall be able to show most favourable results. I should have been glad for this information to have accompanied my works, but I find I cannot postpone them longer for that purpose, as parties have already taken advantage of the delay occasioned by my illness at the time of, and since, my arrival home to collect what scraps of information they could obtain, with the intention of publishing them as my travels. I leave the reward of such conduct to a discriminating public; I shall not fail to carry out my intention with regard to a Telegraph line; and should I have no opportunity of submitting it to the public, I shall take care to advance the matter in such channels as may be most likely to lead to a successful issue. I beg reference

---

[6] Journal 1861 to 1862.) That was one of the hottest winds I ever experienced. I had the horses brought up at 7 o'clock, intending to proceed, but seeing there was a very hot wind coming on, I had them turned out again. It was well I did so, for before 10 o'clock all the horses were in small groups under the trees, and the men lying under the shade of blankets unable to do anything, so overpowering was the heat." Unfortunately, Mr. Stuart had no thermometer.

to my map accompanying this work, which will at once show the favourable geographical situation of the Adelaide River for a settlement, and the short and safe route it opens up for communication and trading with India: indeed when I look upon the present system of shipping to that important empire, I cannot over-estimate the advantages that such an extended intercourse would create."

Mr. Stuart is also very anxious for the formation of a new colony on the scene of his discoveries on the River Adelaide, and would fain have been one of the first pioneers of such an enterprise, but his health has been so much shattered by his last journey that he can only now hope to see younger men follow in the path which he had made his own. He writes as follows:—

"Judging from the experience I have had in travelling through the Continent of Australia for the last twenty-two years, and also from the description that other explorers have given of the different portions they have examined in their journeys, I have no hesitation in saying, that the country that I have discovered on and around the banks of the Adelaide River is more favourable than any other part of the continent for the formation of a new colony. The soil is generally of the richest nature ever formed for the benefit of mankind: black and alluvial, and capable of producing anything that could be desired, and watered by one of the finest rivers in Australia. This river was found by Lieutenant Helpman to be about four to seven fathoms deep at the mouth, and at one hundred and twenty miles up (the furthest point he reached) it was found to be about seven fathoms deep and nearly one hundred yards broad, with a clear passage all the way up. I struck it about this point, and followed it down, encamping fifteen miles from its mouth, and found the water perfectly fresh, and the river broader and apparently very deep; the country around most excellent, abundantly supplied with fresh water, running in many flowing streams into the Adelaide River, the grass in many places growing six feet high, and the herbage very close—a thing seldom seen in a new country. The timber is chiefly composed of stringy-bark, gum, myall, casurina, pine, and many other descriptions of large timber, all of which will be most useful to new colonists. There is also a plentiful supply of stone in the low rises suitable for building purposes, and any quantity of bamboo can be obtained from the river from two to fifty feet long. I measured one fifteen inches in circumference, and saw many larger. The river abounds in fish and waterfowl of all descriptions. On my arrival from the coast I kept more to the eastward of my north course, with the intention of seeing further into the country. I crossed the sources of the running streams before alluded to, and had great difficulty in getting more to the west. They take their rise from large bodies of springs coming from extensive grassy plains, which proves there must be a very considerable underground drainage, as there are no hills of sufficient elevation to cause the supply of water in these streams. I feel confident that, if a new settlement is formed in this splendid country, in a few years it will become one of the brightest gems in the British Crown. To South Australia and some of the more remote Australian colonies the benefits to be derived from the formation of such a colony would be equally advantageous, creating an outlet for their surplus beef and mutton, which would be eagerly consumed by the races in the Indian Islands, and payment made by the shipment of their useful ponies, and the other valuable products of those islands; indeed I see one of the finest openings I am aware of for trading between these islands and a colony formed where proposed."

Mr. Stuart was accompanied on his last journey by Mr. Waterhouse, a clever naturalist, whose report to the Commissioner of Crown Lands of South Australia, although too long for insertion here, is full of most interesting information. Unfortunately, the interests of geographical science were apparently lost sight of in the hurry to effect the grand object of the expedition, namely, to cross from sea to sea. Thermometers were forgotten; two mounted maps of the country from Chambers Creek to Newcastle Water, in a tin case, never came to hand, and the expedition was provided with no means of estimating even the approximate height of the elevated land or of the mountains in the interior. As Mr. Waterhouse remarks: "The thermometers were much needed, as it would have been very desirable to have kept a register of the temperature, and to have tested occasionally the degree of heat at which water boiled on the high table lands. The loss of the maps prevented my marking down at the time on the maps the physical features of the country, and the distribution of its fauna and flora."

Mr. Waterhouse divides the country into three divisions. The first, which extends from Goolong Springs to a little north of the Gap in Hanson Range, latitude 27 degrees 18 minutes 23 seconds, may be called the spring and saltbush country. The second division commences north of the Gap in Hanson Range, and extends to the southern side of Newcastle Water, latitude 17 degrees 36 minutes 29 seconds. It is marked by great scarcity of water—in fact, there are few places where water can be relied on as permanent—and also by the presence of the porcupine grass (Triodia pungens of Gregory, and Spinifex of Stuart), which is the prevailing flora. The third division commences from the north end of Newcastle Water, latitude 17 degrees 16 minutes 20 seconds, and extends to Van Diemen Gulf, latitude 12 degrees 12 minutes 30 seconds; it comprises a large part of Sturt Plains, with soil formed of a fine lacustrine deposit, the valleys of the Roper filled with a luxuriant tropical vegetation, and thence to the Adelaide River and the sea-coast.

On visiting Hergott Springs, Mr. Waterhouse learnt that Mr. Burtt, whose station[7] is only a few miles distant, in opening these springs discovered some fossil bones, casts of which were forwarded to Professor Owen, who pronounced them to be the remains of a gigantic extinct marsupial, named Diprotodon Australis. Bones of this animal have also been found in a newer tertiary formation in New South Wales. Mr. Waterhouse considers that a great tertiary drift extends over this part of the country, obscuring and concealing at no great depth below the surface many springs, which may hereafter be

---

[7] Hergott Springs were only discovered and named by Stuart three years before, yet we now find a station close by them. The explorer is not far ahead of his fellow-colonists, as is well remarked by the Edinburgh Review for July, 1862: "Australian occupation has kept close on the heels of Australian discovery."

discovered as the country becomes better known.

The Louden Spa is a hot spring arising out of a small hillock, and proceeds from the fissures of volcanic rock. This water is medicinal, but not disagreeable to the taste: the damper made with it was very light, and tasted like soda-bread.

In his remarks on the second division Mr. Waterhouse states much that is valuable. He estimates the height of Mount Hay at two thousand feet, regarding it as the highest point of the McDonnell Range, which is the natural centre of this part of the continent. Mr. Waterhouse only saw Chambers Pillar from a distance, but he had an opportunity of examining a smaller hill of the same character, and found it to be composed of a soft loose argillaceous rock, at the top of which was a thin stratum of a hard siliceous rock, much broken up. "The isolated hills appear to have been at some remote period connected, but from the soft and loose nature of the lower rock meeting with the action of water, had arisen a succession of landslips. These have been washed away and others have followed in their turn; the upper rock, from being undermined, has fallen down and broken up, supplying the peculiar siliceous stones so widely distributed on parts of the surface of the country."

The vegetation of this district is poor; the myall is scarce, but the mulga (Acacia aneura) generally plentiful. Both these shrubs are species of acacia, the myall being of much larger growth and longer lived than the mulga. Nutritious grass is seldom found except in the immediate vicinity of the creeks, and the scrubs are very extensive.

Mr. Waterhouse collected a great number of specimens of natural history, but, from want of the convenience for carrying them, many of the more delicate objects were broken.

In the Appendix will be found some remarks by Mr. John Gould, F.R.S., etc., on the birds collected by Mr. Waterhouse during Mr. Stuart's expedition, including a description of a new and beautiful parrakeet. There are also descriptions of new species of Freshwater Shells from the same expedition, by Mr. Arthur Adams, F.L.S., and Mr. G. French Angas, to the skill of which latter gentleman this work is indebted for its admirable illustrations.

Dr. Muller, the Government Botanist, Director of the Botanic Garden at Melbourne, in his report to both Houses of the Legislature of Victoria, April 15th, 1863, says, "A series of all the plants collected during Mr. J.M. Stuart's last expedition was presented by the Hon. H. Strangways, Commissioner of Crown Lands for South Australia, and those of the former expeditions of that highly distinguished explorer, by the late J. Chambers, Esquire, of North Adelaide." Of this collection, Dr. Muller has furnished a systematic enumeration, which will be found in the Appendix. This enumeration must not, however, be accepted as final, for Dr. Muller has forwarded all the specimens to England for the inspection of Mr. Bentham, the learned President of the Linnaean Society of London, who is now elaborating his great and exhaustive work on the Flora of Australia, the second volume of which will shortly be before the public.

WILLIAM HARDMAN.

CONTENTS

JOURNAL OF MR. STUART'S EXPEDITION TO THE NORTH WEST. MAY TO SEPTEMBER, 1858

On the 14th of May, 1858, Mr. Stuart started from Oratunga (the head station of Mr. John Chambers), accompanied by Mr. Barker, with six horses, and all that was requisite (with one important exception, as will be seen hereafter), for an excursion to the north-west of Swinden's Country. They arrived at Aroona the same evening. On the following day (the 15th) they made Morleeanna Creek, and reached Ootaina on the 16th, about 7 p.m. Here they remained for a couple of days, as sufficient rain had not fallen to enable them to proceed. On the afternoon of the 19th they arrived at Mr. Sleep's, who informed them that Mr. M. Campbell had returned from the West, being hard pushed for water; very little rain having fallen to the west. The next day (20th) Mr. Stuart arrived at Mr. Louden's, but, in consequence of some difficulties about the horses, he returned to Ootaina. Various preparations, combined with want of rain, compelled him to delay his start until the 10th of June. Here the journal commences:—

Thursday, 10th June, 1858. Started from Ootaina at 1 p.m. for Beda. Camped on the plain, about thirteen miles from Mount Eyre.

Friday, 11th June, West Plain. Made Mudleealpa at 11 a.m. The horses would not drink the water. Proceeded for about five miles towards Beda. The plains are fearfully dry; they have the appearance as if no rain had fallen here for a long time, and I am very much afraid there will be no water at Beda. If such should be the case, the horses will suffer too much in the beginning of their journey to be without a drink to-night. I think it will be best to return to Mudleealpa, leave our saddles, rations, etc. there, and drive the horses back to water. I sent Mr. Forster back with them, telling him if he can find no water between this and Mr. Sleep's, to take them there, remain for the night, give them a drink in the morning, and return; we shall then be able to make a fresh start to-morrow. Bearings: Mount Arden, 154 degrees 30 minutes; Mount Eyre, 77 degrees 30 minutes; Beda Hill, 272 degrees; Mount Elder, 64 degrees 50 minutes; Dutchman's Stern, 162 degrees 15 minutes.

Saturday, 12th June, Mudleealpa. In examining the creek a little higher up, we found another well. By cleaning it out, the water is drinkable. The horses did not arrive until it was too late to start, and having water here now, that they can drink, we camped here another night.

Sunday, 13th June, Mudleealpa. Started for Beda. Some of the horses would not drink the water, and others drank very little: they will be glad to drink far worse than this before they come back, or I am much mistaken. Arrived at Beda at sundown. I was right in my opinion; no fresh water to be found; nothing but salt, salter than the sea. I can see nothing of Mr. Babbage's[8] encampment; he

[8] It will probably be recollected that Mr. Babbage was sent out by the Government to make a north-west course through the continent, but, when at the Elizabeth, he made an unaccountable detour, and found himself at Port Augusta, his original starting-point. On my return from this journey he called on me at Mount Arden, when I furnished him

must be higher up the creek. All the country we have come over to-day is very dry.

Monday, 14th June, Beda. This morning we have searched all round, but can find no fresh water, although there are numerous places that would retain water if any quantity had fallen. Mr. Forster, whom I had sent up the creek to Mr. Babbage's, to inquire if there was any water at Pernatta, has returned with the information that Mr. B. was up there with all his horses, and that there was still a little water, but not much. Started at 11.30 a.m. for that place; camped in the sand hills one hour after dark. Here we found some pig-faces[9] which the horses eat freely. There is a great deal of moisture in them, and they are a first-rate thing for thirsty horses; besides, they have a powerful diuretic effect. I was unable to fix Beda Hill, all my time being taken up in looking for water, but I hope to get its position at Pernatta. The country was very heavy—sand hills.

Tuesday, 15th June, Sand Hills. Started at break of day for Pernatta. About 10 a.m. met Mr. Babbage's two men returning with some of the horses for rations. They informed me that the water was nearly all gone, but that there was plenty in the Elizabeth, nineteen miles from Pernatta. I intended to keep on the track, but our black insisted that Pernatta lay through a gap, and not round the bluff. I allowed him to have his own way. Our route was through a very stony saddle. When there we saw a gum creek, and made for it; when we arrived at the creek he told us that was Pernatta. We looked for water, and found a little hole, which, to our great disappointment, contained salt water. Could see nothing of Mr. Babbage's camp. I then asked our black where there was another water; he said, "Down the creek," which we followed. He took us to five or six water holes, with native names, every one dry. The last one he called Yolticourie. It being now within an hour of sundown, I would follow him no longer, but unsaddled, and told Mr. Forster to take the black and the horses, and to steer for the bluff; if he found no water between, to intersect Mr. Babbage's tracks, and follow them up and get water. I remained with our provisions. The black fellow evidently does not know the country. I am sorry that I have taken him with me. I think I shall send him back; he is of little use in assisting to get the horses in the morning.

Wednesday, 16th June, Yolticourie. The horses have returned; they found no water last night; they were obliged to camp for the night, it being so dark, but they found Mr. Babbage's camp very early. The horses drank all the water. I was wrong in blaming the black fellow; he took us to the RIGHT Pernatta. It is another water that Mr. B. is encamped at. He moves to-day for the Elizabeth, which I also will do. He found the remains of poor Coulthard yesterday. We

---

with such information as he required, and he again started, and made Chambers' Creek, which I had previously found and named after my old friend, Mr. James Chambers, but which he called Stuart's Creek in acknowledgment of my information, etc. J. McD. Stuart.

[9] These pig-faces belong to the Mesembryaceae, of which the common ice-plant of our gardens is an example

must have passed quite close to them in our search for water. He has sent for me to come and assist at the burial. It being so late in the day (12 o'clock), and the horses requiring more water, and he having four men besides himself, I do not see that I can be of any use, and it might cause me to lose another day, and the horses to be another night without water, which would be an injury to them, they not having had sufficient this morning. Mr. B. also sent to say that he would accompany me to the Elizabeth. I have delayed an hour for him, and he has not yet made his appearance; it being now 1 o'clock, and having to travel seventeen miles, I can wait no longer. Started for Bottle Hill; arrived on the south side of the hill an hour and a half before sundown, found some water and plenty of grass; encamped for the night. Distance to-day, seventeen miles. The former part of the journey was over very stony country; the latter part very heavy sand hills.

Thursday, 17th June, Bottle Hill. Got on the top of Bottle Hill to take bearings, but was disappointed; could see no hill except one, which was either Mount Deception or Mount North-west; the bearing was 51 degrees 30 minutes. There is a small cone of stones on the top, and a flat stone on the top of it, with the names of Louden and Burtt. From here I saw the gum trees in the Elizabeth; course to them 325 degrees 30 minutes, seven miles to the creek. The country from the hill here is of the very worst description—nothing but sand and salt bush.

Friday, 18th June, The Elizabeth. We must rest our horses to-day, they have not yet recovered from their long thirst. I am quite disappointed with this creek and the surrounding country. The water is not permanent, it is only rain water; since we arrived yesterday it has shrunk a great deal. There are small plains on each side from a quarter to half a mile broad with salt bush; the hills are very stony with a little salt bush, and destitute of timber, except the few gum-trees in the creek and the mulga bushes in the sand hills.

Saturday, 19th June, The Elizabeth. The sky was quite overcast with cloud during the night, and a few drops of rain fell, but of no consequence. Started at 9.30 a.m., on a bearing of 308 degrees for six miles; changed the bearing to 355 degrees for one mile and a half; next bearing 328 degrees for four miles, to the north side of a dry swamp; next bearing 4 degrees for ten miles and a half; next bearing 350 degrees for four miles to a sand hill. Camped. Distance to-day, twenty-five miles, over a very bad country, with large fragments of a hard flinty stone covering the surface. Salt bush with small sand hills. No water.

Sunday, 20th June, Sand Hill. Started at 9 a.m., on a course of 25 degrees for sixteen miles. At 1 p.m., came upon a creek, in which I thought there might be water; examined it and found two water holes, with plenty of grass upon their banks. The water is not permanent. Our course to-day has been across stony plains (covered on the surface with fragments resembling hard white quartz), with sand hills about two miles broad dividing them. The black did not know of this water; I am very doubtful of his knowing anything of the country. The stony plains are surrounded by high heavy sand hills, especially to the west and north-west; I dare not attempt to get through them without rain. They are much higher than the country that I am travelling through. It seems as if there had been no rain for twelve months, every thing is so dried and parched up. On further examination of the creek we have found a large hole of clear water, with rushes growing round it; I almost think it is permanent, and intend to run the risk of falling back upon it should I be forced to retreat and wait for rain. The creek seems to drain the large stony plains that we crossed; the water is three and a half feet deep, ten yards wide, by forty yards long.

Monday, 21st June, Water Creek. Started at 9.30 a.m. on a course of 25 degrees. At a mile passed a small table-topped hill to the west of our line; at three miles and a half crossed the creek; at four miles passed another table-topped hill connected with the low range to the east, and passed the first ironstone hill; at seven miles changed to 55 degrees; at eight miles halted at a large permanent water hole (Andamoka). I can with safety say that this is permanent; it is a splendid water hole, nearly as large as the one at the mouth of the gorge in the John. The low range to the east of our course, and running nearly parallel with it, is composed of conglomerate, quartz, and a little ironstone. Part of to-day's journey was over low undulating sandy and very well grassed country. There seems to have been a little rain here lately; the grass is springing beautifully. At eleven miles we came upon a salt lagoon (Wealaroo) two miles long by one broad. From the north end of it, on a bearing of 55 degrees, one mile and a half will strike Andamoka. I think we have now left the western sand hills behind us; and now that we have permanent water to fall back on, I shall strike into the north-west to-morrow. The distance travelled to-day was fifteen miles. The country around this water consists of bold stony rises and sand, with salt bush and grass; no timber except mulga and a few myall bushes in the creek. On an examination of the creek, we have found salt water above and below this hole. In one place above there are cakes of salt one inch and a half thick, a convincing proof that this is supplied by springs.

Tuesday, 22nd June, Andamoka. Started on a bearing of 342 degrees. At seven miles and a half, crossed a low stony range running east-north-east and west-south-west, which turned out to be table land, with sand hills crossing our line, bearing to a high range east of us 93 degrees 30 minutes. About eight miles in the same direction there is the appearance of a long salt lake. At nine miles and a half, on a sand hill, I obtained the following bearings: Mount North-west, 60 degrees 30 minutes; Mount Deception, 95 degrees. At eleven miles and a half passed a large reedy swamp on our left, dry. At seventeen miles sand hills ceased. At eighteen miles and a half the sand hills again commenced, and we changed our course to north for three miles. Camped for the night at a creek of permanent water, very good. The last four miles of to-day's journey have been over very stony rises with salt bush and a little grass.

Wednesday, 23rd June, Permanent Water Creek. The horses had strayed so far that we did not get a start until 10 a.m. Bearing to-day, 318 degrees. At two miles crossed a tea-tree creek, in which there is water, coming from the stony rises, and running to the north of east. At six miles the sand hills again commence. To this place we have come over a stony plain, covered on the surface with fragments of limestone, quartz, and ironstone, with salt bush and grass. In a watery season it must be well covered with grass; the old grass is lying between the salt bushes. We have a view of part of the lake (Torrens) bearing north-east about fifteen or twenty miles from us; to the west again the stony rises, apparently more open. At ten miles, in the sand hills, we have again a view of Flinders range. The bearings are: Mount North-west, 78 degrees 35 minutes; Mount Deception, 107 degrees. At fourteen and a half miles we found a clay-pan of water, with beautiful green feed for the horses. As we don't know when we shall find more water, and as Forster has a damper to bake, I decide to camp for the rest of the day. Our route has lain over heavy sand hills for the last eight miles.

Thursday, 24th June, Sand Hills. At 8.30 we left on a course of 340 degrees, commencing with about two miles of rather heavy sand hills. At eight miles these sand hills diminished, and the valleys between them became much wider—both sand hills and valleys being well covered with grass and salt bush, with courses of lime and ironstone cropping out and running east and west. At twelve miles changed our course to 79 degrees, to examine a gum creek (Yarraout), which we ran down for water, but did not obtain it before four miles, when we found a small hole of rain water. This creek seems to be a hunting-ground of the natives, as we saw a great many summer worleys on its banks. They had evidently been here to-day, for, a little above where we first struck the creek, we saw some smoke, but on following it up, we found they had gone; most likely they had seen us and run away. The latter part of our journey to-day was over a stony plain, bounded on the west by the stony table land with the sand hills on the top. All this country seems to have been under water, and is most likely the bed of Lake Torrens, or Captain Sturt's inland sea. In travelling over the plains, one is reminded of going over a rough, gravelly beach; the stones are all rounded and smooth. Distance to-day, thirty miles.

Friday, 25th June, Yarraout Gum Creek. Started at 9.40 from the point where we first struck the creek last night, bearing 20 degrees for two miles, thence 61 degrees for one mile to a high sand hill, thence 39 degrees for one mile to a stony rise. My doubt of the black fellow's knowledge of the country is now confirmed; he seems to be quite lost, and knows nothing of the country, except what he has heard other blacks relate; he is quite bewildered and points all round when I ask him the direction of Wingillpin. I have determined to push into the westward, keeping a little north of west. Bearing 292 degrees for five miles, sand hills; thence 327 degrees to a table-hill nine miles. Camped without water. Our route to-day has been through sand hills, with a few miles of stones and dry reedy swamp, all well grassed, but no water. We came across some natives, who kept a long distance off. I sent our black up to them, to ask in which direction Wingillpin lay. They pointed to the course I was then steering, and said, "Five sleeps." They would not come near to us. About three-quarters of an hour afterwards I came suddenly upon another native, who was hunting in the sand hills. My attention being engaged in keeping the bearing, I did not observe him until he moved, but I pulled up at once, lest he should run away, and called to him. What he imagined I was I do not know; but when he turned round and saw me, I never beheld a finer picture of astonishment and fear. He was a fine muscular fellow, about six feet in height, and stood as if riveted to the spot, with his mouth wide open, and his eyes staring. I sent our black forward to speak with him, but omitted to tell him to dismount. The terrified native remained motionless, allowing our black to ride within a few yards of him, when, in an instant, he threw down his waddies, and jumped up into a mulga bush as high as he could, one foot being about three feet from the ground, and the other about two feet higher, and kept waving us off with his hand as we advanced. I expected every moment to see the bush break with his weight. When close under the bush, I told our black to inquire if he were a Wingillpin native. He was so frightened he could not utter a word, and trembled from head to foot. We then asked him where Wingillpin was. He mustered courage to let go one hand, and emphatically snapping his fingers in a north-west direction, again waved us off. I take this emphatic snapping of his fingers to mean a long distance. Probably this Wingillpin may be Cooper's Creek. We then left him, and proceeded on our way through the sand hills. About an hour before sunset, we came in full sight of a number of tent and table-topped hills to the north-west, the stony table land being to the south of us, and the dip of the country still towards Lake Torrens. I shall keep a little more to the west to-morrow if possible, to get the fall of the country the other way. The horses' shoes have been worn quite thin by the stones, and will not last above a day or two. Nay, some of the poor animals are already shoeless. It is most unfortunate that we did not bring another set with us. Distance to-day, twenty-four miles.

Saturday, 26th June, Edge of Plain. Started at 9.30 a.m., on a bearing of 314 degrees 30 minutes, over an undulating plain, with low sand hills and wide valleys, with plenty of grass and salt bush. After ten miles the sand hills ceased, and at thirteen miles we reached the point of the stony table land. Here we saw, to the north-north-west, what was apparently a large gum creek, running north-east and south-west. Changing our bearing to 285 degrees, after seven miles of very bad stony plain, thinly covered with salt bush and grass, we came upon the creek, and found long reaches of permanent water, divided here and there by only a few yards of rocks, and bordered by reeds and rushes. The water hole, by which we camped, is from forty to fifty feet wide, and half a mile in length; the water is excellent, and I could see small fish in it about two inches long. About ten miles down the creek the country seems to be more open, and the gum-trees much larger, and in a distant bend of the creek I can perceive a large body of water. The first of the seven or eight tent-like hills that were to the east of our route to-day presents a somewhat remarkable appearance. Of a conical form, it comes to a point like a Chinaman's hat, and is encircled near the top by a black ring, while some rocks resembling a white tower crown the summit. Distance to-day, twenty miles.

Sunday, 27th June, Large Water Creek. Cloudy morning, with prospect of rain. A swan visited the water hole last night, and to-day we have seen both the mountain duck and the large black duck. Having a shoe to fix upon Jersey, and my courses to map down, we did not get a start until 10 o'clock, and we were obliged to stop early in consequence of the grey mare getting so lame that we were unable to proceed. We had an old shoe or two, and Mr. Forster managed to get one on the mare. We started to-day on a bearing of 270 degrees for eight miles to a low flat-topped hill, when we changed to 220 degrees for five miles to a gum creek with rain water. About five miles to the north of our line there are flat-topped ranges, running north-east. The main creek runs on the south side of this course, and nearly parallel to it. Further to the south, at a distance of about ten miles, is still the stony table land with the sand hills. The country is fearfully stony, but improves a little in grass as we get west. It seems to be well watered. Distance to-day, about twelve miles.

Monday, 28th June, Gum Creek. There has been a little rain during the night, and it is still coming down. As I am so far north, I regret that I am unable to go a little further, fearing the lameness of the horses from the stony nature of the country. I intend to follow the creek up, if it comes from the west, or a little to the north of west, to see if I cannot make the fall of the country to the south-west, and get on a better road for the horses. We started on a bearing of 305 degrees, but after a mile and a half, finding the creek wind too much to the north, we changed our course to 287 degrees for five miles to a small flat-topped hill. Changed our bearing again to 281 degrees for twenty-two miles to a tent hill, on the south side of which we camped. This part of the country is very stony and bad, with salt bush and very little grass. It has evidently been the course of a large water at some time, and reminded me of the stony desert of Captain Sturt. Bleak, barren, and desolate, it grows no timber, so that we scarcely can find sufficient wood to boil our quart pot. The rain, which poured down upon us all day, so softened the ground that the horses could tread the stones into it, and we got along much better than we expected. Distance to-day, twenty-eight miles and a half.

Tuesday, 29th June, South Side of Tent Hill. Started at 8.30 a.m. on a bearing of 305 degrees. At eight miles crossed a gum creek, with polyganum, running to the north. At twelve miles crossed another, trending in the same direction. These creeks are wide and formed into numerous channels. I expected to have done thirty miles to-day, but am disappointed, for we were obliged to halt early, after having gone only eighteen miles, as my horse was quite lame. How much do we feel the want of another set of horse-shoes! We have, however, still got an old shoe left, which is put on this afternoon. It had continued raining all last night, but not heavily, and cleared off in the morning shortly after we started. Our travelling to-day has been still very stony, over stony rises; the stony table land that has been all along on our left is now trending more to the south-west. The country is more open: in looking at it from one of the rises it has the appearance of an immense plain, studded with isolated flat-topped hills. The last eight miles is better grassed and has more salt bush. Camped on a small creek in the stony rises. Distance to-day, eighteen miles.

Wednesday, 30th June, Stony Rises. We had a little rain in the former part of the night, and a very heavy dew in the morning. Started at 9.30 a.m., bearing 305 degrees; at five miles crossed the upper part of a gum creek, and at twelve miles ascended a high flat-topped hill, commanding a view of an immense stony plain, but it is so hazy that we can see nothing beyond ten miles. From this hill we changed our course to 309 degrees to a saddle in the next range. At four miles halted at a gum creek, with plenty of green feed. Made a very short journey to-day in consequence of the horses being quite lame. In addition to their want of shoes, a stiff, tenacious brown clay adhered to the hoof, and picked up the small round stones, which pressed on the frog of the foot. These pebbles were as firmly packed as if they had been put in with cement, so that we had hard work to keep the hoofs clear. Distance travelled, sixteen miles. Weather showery.

Thursday, 1st July, Gum Greek. The horses have had such poor food for the last week that I shall rest them to-day. About half a mile below us there is a large water hole a quarter of a mile long, with a number of black ducks upon it, but they are very shy. It rained very heavily and without intermission all last night and to-day. This creek is visited by a great many natives. We saw them making away as we approached.

Friday, 2nd July, Same Place. The creek came down last night: it is now a sheet of water two hundred yards broad. Started at 8.45 a.m. over a stony plain on a bearing of 309 degrees, to the saddle in the range. I ascended one of the highest hills in this range, but the day was too dull to see far. I could, however, distinguish what appeared to be a wooded country[10] in the distance, from south-west to north-east. Observing that the country a little more to the north was less stony, I changed our course to a bearing of 344 degrees, over a plain thinly covered with gravelly stones, consisting of quartz, ironstone, and a dark reddish-brown stone, with a good deal of gypsum cropping out. The soil is of a light-brown colour, with plenty of dry grass upon it, and very little salt bush. In the spring time it must look beautiful. The country was so boggy from the heavy rains, that for the sake of my horse I was obliged to stop early. Camped at a gum creek coming from the south-west, and running a little to the east of north. Distance to-day, eighteen miles.

Sunday, 4th July, Same Place. Not the slightest appearance of a change. It rained in torrents all night and all day, though at sundown it seemed to be breaking a little. The creek came down in the forenoon, overflowed its banks, and left us on an island before we knew what we were about. We were obliged to seek a higher place. Not content with depriving us of our first worley, it has now forced us to retreat to a bare hill, without any protection from the weather. The rain has come from the north-east.

Monday, 5th July, Same Place. The rain lasted the greater part of the night, but became light before morning. Started at 12.30 on a bearing of 312 degrees for eleven miles to some sand hills. A fearfully hard day's work for the poor horses over a stony plain, sinking up to their knees in mud, until at eight miles we crossed a reedy swamp two miles in breadth, and how many in length I know

[10] This "wooded country" afterwards turned out to be sand hills, with scrub.

not, for it seemed all one sheet of water: it took our horses up to their bellies.

Tuesday, 6th July, Sand Hills. All our rations and everything we have got being perfectly saturated with wet, I have made up my mind to stop and put them to rights; if we neglect them it will soon be all over with us. This was a beautiful day, not a cloud to be seen. There are a great many natives' tracks in these sand hills, and plenty of grass.

Wednesday, 7th July, Sand Hills. Heavy dew last night. Started on a bearing of 312 degrees at 9 a.m. At eleven miles the sand hills cease, and stony plain commences. The sand hills were well grassed: also the stony plain. Dip of the country still north-east. We crossed two watercourses—one at this side of the plain, and the other two miles back, broad and shallow. I could see gum-trees on the latter about two miles to the north-east as if it formed itself into a deeper channel. Travelling very heavy. Distance to-day, twenty-five miles.

Thursday, 8th July, Sand Hills. A very heavy dew again last night. Started at 9 a.m. At one mile we came on yesterday's course; could see nothing; changed the bearing to 272 degrees. At seven miles crossed a creek running north and a little west, the water being up to our saddle-flaps. At twelve miles the sand hills ceased, and we came upon an elevated plain, of a light-brown soil, with fragments of stone on the surface. At twenty-five miles, in the middle of this plain, we camped, without wood, and in sight of a large range in the far distance to the west. Distance to-day, twenty-five miles.

Friday, 9th July, Large Plain. Left our camp at 8.50 a.m. on the same bearing as yesterday, 272 degrees. At one mile and half came upon a creek of water, seemingly permanent. Judging from the immense quantity of dry grass that is strewn over the plain, this must be a beautiful country in spring. The dip of the country is to the north and west. Our horses are all very lame for want of shoes, and the boggy state of the soil to-day has tried them severely. If the country does not become less stony, I shall be compelled to leave some of them behind. We camped on a gum creek about three miles to the west of the range. My only hope now of cutting Cooper's Creek is on the other side of the range. The plain we crossed to-day resembles those of the Cooper, also the grasses; if it is not there, it must run to the north-west, and form the Glenelg of Captain Grey. Distance to-day, twenty-one miles.

Saturday, 10th July, Gum Creek, West End of Large Stony Plain. Rested the horses to-day. This evening we were surprised to hear a dog barking[11] at the grey mare; its colour was black and tan.

Sunday, 11th July, Same Place. This morning the sun rose at 62 degrees. Bearing to-day, 272 degrees, so as to round the point of range, which seems to have a little mallee in the gullies on this side, and some trees on the west side. Started at 8.30 a.m., and at four miles ascended the highest point of the range. The view to the north-east is over an immense stony plain with broken hills in the distance. To the north is also the plain, with table-hills in the far distance. To

---

[11] It is commonly supposed that the native dingo or wild dog does not bark. This is an error. The dog in this instance being black and tan, was probably a hybrid. (See below.)

the north-west is the termination of the range running north-east and south-west, distant about ten miles; about half-way between is a gum creek running to north-east. To the west is the same range, and a number of conical hills between. Changed our bearing to 220 degrees in order to break through the range. This range is very stony, composed of a hard milky-white flint stone, and white and yellow chalky substance, with a gradual descent on the other side to the south, which is the finest salt-bush country that I have seen, with a great quantity of grass upon it. The grey mare has been very bad; her belly was very much swollen, but this morning she seemed better. Towards afternoon, however, she fagged very much, which caused me to stop so soon. I am almost afraid that I shall lose her. I shall see how she is in the morning, and, if she is no better, I will endeavour to get her on to some permanent water or creek running to the south. I think we have now made the dip of the country to the south, but the mirage is so powerful that little bushes appear like great gum-trees, which makes it very difficult to judge what is before us; it is almost as bad as travelling in the dark. I never saw it so bright nor so continuous as it is now; one would think that the whole country was under water. Camped without water. No timber as yet on this side of the range, except a few bushes in the creek. A good deal of rain has fallen here lately, and the vegetation is looking fresh.

Monday, 12th July, Large Salt-Bush and Grass Plain. The mare seems a little better this morning, and I shall be able to make a short journey. There was a very heavy white frost during the night, and it was bitterly cold. Not a hill to be seen either to the south-west or west—nothing but plain. Left our camp at 8.30 a.m. on a bearing of 220 degrees; at two miles and a half changed to 112 degrees for three miles to a small creek running south with plenty of feed and water. We found our horses very much done up this morning; they could scarcely travel over the stones, which caused me to alter my course to the eastward, where I found the travelling generally better. All the horses are now so lame that I shall require to rest them before I can proceed. They will not walk above two miles an hour among the stones. The stony plain seems to continue a long way to the south-west, but the country being undulating and the mirage so strong, I cannot say precisely. I intend to see where this creek will lead me to, for I cannot face the stones again. Our distance to-day, five miles and a half.

Tuesday, 13th July, Mulga Creek. Went to the highest point on the stony range east of us, but could only see a very short distance. There are a number of creeks on the eastern side running into this one. The range is low and very stony, composed of flints and pebbles of all colours. No timber.

Wednesday, 14th July, Same Place. During the night it became very cloudy, and I was afraid we were going to have more rain, but it has ended in a light shower, and cleared off this morning. I shall follow down the creek and see what it leads to. The grey mare still seems very bad, and I must make short journeys until she gets a little better. Started at 8.30 a.m., bearing 180 degrees for eight miles to Large Mulga Creek, thence 192 degrees for four miles. The country to-day is good on both sides of the creek, a good salt-bush country with plenty of grass, but rather stony. The gum trees are becoming a little larger on the creek, which at present is formed into a great many channels. The timber consists of mulga and dwarf gum, with saplings. There is plenty of water in the creek at present, from the late rain, but I see nothing to indicate its becoming permanent. Distance to-day, twelve miles.

Thursday, 15th July, Mulga and Gum Creek. Left the camp at 9 a.m. on a bearing of 190 degrees for two miles, thence 230 degrees for one mile and a half, thence 250 degrees for four miles and a half, thence 286 degrees for two miles, thence 290 degrees for one mile, thence 270 degrees for five miles, thence 320 degrees for one mile, to camp at some mallee. The country on both sides of the creek is good, but subject to be flooded; the width of the plain is about fifteen miles, bounded on the south side by bare stony rises, and on the north by scrubby rises. The creek spreads itself all over the plain, which seems to be very extensive. It has been excessively cold to-day: wind from the west. Distance to-day, seventeen miles.

Friday, 16th July, Large Plain, Mulga and Gum Creek. Left the camp at 9 a.m., on a bearing of 270 degrees for nine miles. The first six miles was a continuation of the creek and plain; it then turned to the north-west and the sand hills commenced. At nine miles we had a good view of the surrounding country, from the east to the north-west. To the west we could see the range that we crossed on the 11th instant trending away to the north-west as far as the eye could reach, apparently a sandy and scrubby country with small patches of open ground intervening. There also appeared to be a gum creek, about five miles west of this point. Seeing there was no hope for anything to the west for a long distance, I changed my course to the south on a bearing of 190 degrees to cross the stony rise, keeping on the sand hills for the benefit of the horses' feet. At five miles found that the sandy country swept round the stony rise, the country still having the appearance of scrub and sand hills all round. I altered my course to south-east to 132 degrees for fourteen miles; on this course we have ridden over a scrubby plain of a light sandy soil, most beautifully grassed but dry, the young feed not having sprung. We have not seen a drop of water on the surface; the ground evidently absorbs all that falls; the scrub is principally the mulga and hakea bushes and acacia, with a few other small bushes, but very little salt bush. Camped to-night without water. The grey mare appears to be getting round again; it seems to have been an affection of the chest, and has now fallen down into the left knee, which has become very much swollen, but it seems to have relieved her chest; she now feeds as well as ever. Distance to-day, twenty-eight miles.

Saturday, 17th July, Scrub and Sandy Plain without Water. Started at 8.10 a.m. on the same course, 132 degrees. At two miles and a half, rain water; at seven miles crossed a stunted gum creek running towards the south-west; at twenty-five miles came upon a little rain water. Camped. The plain still continues with very low rises at intervals; the scrub is much thicker and the greater part of it dead, which makes it very difficult to travel through. The grass is not so plentiful, and it is more sandy. The creek that we crossed at seven miles was running; it had salt tea-tree on its banks, and seems likely to have some permanent water either above or below. I did not examine it, because, the surrounding country being so sandy and scrubby, it will be of little use. Distance to-day, twenty-five miles.

Sunday, 18th July, Dense Scrubby Plain. Rain Water. Left at 9.15 a.m. on the same bearing, 132 degrees. We saw some native worleys, and the tracks of a number of natives having passed this place a day or two ago, going to the south-west. Distance to-day, twenty miles. Had to halt early in consequence of grey mare being done up and unable to proceed. The first part of the day's journey the scrub became more open and splendidly grassed, the latter part was fearfully thick, it is composed of mulga, dead and alive, and a few hakea and other bushes, with salt bush and plenty of grass of two or three different sorts. We have a view of rising ground a little to the north of our line, about from fifteen to twenty miles distant. To-morrow I shall alter my course to strike the highest point; it is a range, and seems to be wooded. I suppose it is the same range that we crossed on the 11th instant. It is very cloudy, and seems as if it will rain. Distance to-day, twenty miles.

Monday, 19th July, Dense Scrubby Plain. Started at 9.15 a.m. on a bearing of 120 degrees to the highest point of the range. A slight shower fell early this morning; it still looks very cloudy. We could only accomplish ten miles to-day in consequence of the grey mare being unable to proceed farther; if I can get her on to permanent water I shall leave her; she only keeps me back, and endangers the other horses. I shall be very sorry to do so, for she is a great favourite. We are now camped at a place where there are five or six small watercourses; if we can find water I shall give her until to-morrow to rest. The country that we have come over to-day is most splendidly grassed, of a red light sandy soil, but good; the mulga bushes in some places grow thick, and a great many are very tall. Forster caught an opossum—the first that we have seen; we intend making a dinner from him to-day. This is the first game we have been able to secure, except two small ducks we had at the beginning of our journey. We have found water a little way down the valley, which I think will become a large creek further to the south-west. We are again in the country of the kangaroos. Distance to-day, ten miles.

Tuesday, 20th July, Grassy Valley. We had another shower this morning. I must try and make the hills to-day if I can. Started at 10.10 a.m. on the same bearing as yesterday, 120 degrees, and at four miles ascended the peak on the range. I see around me a scrubby country, with open patches, and here and there in the far distance what appear to be belts of mulga. Four miles beyond this hill we halted at some rain water. We have seen three or four kangaroos to-day; they were the red sort with white breasts. Distance travelled, eight miles.

Wednesday 21st July, Grass and Salt-Bush Plains. Left the camp at 9 a.m. on a bearing of 97 degrees. Camped at some rain water in a clay-pan. At twelve miles there is low rising ground running north-west and south-east, which divides the two plains; there are no creeks, but the dip of the country is to the south-west. This is as fine a salt-bush and grass country as I have seen. It is a pity there is no permanent water. Distance to-day, twenty miles.

Thursday, 22nd July, Open, Good Country. Started at 9 a.m. on the same course as yesterday, 97 degrees. At ten miles crossed a small watercourse running to the south-south-west; at sixteen miles came through the saddle of a low range running north-west and south-east composed of limestone; it forms one of the boundaries of a large plain, which seems well adapted for pastoral purposes; it is well grassed, with salt bush, although we could find no permanent water. I think I can see a gum creek to the east of us, but the mirage is so powerful that I am not quite certain. Distance to-day, twenty miles.

Friday, 23rd July, Large East Plain. Started at 9.10 a.m. on a bearing of 82 degrees, and at four miles ascended an isolated hill, but can see nothing of the gum creek. Changed our course to 122 degrees, and at four miles crossed a mulga creek running to the east. Camped on the south-east side of a flat-topped hill, which, although the highest I have yet seen, enabled me to see nothing but the range to the north-east, and a high conical hill about ten miles south-west, connected with the ranges. The country is without timber except a few mulga bushes at intervals. Distance to-day, twenty-one miles.

Saturday, 24th July, South-east Side of Flat-topped Hill. Left at 8.10 a.m. on the same course, 122 degrees, over an undulating stony plain, with narrow sand hills at intervals, and a number of lagoons containing rain water, where we camped. I intend to move to-morrow to another large lagoon that we have seen from a small rise, and rest the horses there; they have had a very severe day of it, and feel the want of shoes very much. The stones are mostly white quartz and ironstone, small and water-washed. I conclude they have come from the hills that are to the south-west. Distance to-day, twenty-four miles.

Sunday, 25th July, A Lagoon of Rain Water. Finding that we have sand hills to cross, and being anxious to meet with the gum creek that the blacks have talked about, I have determined to proceed to-day, but if I do not find it on this course I shall turn to the south. Started at eight a.m. on a bearing of 122 degrees. At five miles, one mile to the south is a large reedy swamp. At fourteen miles changed the bearing to 135 degrees to the head of a swamp, two miles and a half, found it dry, a large clay-pan about three miles in circumference. I am obliged to halt, the horses are very tired and want rest; and there being plenty of beautiful green feed about, I have halted without water. Our journey has been through a very thick mulga scrub and sand hills, very heavy travelling. The trees in the scrub are of a different description to any that I have seen; they grow high and very crooked, without branches until near the top, and with a rough, ragged bark; seven or eight seem to spring from one root. The wood is very tough and heavy, and burns a long time, giving out a glowing heat. The leaves resemble the mulga, but are of a darker colour and smaller size. The native name is Moratchee. Shot a wallaby, and had him for dinner. They are very wild, no getting within shot of them, which is unfortunate, as our provisions are getting rather short. From the number of native tracks about, this would seem to be their season for hunting in the sand hills, which accounts for everything being so wild. We saw five turkeys yesterday, but could not get within shot of them. All the water seems to drain into the reedy swamp and clay-pans. I shall go no further to the east on this course, for I can see no inducement. I shall go south to-morrow, and see what that produces; if I cross no large creek within forty-five miles in that direction, I shall then direct my course for the north-west of Fowler's Bay to see what is there. Distance to-day, sixteen miles.

Monday, 27th July, Sand Hills and Dense Scrub. Left our camp at 9.20 a.m. on a southerly course, 182 degrees. At thirteen miles we camped at some rain water to give the horses a little rest. We have come through a very thick scrub of mulga, with broken sand hills and a few low rises of lime and ironstone. We have seen two or three pines for the first time, and a few black oaks. No appearance of a change of country. From a high sand ridge I could see a long way to the north-east, seemingly all a dense scrub. The grey mare is unwell again. Distance to-day, thirteen miles.

Tuesday, 27th July, Sandy Undulations. Started at 9 a.m. on the same bearing as yesterday, 182 degrees. At twenty-one miles changed our course to 235 degrees to some gum-trees. The first part of our journey the scrub became lower and more open, with limestone and sand rises at intervals, and with a good deal of grass in places. The last ten miles the mulga scrub was so dense that it was with difficulty we managed to get through. We have seen no water on this day's route, except that in the lagoon we are now camped at, and which is as salt as the sea. There is another large lagoon about a mile to the westward of us, which I will examine to-morrow to see if it gives rise to any creek. Distance to-day, twenty-two miles.

Wednesday, 28th July, Sand Hills. Started at 9 a.m. on a bearing of 283 degrees for two miles to examine the other lagoon, which is about three miles long, water salt. Changed our course to 182 degrees for ten miles to a large lake crossing our course. Changed our bearing to 240 degrees, and at four miles changed to 270 degrees, crossing some horse-tracks going towards the large lake. This seems to be a country of salt lagoons, for we passed three, and have seen a great many more. The large one that crossed our south course is evidently the head of Lake Gairdner. I could see it winding away in that direction. We have now got upon a plain slightly undulating with thick scrub and the unceasing mulga, intermixed with a few black oaks; no signs of water, no creeks. I intend to proceed north of west to intersect any creek or country that may come from the good country that we found on our south-east course, and the land of kangaroos; there is no hope of anything here. Camped without water. Distance to-day, twenty miles.

Thursday, 29th July, Mulga Plain, West of Lake Gairdner. Our course to-day is 310 degrees. Left our camp at 8.30, and accomplished twenty miles of the same scrubby plain, slightly undulating. Plenty of grass, but no water. Same description of country as on the 18th instant.

Friday, 30th July, Mulga Plain. Started at 7.35 on same course, 310 degrees. The scrub is so dense that I cannot see above one hundred yards ahead, and sometimes not that. During the night some swans and two ducks flew over, apparently from Lake Gairdner, and going in our direction. At ten miles, having met with some rain water, we halted, for the horses had been three nights without it. I have given them the rest of the day to drink their fill. This seems to be a continuation of the stony plain we crossed on our south-eastern line. The country appears open to the south, but no sign of any permanent water. Forster bakes the last of our flour this afternoon—the last of our provisions. Distance to-day, ten miles.

Saturday, 31st July, South Stony Plain. Left at 8.30 on the same bearing, 310 degrees. At ten miles we ascended a low range running north and south. We did not see a drop of water all day. Our course was over a gradually rising plain, well grassed at intervals, with plenty of salt bush, and with stone on the surface, composed of quartz, ironstone, and the hard white flinty stone so frequently met with. The scrub has nearly ceased. The dip of the country is south. During the night we again heard a dog barking at one of the horses, and during the day we saw two kangaroos. At ten miles we crossed a valley, through which water has been flowing to the south-south-west. Camped without water. Distance to-day, fifteen miles.

Sunday, 1st August, Stony Plain Valley. Left at 8.45 on the same bearing, 310 degrees. My reason for keeping this bearing is that there seems to have been very little rain to the south of us, and I am unwilling to get too far away from where it has fallen, in case I have to put to my former line for it. If I should meet with it to-day I shall turn south-west or west. This country is very dry, and absorbs all that falls. It is of a bright red soil, mixed with sand and, in some places, lime. At ten miles I am obliged to stop, in consequence of the grey mare being quite done up; the stones play the mischief with her. I have great doubts of her living through the journey. Distance to-day, ten miles.

Monday, 2nd August, Salt Bush—a Stony Plain. We had a little rain during the night. Started at 9 on a bearing of 315 degrees. At three miles changed our course to 230 degrees. The last three miles of this day's journey were through rather a thick scrub, but well grassed, with few stones. The former part was through a very well-grassed country, with a little salt bush and low scrub. Saw a number of kangaroos, but they were too wild to get near them. Distance to-day, twenty miles.

Tuesday, 3rd August, Good Country. It has rained during the whole night, and is likely to do so to-day. Started at 9, on the same course as yesterday, 230 degrees. The first portion of our journey was over six miles of splendid alluvial country, covered with grass—partly spear grass—with a little salt bush intermixed with it, also a few mulga bushes at intervals; no other timber. It is a most beautiful open piece of country, and looks much better than the Adelaide plains did at the commencement of the colony. Four miles further it was not so good; the soil became a little lighter, with more salt bush, and a little scrub. The last eleven miles the soil is good, with grass and salt bush in abundance, but much thicker with mulga and other low scrubs. It seems to be a continuation of the same scrub that we passed over on the 19th ultimo, and I observe that the ants build their habitations in the same style as they did there. They are about one foot in diameter at the base, and formed in the shape of a cone, and are supported by the dead root of a mulga. Others, however, stand from eighteen inches to three feet in height, built of clay, and on the surface. The kangaroo and emu inhabit the country. We have also found a number of places where the natives have been encamped. They seem to be numerous, judging from the number of places where they have had their fires; but we have not seen any of them. We have had it raining nearly all day, and it still looks bad. Our black fellow left us during the night; he seemed to be very much frightened of the other natives. He knows nothing of the country, and if he follows our tracks back, I don't envy him his walk. He was of very little use to us, and I wish I had sent him off before, but I thought he might be useful in conversing with the other natives when we should meet them. He was of no other use than for tracking and assisting in getting the horses in the morning, for I have given them every advantage—they have been seldom hobbled. There are three small valleys on our line in which water seems to have run at some former period. We have crossed no course of rocks of any description since our northern line; from which I am of opinion that the drainage is underneath, so that there ought to be numerous springs near the sea-coast. Camped without water. Distance to-day, twenty miles.

Wednesday, 4th August, Scrubby Good Country. Started at 8 on the same bearing as yesterday, 230 degrees. At thirteen miles ascended a low red granite range in which there is water. Changed our bearing to 209 degrees to a hill on the opposite range; when I returned I found the grey mare so done up that she is unable to proceed. I should not like to leave her, but I cannot delay longer with her. For about half a mile under the range where we are now camped is beautiful feed up to the horses' knees. Six cockatoos passed over to another range. We have also found a small running stream where I shall leave the mare to-morrow; I will make an attempt to regain her as I return.

Thursday, 5th August, Granite Range. Started at 8 on the same bearing for the hill on the opposite range. At six miles another low granite range with water, where we left the mare. At twelve miles went to the highest point of the range composed of hard flinty quartz and ironstone. We had a good view of the surrounding country, which was generally low and undulating, with salt lakes crossing at about ten miles. This region appears to be dotted with the lagoons from nearly the foot of the range. Changed our bearing to 268 degrees for nine miles. Camped under a range of low hills with good feed for the horses. On our west course we crossed a plain of red light soil, with abundance of grass and a little salt bush with a very thick scrub close to the range, but as we advanced it became more open, and the scrub lower. Shot a wallaby and had him for supper. Distance to-day, twenty-five miles.

Friday, 6th August, Under the Low Range. Left at 8.30 a.m. on a bearing of 239 degrees to avoid the stones on the hills. At five miles and a half got some rain water; at nine miles changed our bearing to 255 degrees; at fifteen miles camped among the sand hills. Shot another wallaby. The timber about here is very large, consisting of black oaks, mallee, mulga, the native peach, the nut, and numerous low scrubs. The grass is good in some places. The mountain that I am steering for is further off than I anticipated; we got sight of it a short time before we halted; it seems to be very high, and I expect something good will be the result of our visit to it to-morrow. The hills that we were camped under last night are composed of quartz, and are connected with the range that we were on running to the south-west. Distance to-day, twenty six miles.

Saturday, 7th August, Sand Hills going to the High Mount. Left at 8.30 a.m. on the same bearing, 255 degrees, for eighteen miles to the foot of the mountain. At fifteen miles camped under the highest point, which is composed of quartz rock. The journey to-day has been through horrid dense scrub and heavy sand hills, to the foot of the hill, which I have named Mount Finke. It is as high as Mount Arden; I have not light to get on the top of it to-night. Very little rain has fallen here, and we have been without water for the last two nights: the country is of such a light sandy soil that it will not retain it. I almost give up hopes of a good country; this is very disheartening after all that I have done to find it. If I see nothing from the top of the mount to-morrow, I must turn down to Fowler's Bay for water for the horses. As I could not remain quiet, I got on one of the lower spurs of Mount Finke to see what was before me. The prospect is gloomy in the extreme! I could see a long distance, but nothing met the eye save A DENSE SCRUB AS BLACK AND DISMAL AS MIDNIGHT. On my return I found that Forster had succeeded in finding water by digging in the creek. Distance to-day, twenty miles.

Sunday, 8th August, Mount Finke. At dawn of day I ascended the mountain, but was unable to see much more than I did last night, in consequence of there being a mist all round. No high rising ground is to be seen in any direction. A FEARFUL COUNTRY. Left the mount at 9.30 a.m. on a bearing of 270 degrees. At eighteen miles halted to give the horses some food, as they were obliged to be tied up all last night, there not being any feed for them, and the scrub very dense. The horse Blower seems to be very unwell; he has lain down twice this morning, and an hour's rest will do him good. After leaving the mount we have a thick mallee and mulga scrub to go through with spinifex. At ten miles changed our bearing to 190 degrees; at eight miles camped. The whole of our journey to day has been through a dreadful desert of sand hills and spinifex. In the last eight miles we have not seen a mouthful for the horses to eat and not a drop of water; it is even WORSE than Captain Sturt's desert, where there was a little salt bush; but here there is not a vestige. Distance to-day, twenty-five miles.

Monday, 9th August, Desert. Started at 8.30 on the same bearing, 190 degrees. At five miles there is a change in the country; the spinifex has suddenly ceased and low scrub taken its place; the sand ridges are spread and the valley wider. At seven miles discovered some rock water in the middle of a valley with plenty of salt bush and green grass, first rate for the horses, which have had nothing to eat for two nights. I shall give them the rest of the day to recover. They were beginning to be very much done up, and it was with difficulty we could get them to face the spinifex. Shot a pigeon and had him for supper. We have seen where a horse has been a long time ago. Distance to-day, seven miles.

Tuesday, 10th August, Rock Water. Started at 8.30 on a bearing of 180 degrees. Camped at eighteen miles without water, and a very little food for the horses, only a little salt bush. The appearance of a change from the dreary desert lasted only for about one mile from where we camped last night; it then became even worse than before—the sand hills higher, steeper and closer together, the spinifex thicker and higher; we got the horses through it with difficulty. It rained all last night and all day. There is some rising ground to the west. Distance to-day, eighteen miles.

Wednesday, 11th August, Dense Scrub. Left our camp at 8 on the same bearing, 180 degrees. At 9 obliged to halt for the remainder of the day, the horses being too tired to proceed further; the fearful sand hills are very trying for them. To-day's few miles have been through the same DREARY, DREADFUL, DISMAL DESERT of heavy sand hills and spinifex with mallee very dense, scarcely a mouthful for the horses to eat. When will it have an end? We again saw the rising ground a little to the north of west of us; I should have gone and examined it, but our small remaining quantity of provisions being nearly exhausted, I could not venture; my object now being to make Fowler's Bay for water for our horses, and thence to Streaky Bay, to endeavour to get some provisions there to carry us home. We have now travelled considerably upwards of a thousand miles, and in that journey my horses have had only four clear days to themselves; they have done most excellently well. No water.

Thursday, 12th August. Dense Scrub. Left at 8.25 on a bearing of 165 degrees. Camped at ten miles; the horses done up. The same dreary desert. No water.

Friday, 13th August. Dense Scrub. The horses look very bad this morning. I hope we shall be able to make the sea-coast to-day. Started at 8.30 on the same bearing, 165 degrees, but was unable to get more than ten miles out of the horses; Bonney is nearly done up, and there is no water for the poor animals. I hope I shall not be obliged to leave the poor old horse behind, but I very much fear that I shall have to do so if nothing turns up to-morrow. The country is still the same. This is dreadful work!

Saturday, 14th August, Dense Scrub. Started at 8.15 on the same bearing, 165 degrees. At ten miles came upon some green feed for the horses, and gave them the benefit of it for the rest of the day. Bonney still very bad. For the last two miles we have had no sand hills, but very dense mallee and tea-tree, with a light sandy soil with a little limestone, also salt bush and pig-face in abundance. No water.

Sunday, 15th August, Dense Mallee Scrub. Started at 8.45 on same bearing, 165 degrees. At two miles and a half changed our course to 225 degrees, having found some fresh horse-tracks; at seven miles camped for the remainder of the day to recruit the horses, having come upon some new green grass. Distance actually travelled, fifteen miles.

Monday, 16th August, Dense Mallee Scrub. Started at 9 on a course of 205 degrees. Twelve miles to Miller's Water. I intended to have given the horses two days' rest here, but there is not sufficient water; there are only three holes in the limestone rock, and the thirsty animals have nearly drunk it all: there will not be enough for them in the morning. The country that we have come through yesterday and to-day resembles the scrub between Franklin Harbour and Port Lincoln—mallee with grassy plains occasionally—only the mallee is larger, and the plains are met with at shorter intervals, more numerous and of larger extent. The soil is good but light, being produced by decomposed limestone, of which the low range to the north-west is composed. I am unable to go to Fowler's Bay as I intended; our provisions are exhausted, and the horses unable to do the journey. I must now shape my course for Streaky Bay to get something to eat.

Tuesday, 17th August, Miller's Water. Watered our horses from a waterproof with a quart pot. Started at 9.15, our course 160 degrees, six miles to Bectimah Gaip. For the first three miles the grassy plains are very good, and seem to run a considerable distance between belts of large mallee, in some places wider than in others, and seem to be connected by small gaips; I think water could be easily obtained by digging. The last three miles to the coast is very dense small mallee. Actual distance, twelve miles. I intend to give the horses a rest to-morrow. I regret exceedingly that I was unable to make Fowler's Bay. It is with difficulty that I have been able to save Bonney; he is still very weak and unable to do a day's journey; we can scarcely get him to do the short journeys we have been doing lately. For upwards of a month we have been existing upon two pounds and a half of flour cake daily, without animal food. Since we commenced the journey, all the animal food we have been able to obtain has been four wallabies, one opossum, one small duck, one pigeon, and latterly a few kangaroo mice, which were very welcome; we were anxious to find more, but we soon got out of their country.

These kangaroo mice are elegant little animals, about four inches in length, and resemble the kangaroo in shape, with a long tail terminating with a sort of brush. Their habitations are of a conical form, built with twigs and rotten wood, about six feet in diameter at the base, and rising to a height of three or four feet. When the natives discover one of these nests they surround it, treading firmly round the base in order to secure any outlet; they then remove the top of the cone, and, as the mice endeavour to escape, they kill them with the waddies which they use with such unfailing skill. When the nest is found by only a few natives, they set fire to the top of the cone, and thus secure the little animals with ease. For the last month we have been reduced to one meal a-day, and that a very small one, which has exhausted us both very much and made us almost incapable of exertion. We have now only TWO meals left to take us to Streaky Bay, which is distant from this place ONE HUNDRED MILES. We have been forced to boil the tops of the pigface, to satisfy the wants of nature. Being short of water, we boiled them in their own juice. To a hungry man they were very palatable, and, had they been boiled in fresh water, would have made a good vegetable. Yesterday we obtained a few sow-thistles, which we boiled, and found to be very good.

Wednesday, 18th August, Bectimah Gaip. Rested the horses and obtained a few shell-fish from the beach: there are very few, which was a disappointment to us.

Thursday, 19th August, Bectimah Gaip. Started at 8 a.m. for Streaky Bay. I managed to get thirty miles to-day, which is a great help. I only hope that Mr. Gibson is at Streaky Bay, so that we may be able to get something to eat; we must endure three days' more starving before we shall be able to reach there.

Friday, 20th August, Smoky Bay. Started at 7.15. Mallee scrub in some places very dense, in others open, with good grassy plains at intervals, in which I think water could be had by digging; very few birds about, and those small. At twenty-five miles we got some rock water. Distance to-day, thirty-five miles.

Saturday, 21st August, Small Grassy Plains. Started at 7.30 on a south-easterly course. Got a little water in the limestone rock for our horses. Camped on the shore at Streaky Bay at sundown. The last sixteen miles were through very dense scrub; the former part through scrub with good grassy plains at intervals. Distance, thirty-eight miles.

Sunday, 22nd August, On the Shore at Streaky Bay. Started at 11 a.m. to make Mr. Gibson's station. The horses did not arrive until 10.30, as they had gone back on their tracks of yesterday. During the time Forster was after them, I managed to shoot a crow, and cooked him in the ashes. We had him for breakfast—the first food we have had for the last three days; it was very agreeable to taste and stomach, for we were beginning to feel the cravings of nature rather severely. I hope Mr. Gibson will be at the Depot; it will be a fine trouble if he is not, and we have to travel two hundred and forty miles on the chance of shooting something. Twenty-four miles to Mr. Gibson's station, where we were received and treated with great kindness, for which we were very thankful. We enjoyed a good supper, which, after three days' fasting, as may readily be imagined, was quite a treat.

Monday, 23rd August, Mr. Gibson's Station. Both Forster and myself felt very unwell, especially Forster, who is very bad; the sudden change from a state of starvation to plenty of good and wholesome food has been the cause. I am suffering chiefly from weakness and a very severe pain between the shoulder-blades, which I have felt for some weeks back. It is a dreadful pain, and nearly incapacitated me from sitting in the saddle all day yesterday; I thought I should not have been able to reach here, I was so very bad with it. I have been obliged to send down to the next station, about thirty miles distant, to try and get some horseshoes. I must rest here a few days to recover.

Tuesday, 24th August, Mr. Gibson's Station. Forster appears to be a little better this morning, but very weak; I also feel a little better this morning from yesterday's rest.

Wednesday, 25th August, Mr. Gibson's Station. I have succeeded in getting some shoes for the horses from Mr. Miller, to whom I am deeply indebted for his kindness in allowing me to have them.

Thursday, 26th August, Mr. Gibson's Station. Shoeing the horses and preparing for a start at the beginning of next week.

Friday, 27th August, Mr. Gibson's Station. At the same thing. Improving in health and strength.

Saturday, 28th August, Mr. Gibson's Station. I have been very unwell all night.

Sunday, 29th August, Mr. Gibson's Station. Still very ill; unable to do anything.

Monday, 30th August, Mr. Gibson's Station. The same.

Tuesday, 31st August, Mr. Gibson's Station. I had a dreadful night of it; seized with cramp in the stomach, and thought I should never see morning; no medicine to relieve me. I intended to have started to-day, but am quite unable to do so.

Wednesday, 1st September, Mr. Gibson's Station. Can stay no longer; made a start to-day, and got as far as one of Mr. Gibson's out-stations, twenty-five miles. Quite done up.

Thursday, 2nd September, One of Mr. Gibson's Out-Stations. Raining this morning; unable to proceed. Very unwell.

Friday, 3rd September, Same Place. Feel better this morning. Started at 8.30 for Parla. I am unable to make any attempt to recover the grey mare. Made Parla at 1 p.m.; camped at ten miles beyond. Distance to-day, twenty-five miles.

Saturday, 4th September, Ten Miles beyond Parla. Started at 8.15 on an east bearing twenty-three miles to Rock Water. Camped. Very poor country. The granite range that Mr. Hack has laid down on his chart, I cannot find. I have come east from Parla, and ought to have crossed about the middle of it.

Sunday, 5th September, Rock Water. I shall shape my course for the Freeling range, and see what that is made of. Started at 7.30 on a bearing of 84 degrees twenty-two miles. Rock water with plenty of grass. Gave the horses the rest of the day.

Monday, 6th September, South of Mount Sturt. Started at 8.15 on a bearing of 84 degrees for twenty-five miles. Changed the bearing to 60 degrees for three miles to a fine plain covered with grass. Halted. No water. There are some high hills to the east-north-east, to which I have now changed my course, and which I conclude to be the Freeling range. Our journey to-day has been through very scrubby and sandy country, especially the last fifteen miles. At six miles south there is a high table-topped hill, which I think is granite. I intended going down to it, but the country, so far as I could see, was apparently not good, and, having crossed the tracks of some horses going towards it, and being very unwell myself, I thought it would be useless my going. Distance to-day, thirty-eight miles. No water.

Tuesday, 7th September, Freeling Range. Started for the range at 8 on a bearing of 60 degrees. At eleven miles ascended the south-west hill of Freeling range, Mount Sturt bearing 266 degrees. Changed the bearing to 96 degrees to a stony hill of granite. Found a little water, and halted for the remainder of the day. Distance, fifteen miles.

Wednesday, 8th September, Freeling Range. Started at 7.30 for Separation Camp, bearing 72 degrees. Halted at thirty-three miles. The first twenty-five miles were mallee scrub with patches of grass; the last eight miles were over elevated table land, salt bush, and a little grass with a few patches of scrub, the soil being red, with a few fragments of quartz and ironstone on the surface. No water.

Thursday, 9th September, Salt-Bush Country. Started at 9.15 on the same bearing, 72 degrees, fourteen miles; changed to 160 degrees (1.30 p.m.) two miles and a half; thence 80 degrees three miles to a small creek, where we can obtain water by digging in the sand. Camped. Distance to-day, twenty miles. Did not see Separation Camp; it is wrongly placed on the map.

Friday, 10th September, Small Creek. Started at 9 on a bearing of 110 degrees for Cooroona; at seventeen miles made Cooroona. Camped fifteen miles beyond.

Saturday, 11th September. Arrived at Mr. Thompson's station, Mount Arden.

I cannot conclude this narrative of my first journey, without acknowledging that it was with the advice and assistance of my friend Mr. Finke SOLELY, that I undertook this exploration of the country. I therefore look upon him as the original pioneer (if I may be allowed so to express myself) of all my subsequent expeditions, in which our friend Mr. Chambers afterwards joined.

JOURNAL OF MR. STUART'S SECOND EXPEDITION (IN THE VICINITY OF LAKE TORRENS). APRIL TO JULY, 1859

SATURDAY, 2nd April, 1859. Started from Mr. Glen's for St. A'Becket's Pool, where we camped. This water hole is a large one, and likely to last a long time. The country around is good—a large salt bush and grassy plain, with upwards of 300 cattle feeding upon it. Found the native cucumber growing.

Sunday, 3rd April. Shortly after sunrise started from St. A'Becket's Pool, over low sand hills with large valleys between, well grassed, as described by Mr. Parry. Camped about two miles to the north-east of it, in a polyganum and grassy valley.

Monday, 4th April. The saddles injuring our horses' backs, we must stop and repair them. Herrgott and I rode to Shamrock Pool. There is still water there. It may last about a month, but it is not permanent.

Tuesday, 5th April. The horses could not be found before noon. One of them has lost a shoe, which will require to be put on. It is too late to start to-day for St. Francis' Ponds, the distance being thirty-two miles, and no water between. I deem it advisable to remain until to-morrow.

Wednesday, 6th April. Started on a bearing of 330 degrees, and at six miles came upon a gum creek, with abundance of water, which I believe is permanent. For fifty yards on each side of the creek there is a great quantity of polyganum and other water-bushes. On the water there are a great many ducks, cranes, and water-hens. The water hole is upwards of three-quarters of a mile long; at the broadest place it is fifty yards in breadth. There are two trees marked "J.G. and W. Latitude, 30 degrees 4 minutes 1 second." At one mile struck Mr. Parry's tracks; had a view of the country on the bearing that I intended to steer; saw that it would lead me into a very rough country, therefore followed his tracks to where he had camped. Camped south of Mount Delusion, without water. I do not doubt that there is water further down the creek to the eastward.

Thursday, 7th April. Went to the top of Mount Delusion and took bearings. Had some difficulty in finding St. Francis' Ponds. Towards sunset we found them, and, to our great disappointment, quite dry; all the water had disappeared, except a little in one of the creeks, which was salter than the sea, and of no use to us. There seems to have been no rain here this season; I have searched the country all round, but can see no sign of water. I must return to-morrow morning to the creek that I passed yesterday. The horses have now been two nights without water; they appear to feel it very much.

Friday, 8th April. Started back on a straight line, 6.40, for the gum creek, and arrived at 1.40 p.m., the horses being so much done up that I must give them two days' rest. I expect they will endure it better next time; they now know what it is to be without. In our course we crossed the middle of Mr. Parry's dry lake. It can be crossed at any time, for there are large courses of slate running through it in a north and south direction, level with the bed of the lake. The country around St. Francis' Ponds is as Mr. Parry describes it, with the exception of the water, which is gone. There is a great deal of Cooper's Creek grass growing in places. It is my intention to start with one man (as soon as the horses recover), and endeavour to find water nearer Mount North-west range. If I can find water east or west of St. Francis I shall then be able to make the Finniss Spring.

Saturday, 9th April. Resting the horses.

Sunday, 10th April. I intended to have gone to the north to-day to search for water, but I am so unwell from the effects of the water of this creek that I am unable to do so. I have been very ill all yesterday and all night, but I hope I shall be right to-morrow.

Monday, 11th April. I am unable to go and search for water, being too weak and not able to ride. I have sent Herrgott and Muller to find St. Stephen's Ponds, and see if there is water; they are to return by the foot of the range and endeavour to find water there also. I have been very ill indeed during the night; I have had no sleep for the last two nights, and I am so weak that I am scarcely able to move.

Tuesday, 12th April. Feel a little better this morning, but still very unwell.

Wednesday, 13th April. I feel a good deal better. I hope by to-morrow I shall be all right again. Herrgott did not return until noon to-day. He reports that there is no water in St. Stephen's Ponds, which I expected; but he also states that he has found a batch of springs three miles on this side of the ponds, with abundance of water. They are twelve in number. I shall go to-morrow with the party to them. I am very glad he has found them. There will now be no difficulty in taking stock to Chambers Creek. From this camp to the springs will be the longest journey to be encountered in a season like this, in which so little rain has fallen. After rain has fallen there will be no difficulty at all. The native cucumber grows about here.

Thursday, 14th April. Started at 8.10. The country travelled over was fine salt-bush country, but there was no water on our course, although we disturbed numerous pigeons and other birds. There are three table-topped hills to the east of the end of our north line; I think they are those within a short distance of which Major Warburton mentions that he found water. It would take me too much to the east of my course to examine them at present. I should have gone that way if Herrgott had not found those twelve springs, which we hope to make early to-morrow morning, and then proceed to the Finniss Springs. Camped on the east side of Decoy Hill, without water.

Friday, 15th April, East Side of Decoy Hill. At daybreak despatched Campbell for the horses. At 7.30 he returned with only five, and said that he found them on the track, going back for the water from which we have come, and that the others had left the tracks and gone west towards the hills. I immediately despatched Muller on horseback to track and bring them back, and I sent the others by Herrgott to get water at the springs. Sundown: no appearance of the horses. They must have gone back. If they have, it will be the middle of the night before Muller can be here. It is vexing to be delayed thus with the brutes.

Saturday, 16th April, Same Place. Muller and the horses have not yet come. I must go to the top of Decoy Hill to take some bearings. At 9.30 returned to the camp, and found Muller had just returned, but no horses; he had followed upon their tracks until they crossed a stony hill, where he lost them, and, on purpose to find them again, he tied the mare to a bush; she broke loose, and would not allow him to catch her until she got to the water. It was then sundown; he remained there during the greater part of the night to see if the others would come in: they did not, and he therefore came up to inform me of what had occurred. He was without fire, blankets, or anything to eat. I did not pity him; he ought to have been more careful. I had several times warned him not to leave the mare insecurely tied, or she would be off. I gave him a fresh horse, and sent him and Campbell off to follow them up to wherever they go, and not to come back without them. It is most dreadfully annoying to be kept back in this manner, all through the carelessness of one man: he must have been quite close to them when the mare got away. They were short hobbled, and I had looked at them at half-past two in the morning, to see if they were all right, and found them feeding quietly, so that they could not have gone far. Sundown: no appearance of the horses. I feel much better to-day.

Sunday, 17th April, Same Place. Still neither horses nor men. At 1.30 they arrived; my men had gone over to the range, and had searched every creek, but without success. When found, the runaway animals were standing on a rise looking very miserable and at a loss what to do; they had skirted the hill as far down as Mount Delusion. The men took them to the last water, remained there through the night, and left for this place this morning. I will give them an hour's rest, and go to the springs to-night. Arrived at the springs at sundown; they are about nine miles from Decoy Hill.

Monday, 18th April, Same Place. Resting horses. I went to the top of Mount Attraction, accompanied by Herrgott, to see what appearance the country had to the north of west. I observed a high red table-topped hill bearing 276 degrees from this point, for which I started in search of water. I had a good view of the country all round; it seems very low to the westward with low ranges and valleys between; plenty of salt bush and grass. There is copper with the ironstone on the top of Mount Attraction; native copper is adhering to the sides of the large pieces of ironstone. No water. Changed our course to north one mile and a half, thence to north-east five miles, thence to the springs, but could neither find water nor Major Warburton's tracks. To-day's journey forty-five miles. Arrived at the springs after dark.

Tuesday, 19th April, Springs. To the south of our tracks yesterday there was the appearance of a gum creek, and I think it advisable to send Herrgott to-day to examine it for water. It would be a great advantage for stock going to the new country. Seen from a little distance these springs, at which we are camped, resemble a salt lagoon covered with salt, which however is not the case; it is the white quartz which gives them that appearance. There are seven small hillocks from which flow the springs; their height above the plain is about eight feet, and they are surrounded with a cake of saltpetre, but the water is very good indeed, and there is an unlimited supply. Herrgott has taken a sketch of them. He has returned from examining the gum creek, but can find no water. I must push on to-morrow for Finniss Springs, and trust to find water on the way.

Wednesday, 20th April, Same Place. Started at 7.30 on a bearing of 275 degrees over a stony, undulating country with plenty of grass and salt bush, but no water. At twenty miles we saw a smoke raised by the blacks to the south of our line, under the range. Camped at 5.15 under a low range about thirty feet high and very perpendicular, running nearly north-east and south-west. Distance to-day, thirty-three miles.

Thursday, 21st April. Started at daybreak this morning. Same course. Cut Major Warburton's tracks at two miles, and changed to his course, 252 degrees. At one mile, saw Finniss Springs a mile and a half to the south of us; went down to them and camped. There is an immense quantity of water flowing from them. I shall raise a large cone of stones upon the hill, which is very prominent and can be seen from a long distance.

Friday, 22nd April, Finniss Springs. Went to the top of Hermit Hill, whence I obtained a very extensive view of Lake Torrens from north-west to north-east. Mount Hermit is surrounded by low hills, and in the far distance there seems to be rising ground. To the south are broken hills, the termination of the Mount North-west range. I shall examine that part of the country to-morrow. Between this and the lake (Eyre) to the north the country is very rough—broken cliffs, with sand; the good country does not extend more than three miles. The springs are very numerous all round this mount, and seem to drain into the lake; they give out an immense quantity of water, and there are many streams of water running from them. The ground is covered round about the springs with a cake of soda and saltpetre. I intended to have moved on to Gregory Creek this afternoon, but took the precaution to send my stockman to see in what state the water was. He reports the water in the creek to be quite salt, and many of the small fish dead; he also found some very perfect fossil shells, the mussel and oyster; they have now become a solid limestone; they were found in a large circular piece of limestone.

Saturday, 23rd April, Finniss Springs. Started at 8 a.m. with Herrgott to examine the country south of this. Between this and the range the land is good in places. It is a little rotten and stony, but the range is a beautiful grass country to the very top. In the creeks the grass and other plants are growing luxuriantly, but we could find no water. I was unable to prosecute the search as far as I wished, in consequence of my horse having lost a shoe and becoming quite lame, which forced me to return to the camp, where we arrived at 9 p.m. The view from a high conical hill of white granite with black spots at the north-west point of the range, is very extensive, except to the south, which is limited. We saw smoke in one of the creeks to the east; but as I was anxious to examine the creek to the south-west, which we saw from the top of the conical hill, I did not go to where the smoke was rising, thinking that the blacks might only be hunting. I therefore crossed the hills to the creek over a good feeding country, timbered with box and gum-trees. We expected to find water in it, from the great number of birds of all descriptions that were flying about; we followed it down, but were unsuccessful, although the birds continued all the way. There must be water about the hills in some place. At sundown, my horse becoming very lame, I was forced reluctantly to return. The flow of the waters is northward into North Lake Torrens. On Monday I shall start again to the south-west, and leave the examination of the range to the south-east until my return.

Sunday, 24th April, Finniss Springs. Latitude, 29 degrees 33 minutes 30 seconds. Rested.

Monday, 25th April, Finniss Springs. As it seemed likely to rain, in which case the country would be very soft, I started at 9.30 on a bearing of 242 degrees for Chambers Creek. After three miles of gravelly soil and scanty feed we came to the banks of the two creeks passed by Major Warburton, splendidly grassed, but the water very salt. They flow into Lake Torrens. After leaving these creeks we had four miles of sand hills, very rich with feed, thence over some stony ground to the creek, all good; my course brought me about three-quarters of a mile to the south of the creek, which I expected. Distance from the springs to this water hole, two miles; this is a very long water hole, with plenty of water in it, and the feed good. We saw some fresh tracks of natives to-day, but did not meet with any of them.

Tuesday, 26th April, Chambers Creek. I intend to remain here to-day to fix this place and examine the country about it. Latitude, 29 degrees 39 minutes 9 seconds. I sent Campbell (my stockman) in one direction, and Muller (the botanist) in another; they report quantities of water, also a great deal of salt water, with plenty of salt for the use of stations, with abundance of feed. The stockman saw numerous fresh tracks, but did not see any natives. The fires were still burning. Muller saw an old man, a woman, and a child. They were very much frightened, and when he approached, they called out "Pompoy!" and moved their hands for him not to come any nearer. As they seemed quite unwilling to hold any conversation, he left them.

Wednesday, 27th April, Chambers Creek. Started at sunrise this morning, accompanied by my botanist. After travelling thirty miles in a fruitless search for water, we camped upon a large stony plain with plenty of vegetation. The horses were very much tired by reason of the heavy sand. We could see no sign of Lake Torrens. Latitude, 29 degrees 53 minutes 58 seconds.

Thursday, 28th April, Large Stony Plain. Saddled by break of day. Changed my course to see if the water is still at Yarra Wirta. In order to avoid the heavy sand hills, which will not do for the horses if there is no water, I steered for the creek, struck it a little to the north of where I crossed it on my former expedition, and followed it down. Passed my former encampment, and found no water there, but on following it down to where I considered it permanent, I found water still there. I shall give the horses the afternoon to recruit, and start early in the morning. Distance to-day, twenty-three miles.

Friday, 29th April, Chambers Creek. Started at sunrise for about a mile to that part of the north shore of the lake opposite to where the Yarra Wirta empties itself into it. The country close to the lake is very stony and scanty of feed; there is some water in it, but it is very salt; a few salt creeks run into it, but no great body of water. I ascended a hill for which I had been steering, and obtained an observation of the sun and bearings. Latitude, 30 degrees 8 minutes 11 seconds. There is no appearance of any lake between this point and Mount Deception; it appears to be a stony plain with some ridges of sand hills. This hill, which I have named Mount Polly, for distinction, is the easternmost of the flat-topped hills on the north side of the lake, and is a spur from the Stuart range. It is very stony, and there is grass nearly to the top; it is very level, and extends for six miles in a north-westerly direction. I saw that there was little prospect of my obtaining water to-night; and knowing that the natives had been seen within a few miles of the camp, I felt anxious about the safety of my party. I determined to proceed towards the camp on a north-westerly course. Arrived at the creek at 11.30 p.m. and found all right; the natives had paid them a visit, as I anticipated, but my people could get no information from them. They were six in number; one was very forward, wishing to examine everything. I had left orders that, if they came, they were not to be allowed to come near the camp, but were to be met a little distance from it. They remained for some time, and then stole off one by one without being perceived, and were out of sight in a moment. The one that remained to the last in his flight did not forget to carry along with him a piece of blanket that had been a saddle-cloth, and which happened to be lying outside the camp.

Saturday, 30th April, Chambers Creek. Sent Muller and my stockman to build a cone of stones upon the highest of the three table-topped hills, for the base line of the survey. They are three remarkable hills close together; two only can be seen coming from the south and from the north-east. Latitude, 29 degrees 40 minutes 27 seconds. From the hill the men saw a number of native fires smoking to the westward on the creek, but have not seen any natives.

Sunday, 1st May, Chambers Creek. This morning we had a heavy dew. Went to the top of the three table-tops, and had a fine view of Mount Hamilton and the lagoon where the springs are, and the other hills; they are the same hills that I saw on my north-west course, when on my last journey.

Monday, 2nd May, Chambers Creek. Sent Muller and Campbell to build a cone of stones on Mount Strangways, which I have fixed as a south point of my base line. The mean of all the observations that I have got to-day makes the latitude to be 29 degrees 39 minutes 15 seconds.

Tuesday, 3rd May, Chambers Creek. Spent the day examining the neighbourhood for water, and in taking numerous bearings.

Wednesday, 4th May, Chambers Creek. I intend to move to-day to the large water holes westward, where I first struck the creek. The horses having strayed a long way off this morning, made it 11 o'clock before we got a start. About four miles from last night's camp the chain of large water holes commences, and continues beyond to-night's camp. They are indeed most splendid water holes—not holes, but very long ponds; they are nearly one continuous sheet of water, and the scenery is beautiful. I am sorry I did not name it a river in my former journal. I must bring my survey up to this night's camp to-morrow. It is very cloudy to-night, with a strong wind from the south-west, from which quarter the clouds are coming. The country is a little stony, but well grassed.

Thursday, 5th May, Chambers Creek. Moved the camp to a better situation. Ascended a hill, got some bearings to fix it, and built a cone of stones upon it. I have had the creek, which joins this, run up for three miles to the sources to-day. There is no more permanent water. There are an immense number of small fish in the ponds, and on the banks there is a shrub growing that tastes and smells like cinnamon; we happened to stir up the sugar in a pannikin of tea with a small twig of the bush, and it left quite the flavour of it in the tea. I have had Herrgott to take sketches of some of the ponds, also of the fish and other remarkable things. It has been rather cloudy to-day, and I could not depend upon my observations. There are numerous tracks of natives about, but we have not seen any of them; we have also found some new plants in the creek.

Friday, 6th May, Chambers Creek. Moved further up the creek on the south side to the last water that we knew of. It is a hole of rain water, very large, and will last a long time, being well sheltered by gum-trees and other shrubs.

Saturday, 7th May, Chambers Creek. Sent Muller to see if there is any more water to the west, and went myself to the top of a small hill, and built a cone of stones to connect this point with the last point. Muller returned after dark, and reported that there was no more permanent water. I shall start to the north to-morrow.

Sunday, 8th May, Chambers Creek. Started to the north over the range, which is rather difficult to get the horses up and down. On the top it is very stony, with salt bush and scanty grass. Crossed the Margaret and a salt creek, in which there is water, some of which is salt and some brackish, but not unfit for the use of cattle. There is abundance of feed all round. We arrived at Hamilton Springs a little before sundown. Distance, twenty-one miles.

Monday, 9th May, Mount Hamilton. Some of the horses require to be shod to-day. I shall also require to build a cone of stones upon Mount Hamilton (the one built by Major Warburton having fallen down), and get an observation of the same. Latitude, 29 degrees 27 minutes 37 seconds. The springs are certainly very remarkable, and Major Warburton gives a very good description of them.

Tuesday, 10th May, Mount Hamilton. Started for the Beresford Springs. Arrived at Mount Hugh at 11 o'clock, seven miles distant from Mount Hamilton, and, as I anticipated, found a number of splendid springs, giving out a fine stream of water, not the least brackish. The hill from which this stream issues is one hundred feet above the level of the plain, the water coming from the very top. My horse got bogged on the top, and I had some difficulty in getting him out, but I did so at last without injuring him. Started from the mount at 12.30, and, after three miles and a half, arrived at Beresford Springs. The Beresford Springs are nothing in comparison to the others; there are only two that are running, but they are very good. The country travelled over to-day has been very well grassed, with salt bush; take it altogether I have not seen better runs in the colony, and in the driest summer the furthest distance from water will not be above five miles at the most, but the feed is so abundant that they would not require to go so far. On that account they will feed double and treble the number of stock that the runs down the country do. At two miles on this side of the Hugh Springs discovered another batch of springs with plenty of water running from them; there are about eight or nine of them very good; those springs have not been visited by Major Warburton. We examined all round, but could find no tracks. I have named them the Elizabeth Springs. There is enough water running to drive a flour-mill in two or three places. They are really remarkable springs—such a height above the level of the plain; I saw them from a hill on Chambers Creek (the Twins). From whence do they derive their supply of water, to cause them to rise to such a height? It must be from some high ranges to the north-west, or a large body of fresh water lying on elevated ground. This is another strange feature of the mysterious interior of Australia. I shall remain here until after 12 to-morrow, to get an observation of the sun to fix this hill. I shall return to Mount Hamilton, and proceed to examine the country west of North Lake Torrens, for one of the east runs, which will complete my survey of them, and I shall despatch thence a messenger to Oratunga.

Wednesday, 11th May, Elizabeth Springs. Latitude, 29 degrees 17 minutes 43 seconds. I omitted to mention yesterday that, two miles before we reached Beresford Hill, we crossed Pasley Ponds and saw one of the Major's camps. The water is brackish, but not bad. The white deposit round these springs, and also round the Elizabeth, is soda. In returning, I examined the Coward Springs; the water is good, and running. There is a plentiful supply. It was dark when I arrived at Mount Hamilton. Saw four natives to-day, but they gave us a wide berth; they do not like to come near us.

Thursday, 12th May, Mount Hamilton. Some of the horses require shoeing, and I wish to get another observation of the sun. I shall remain here to-day, and examine the country to the north-east. About seven miles in that direction is the salt creek of Major Warburton. The country is of a light sandy soil covered with grass.

Friday, May 13th, Mount Hamilton. Started to the eastward, to complete the survey of the runs, and see if there are any more springs. To the south of east, about four miles, we discovered four springs not seen by the Major; there is a plentiful supply of water, and would be more if they were opened. One is choked up with reeds, but the other two are running. Saw some natives; they seemed frightened at first, but were induced to come close up: they were very much amused at our equipments. Two had seen or heard of whites before; they knew the name of horse, but no more; they call water courie, and some of their words very much resemble those of the natives in Port Lincoln. We could make nothing of them—they repeat every word of the question we ask them. They followed us over to the Margaret, and took us to some fresh-water springs in the creek, the water of which is very good. There is a quantity of reeds growing round them, also tea-tree. From this we followed the creek to the north, thence north-east towards the lake, but the water being too brackish, I returned to the springs, the natives walking with us all the time; they seemed very inoffensive. In following down the creek, another native joined us from the creek, carrying a net in which were some small fish; the net was a hoop one, well made.

Saturday, May 14th, The Margaret Creek. The morning very cloudy; every appearance of rain. Saddled and proceeded in search of Emerald Spring, on a north course. At seven miles made Mr. Babbage's old camp on a sand hill. Camped a little way from it. I did not know the position of the spring, but Herrgott informed me that it was three miles to the west. It commenced raining before we started, has rained all the way up, and is still doing so; it is a very light rain, but the wind is very strong and cold from the south-west. Intended to have brought up my plan, but the rain and wind prevent me.

Sunday, 15th May, Mr. Babbage's Old Camp. It cleared off during the night, but the clouds have come up again this morning and look very threatening. Sent Herrgott to find the spring. The wind is still from the same quarter, and too strong for me to do anything to the plan, which is a great annoyance. I will finish the survey of the runs from this place, and send Campbell back to Oratunga with the plan. Herrgott did not return until after sundown: he could not find the spring.

Monday, 16th May, Same Place. Sent Muller to the west; he returned at 10 o'clock, having found the spring about two miles and a half distant from the camp; it is not hot, but a little warmer than milk-warm. There is a good stream running from it, and the water is excellent; to me it has a mineral taste, very good. There were some small fish lying dead on the bank, near the mouth; they seemed to have been left there by the retiring of the flood—they were quite dried up. I intended to have taken some with me, but they were too dry—nothing but skin and bone. The creek empties itself into the lake, about a mile north from where Chambers Creek goes into it.

Tuesday, 17th May, Same Place. Again very cloudy, with a little rain. Busy finishing the survey. Could not obtain an observation of the sun. Wind still very strong.

Wednesday, 18th May, Same Place. Weather clearing up. Engaged with survey.

Thursday, 19th May, Same Place. Finishing tracings, etc.

Friday, 20th May, Same Place. At sunrise started Campbell for Oratunga with tracings, letter, etc., with orders to proceed to Finniss Springs, thence to Herrgott Springs, thence to St. A'Becket's Pool, thence to Mount Glenns, thence to Mount Stuart, and thence to Oratunga, taking six days to perform the journey. Preparing my other plans for a start to-morrow for the north-west, to see what the Davenport range is. Latitude, 29 degrees 23 minutes 20 seconds.

Saturday, 21st May, Same Place. Started at 8 o'clock on a bearing of 310 degrees for the Davenport range. At twenty-two miles changed our course to examine a large lagoon to the south-west of us, bearing 238 degrees. At two miles reached the lagoon, which we examined for springs, but found none. I suppose it receives Major Warburton's salt creek. It is caked with a crust of salt, and is dry; it is seven miles long by three broad, running north-west and south-west. On the south-west side it is bounded by steep cliffs, and high sand hills on the top. Changed to 310 degrees, our original course. Came upon some rain water at four miles, and camped for the night. Distance to-day, twenty-eight miles.

Sunday, 22nd May, Rain Water. Sent Herrgott to examine the south-west side of the lagoon which we passed last night, with orders to overtake me by 11.30, so that I may get an observation of the sun at noon. The horses having strayed some distance during the night, our start was delayed until 9.15. Started on the same bearing as yesterday, 310 degrees. Stopped at 11.20 for Herrgott to come with the instruments, but he did not come up until 1.15, so that I lost my observation. I had told him, if there was no appearance of springs not to go far, but to return immediately; instead of which he went round the lagoon. Camped on a stony rise, with a little wood. Distance to-day, twenty-one miles.

Monday, 23rd May, Stony Rise. Started towards the Davenport range. The sand hills again commenced with beautiful feed upon them—low, with broad valleys; they continued for five miles, when the stony plain again commenced. The highest part of the range seems to be at the north-eastern point, which has the appearance of a detached hill. At three miles and a quarter from the last of the sand hills we saw the Douglas, and changed our bearing to 328 degrees 30 minutes. At one mile and a quarter struck the creek, but found no water in it. There were a number of gums, but not very large, also plenty of myalls there. The bed of the creek is bad, and will not retain water. We followed it down for three miles to see if there was water; but no sign of it, the creek still continuing broad and sandy. I was obliged to return to where I struck it, because it was nearly sundown, and I had found a little rain water about a mile to the south, which would do for the horses in the morning.

Tuesday, 24th May, The Douglas. Herrgott's horse in want of shoes. Could not get a start until late. Found a little more rain water in a clay-pan. If I can find no water near the range, I shall have to fall back upon Strangway Springs. I am anxious to see what is on the other side of the range, or I would run this creek down. There are numerous tracks of natives about the creek; we have also seen three fires three or four days old. Latitude, 28 degrees 45 minutes 4 seconds. Started at 12.30 on a bearing of 313 degrees for the highest point of the range east, over stony table land. The creek runs in the same direction for four miles, it then turns to the westward, and is lost sight of among some hills. At ten miles struck a stony box-tree creek; its bed was sand and gravel, but no water. At 11.30 descended from the table land, and camped at a gum creek at sundown; the bed the same as the last, and no water. There were numerous native foot-tracks here also. I am sorry I could not reach the range to-night, but we had some very bad ground to travel over, and no water.

Wednesday, 25th May, Dry Gum Creek. Examined the creek for water, but found none. Started on the same course as yesterday, 313 degrees, for the north-east highest point, which I suppose to be the Mount Margaret of Major Warburton. Native tracks seen in the creek. There may be water some distance down the creek, but here it is too sandy to retain it. At four miles struck another gum creek in turning round the south side of the range; it was of the same description as the others, too sandy to hold water. Proceeded towards the highest point of the range, and obtained an observation of the sun within a mile and a half of the mount. Left the horses in charge of Muller and ascended the mount, which was very difficult; it took us an hour to go up, and three-quarters of an hour to come down. The hill is composed of a greenish slate, lying horizontally at the base, and courses of quartz and granite, with ironstone; but I can see nothing of Major Warburton's quartz cliffs; they must be more to the south-west. The range has a very peculiar appearance from a short distance off; it seems to be an immense number of rugged conical hills all thrown together. From the top, the view to the north-west was hidden by a higher point of the range. To the north-north-west there is another range, about twenty miles distant, apparently higher than this, running south-west and north-east. To the north is another far-distant range; to the east, broken hill and stony plain, with a number of clay-pans. A number of creeks run to the eastward from this range; they become gum creeks further down, but in and close to the range they have myall bushes, and other shrubs. No water to be obtained in this range. Changed my course to the north-east to examine a white clay-pan that I thought might contain some fresh water. At three miles came upon it, and was very much disappointed to find it salt. This being the second day that the horses have been without water, I must give up the search for springs and return to one mile south of the Douglas, where we had found a little rain water. It being nearly sundown, I made for the last large gum creek, striking it lower down, also cutting the other creeks between, hoping to find water in some, but there was none. Made the large gum creek at 10 o'clock. Camped for the night. Horses very much done up, in consequence of the ground that we have been travelling over being so rotten and stony. The country is not good, nor the range; but at three miles to the east it becomes less stony and better grassed. No water.

Thursday, 26th May, Large Gum Creek. Started at daylight for beyond the Douglas. At 3 o'clock arrived at water. Horses so much done up that I shall require to give them two days' rest, if the water will hold so long, and then I must return to the Strangway Springs, as we know that to be permanent water. There are some heavy clouds coming up from the south-west, which I hope will bring rain.

Friday, 27th May, The Douglas. Rain all gone after a slight shower, which did not assist me much. Very sorry for it.

Saturday, 28th May, The Douglas. Horses looking better this morning, so I will give them this day also. I have sent Muller down the creek to the eastward, to see if there is any water in it. I should have gone again to-day to the Davenport range, to see if I could find the quartz reefs by striking it more to the south-west, but it would be too much for the horses, which are my mainstay, and this water will not last longer than to-day; it is going very fast. I do wish to goodness it would rain, for I do hate going back. Muller returned at sundown. He has been about twelve miles down the creek, but can find no water. It still continued sandy. He shot three new parrots.

Sunday, 29th May, The Douglas. Not being satisfied with my hurried examination of the range, I shall make another attempt to-day, and endeavour to find water. If we do not succeed we must fall back upon the springs. Started on a course of west-north-west. Crossed the Douglas three times. It turned to the south-west, but I continued my course, over low hills and valleys, with plenty of feed, with quartz, ironstone, and granite. At fifteen miles changed a little more to the north towards a rise. The country becomes very broken and rough, but still plenty of grass. At twenty miles crossed the upper part of the gum creek that I camped on on the 25th instant. The banks are nearly perpendicular cliffs of slate. Followed it up for two miles, but no water. I continued my course for the rising ground. At six miles I found that I was getting upon high table land; so, as the sun was nearly down, I returned to the creek, where there is some green feed for the horses, as they will be without water to-night. It was after sundown before I reached the creek and camped. I have named this creek Davenport Creek, after the Honourable Mr. Davenport, M.L.C.

Monday, 30th May, Davenport Creek. Started at sunrise determined to follow down the creek, for I think there must be water somewhere before it enters the plain. The flow is to the east. At five miles came upon a beautiful spring in the bed of the creek, for which I am truly thankful. I have named this The Spring of Hope. It is a little brackish, not from salt, but soda, and runs a good stream of water. I have lived upon far worse water than this: to me it is of the utmost importance, and keeps my retreat open. I can go from here to Adelaide at any time of the year, and in any sort of season. Camped for the rest of the day. Latitude, 28 degrees 33 minutes 34 seconds.

Tuesday, 31st May, The Spring of Hope. Shoeing horses, and repairing various things.

Wednesday, 1st June, The Spring of Hope. Not being satisfied with my hurried view of the salt clay-pan that I visited on the 25th ultimo, I have sent Muller to-day to examine it for springs, before I proceed to the north-west. On a further examination of this water, I find a very large portion of magnesia in it, and also salt, but very little. Muller has returned, having been down the creek, and, as I expected, has found a small spring of very good water on the banks of the salt creek. I expect there will be others. I shall move down there to-morrow and examine it. I expect we have fallen upon the line of springs again, which I hope will continue towards the north. No rain seems to have fallen here for a long time.

Thursday, 2nd June, The Spring of Hope. Started at 9 o'clock for the springs, and arrived there in the afternoon. Travelled over a stony but very good feeding country, which became better as we approached the springs. There is a creek with a large water hole, and around the small hills are numerous springs. On the banks of the creek and round the springs an immense quantity of rushes, bulrushes, and other water-plants are growing. The quantity of land they cover is very great, amounting to several square miles. Some of the springs are choked up, others are running, though not so active as those further to the south. Round about them there is a thin crust of saltpetre, magnesia, and salt. The water of these springs is very good, but that of the creek is a little brackish, but will do very well for cattle. Some of the holes in the creek are rather salt. There is enough of good water for the largest station in the colony. Round the small hill, where I am now camped, there are twelve springs, and the water is first-rate.

I have named them Hawker Springs, after G.C. Hawker, Esquire, M.L.A.[12] The hills are composed of slate, mica, quartz (resembling those of the gold country), and ironstone. Latitude, 28 degrees 24 minutes 17 seconds. One of the horses seems to be very unwell to-day; he has endeavoured to lie down two or three times during the journey, but I hope he will be better by the morning.

Friday, 3rd June, Hawker Springs. I find that the horse is too unwell to proceed. I shall give him another day, for fear I should lose him altogether. I sent Muller to see if there are any springs round the hill about six miles to the east. He states that the creek flows past that hill, and on towards other hills of the same kind. The springs continue to within half a mile of the hill, where he found two large springs running over, covered with long reeds. I do not doubt but that they still continue on towards the lake, (wherever that may be), which I intend to examine on my return.

Saturday, 4th June, Hawker Springs. This morning the horse does not look much better, but still I must push on. Started at 8 towards the highest point of the next range. At one mile struck a gum creek coming from the Davenport range, and running to the north of east; the bed sandy and grassy. At four miles another gum creek of the same description, with the gum-trees stunted. At eight miles and a half struck three creeks joining at about a quarter of a mile to the east; the centre one is gum, and the other two myall. At twelve miles changed my course to 29 degrees to examine three dark-coloured hills, where I think there will be springs. At a mile and a quarter came upon a small batch of springs round the north side of the hills in a broad grassy valley, with plenty of good water. Changed my course again to 318 degrees towards the highest point of the range. At one mile a myall and gum creek; at three miles another gum creek; at seven miles a very large and broad gum creek, spread out into numerous channels. I have not the least doubt but there is water above and below, judging from the number of tracks of natives and emus that have been up and down the creek. As this is the largest creek that I have passed, and is likely to become as good as Chambers Creek, which it very much resembles, I have called it The Blyth, after the Honourable Arthur Blyth. I have named the range to the east The Hanson Range, after the Honourable R.D. Hanson. At nine miles and a half attained the highest point of the range, and built a cone of stones thereon, and have named it Mount Younghusband, after the Honourable William Younghusband. From it I had a good view of the surrounding country, which seems to be plentifully supplied with springs. To the north-west is another isolated range like this; I should think it is about seven hundred feet high. I have named it Mount Kingston, after the Honourable G.S. Kingston, Speaker of the House of Assembly. To the north the broken ranges continue, and in the distance there is a long flat-topped range, broken in some places. It seems to be closing upon my course on the last bearing. I cannot judge of the distance, the mirage being so great. Descended from the mount, and proceeded on a bearing of 336 degrees towards a spring that I saw from the top. As we were rounding

[12] Now the Honourable G.C. Hawker, Speaker of the House of Assembly at Adelaide.

the mount to the east, we found eight springs before we halted, in a distance of three miles; some were running, and others were choked up, but soft and boggy. At dark arrived at another batch of springs—not those that I intended going to—they are on the banks of a small creek, close to and coming from the range; they are not so active as the others, and taste a little brackish; they are coated with soda, saltpetre, and salt. The horse seems to be very ill; he has again attempted to lie down two or three times. I cannot imagine what is the matter with him.

Sunday, 5th June, Mount Younghusband. I must remain where I am to-day; the horse is so bad that he cannot proceed; he neither eats nor drinks. I have sent Muller to the west side of the mount to see the extent of the springs; they are on the banks of a creek which has brackish water in it, large and deep, and a great quantity of rushes. The water comes from the limestone banks which are covered with soda. He rode round the mount: it is all the same, and the feed is splendid right to the top of the mount. It is a wonderful country, scarcely to be believed. I have had one of the springs opened to-day, and the water to-night tastes excellent; it could not be better. Native tracks about; I am surprised we see none of them; we are passing old fires constantly. Latitude, 28 degrees 1 minute 32 seconds.

Monday, 6th June, Mount Younghusband. The horses being some distance off, and my horse requiring a shoe, I was unable to make a start until 10 o'clock, on a bearing of 307 degrees 45 minutes, passing Mount Kingston on the south-west side. At three-quarters of a mile came upon the springs that I intended to have camped at on Saturday night: they are flowing in a stream strong enough to supply any number of cattle. I named them The Barrow Springs, after J.U. Barrow, Esquire, M.L.A. At four miles and a half struck a large broad valley, in which are the largest springs I have yet seen. The flow of water from them is immense, coming in numerous streams, and the country around is beautiful. I have named these The Freeling Springs, after the Honourable Major Freeling, M.L.C. After leaving the springs I ascended a rough stony hill, to have a view of them, but I could not see them all, their extent is so great. They extend to under the Kingston range, and how much further I do not know. From this point I changed my course to 322 degrees. I can just see the top of a distant range, for which I will go on that bearing. At one mile and a half crossed a broad gum salt creek, coming from the west, with a quantity of salt water in it. I have named this Peake Creek, after C.J. Peake, Esquire, M.L.A. After crossing this, we travelled over low rises with quartz, ironstone, and slate; the quartz predominating. Herrgott and Muller, who have both been long in the Victoria gold diggings, say that they have not seen any place that resembles those diggings so much as this does. The country seems as if it were covered with snow, from the quantity of quartz. At eleven miles passed a brackish water creek and salt lagoon; searched for springs but could find none, although reeds and rushes abound, but no water on the surface. I thence proceeded three-quarters of a mile, and struck a gum creek with a number of channels and very long water holes, but the water is brackish; it might do for cattle. This I have named The

Neale, after J.B. Neale, Esquire, M.L.A. I think by following it down, there will be a large quantity of water, and good, and that it will become a very important creek. No person could wish for a better country for feed than that we have passed over to-day; it resembles the country about Chambers Creek.

Tuesday, 7th June, The Neale. At 8 o'clock started on a bearing of 180 degrees for the northernmost of the isolated hills, to see if there are springs round it. At four miles ascended it, but could see no springs. This I have named Mount Harvey, after J. Harvey, Esquire, M.L.A. from Mount Kingston it bears 47 degrees 45 minutes. Thence I started for the other mount, which I have named Mount Dutton, after the Honourable F.T. Dutton; four miles and a half to the top. The Hanson range is closing upon my course, and I think to-morrow's journey will cut it. On the north side are a few springs, some of them a little brackish, and some very good. We cleared out one, and found it very good. Here I camped for the night. From south-west to north-west it seems to be an immense plain, stony on the surface, with salt bush and grass. Mount Dutton is well grassed to the top; it is composed of the same rock as the others.

Wednesday, 8th June, Mount Dutton. at 9.15 started on a course of 310 degrees. At three-quarters of a mile passed another batch of springs, some of them brackish, and some very good indeed. Leaving them we passed over a good feeding country, crossing several gum and myall creeks, one with polyganum, all coming from Hanson range and flowing into the Neale. At nine miles crossed the top of Hanson range. From it I could see, about fifteen miles to the west of north, a high point of this range, which I have named Mount O'Halloran (after the Honourable Major O'Halloran), on the west side of which there appears to be a large creek coming from the north-west. We then proceeded on a course of 324 degrees towards Mount O'Halloran. At four miles and a half struck a large gum creek coming from the range and running for about four miles north-west on our course; examined it for water, but found none. It divides itself into numerous channels, and when full must retain a large quantity of water for a long time. The gum-trees are large and numerous, and numbers of pigeons frequent its banks. At a mile further came upon some rain water in a stony flat, where we camped for the night between low sand rises covered with grass.

Thursday, 9th June, Stony Flat. This country must be examined today for springs. I have therefore sent Muller down the creek to search that, whilst I must remain and get an observation of the sun. My party is far too small to examine the country well. I cannot go myself and leave the camp with the provisions to one man; the natives might attack him, and destroy the lot, there seem to be a great many tracks about. Three o'clock. Muller has returned; he has run the creek down until it joined another very large gum creek coming from the north-west—the one that I saw from the top of the range. The gum-trees were large; from one of them the natives had cut a large sheet of bark, evidently for a canoe. He also saw two large water holes, one hundred yards wide and a quarter of a mile long, with very high and steep banks. It seems to be the same creek as the Neale. Can it be Cooper's Creek? the country very much resembles it. My course will strike it more to the north-west to-morrow.

Friday, 10th June, Same Place. I have been very unwell during the night with cramp in the stomach, but hope I shall get better as I go on. Started at 8 o'clock on a bearing of 32 degrees 4 minutes. At four miles went to the top of Mount O'Halloran. The creek is about three miles to the west; it breaks through the Hanson range. Changed my course to 317 degrees to get away from the stones, which are very rough close to the hill. At six miles changed my course to 270 degrees to examine an isolated hill for springs, but found none. The creek winds round this hill, and spreads out into numerous channels, covering a space of two miles; but there is no water here, nor for three miles further up the creek. We have, however, found some rain water; and, as I feel so unwell that I am unable to ride, I have camped here for the night, and sent Muller to examine the creek for water. He has been unsuccessful.

Saturday, 11th June, Rain Water. I feel a little better this morning. Started at 9.20 on a bearing of 317 degrees. Crossed the creek, which is about a mile wide. For five miles it ran parallel to my course, and then turned more to the west. There is a beautiful plain along the bank, about three miles wide, and completely covered with grass. At nine miles and a half, on a small rise, changed my course to 318 degrees 30 minutes, to a distant hill. Travelled for nine miles and a half over another large and well-grassed plain of the same description; thence over some low stony hills to a myall flat, the soil beautiful, of a red colour, covered with grass; after four miles it became sandy. Camped for the night, after having gone thirty-one miles. The country of to-day surpasses all that I have yet travelled over for the abundance of feed. We have passed a number of native tracks, but only one or two are fresh. We have found no water to-day, except some little rain water, which is nearly all mud. I have no doubt but there is plenty towards the east.

Sunday, 12th June, Myall Flat. I feel still very unwell. We are now come to our last set of shoes for the horses, and, having experienced the misery of being without them in my previous journey, I am, though with great reluctance, forced to turn back. My party is also too small to make a proper examination of such splendid country. Started back, keeping more to the east to examine a high hill in search of water. If I can find water, I shall endeavour to reach the north boundary. At 11.40 arrived at the hill. Latitude, 27 degrees 12 minutes 30 seconds. Can see no appearance of water, although the country seems good all round. Ten degrees to the east of north is a large dark-coloured hill, which I saw from last night's camp, from fifteen to twenty miles distant. I should like to go to it, but can find no water. I have named it Mount Browne, after Mr. J.H. Browne, of Port Gawler, my companion in Captain Sturt's expedition. I dare not risk the horses another night without water, the grass is so very dry; had there been green grass, I would not have hesitated a moment. Turned towards the Neale by a different course to try and find water; was unsuccessful until within an hour of sundown, when we struck some muddy water. As I expected, the horses were very thirsty and devoured the lot. Reached the creek after dark.

Monday, 13th June, The Neale. Found some rain water on the bank of the creek, and, two of the horses requiring shoes, I stopped for the day. At noon sent Muller up the creek to see if he could find any water holes, but he saw none. At six miles another creek coming from the south-west joins this. I am afraid I shall not have enough shoes to carry me into the settled districts. The creek does not seem to have been running for a number of years. The water has, some time or other, been ten feet high. The breadth of the plain where the channels flow is a mile and a half, and the quantity of water must be immense. It drains a very large extent of country. After examining the country during the next two or three days, I shall endeavour to follow this creek down, and learn where it empties itself.

Tuesday, 14th June, The Neale. Started at 9 o'clock. Running the creek down. At eight miles crossed another branch of the creek coming from the south of west. We found no water. At twelve miles changed my bearing to south. At three miles and a half camped at the two water holes that Muller found when I sent him to examine the creek on the 9th instant. I can not with certainty say they are permanent, there are neither reeds nor rushes round them; they are very large and very deep, and, when filled with rain, must hold a large quantity of water for a long time. There are ducks upon them. The water does not taste like rain water, which leads me to think that it may be permanent and supplied by springs from below.

Wednesday, 15th June, Water Holes found by Muller on the 9th. Started at 9.15 a.m. Following the creek down. As we approached Hanson range, where it broke through, we came upon two nice water holes with ducks upon them. They are long, wide, and deep, with clay banks, and about three feet of water in the middle. There are no reeds nor rushes round them, and it is doubtful whether they are permanent. At seven miles and a half the creek winds a little more to the west. Shortly afterwards we struck (in the gap) two very long and large water holes a quarter of a mile long, and between forty and fifty yards wide, and very deep. These I may safely say are permanent. After getting through the range, the creek spreads out over a large plain in numerous courses, bearing towards the south-east. At four miles and a half changed my course. At six miles, going more to the east, changed again, and at eight miles camped for the night, without water. We have found no water since leaving the last water hole, although I do not doubt of there being some. It would have taken us too long a time to examine it more than I have done, my party being so small. We have passed several winter worleys of the natives, built with mud in the shape of a large beehive, with a small hole as the entrance. Numerous tracks all about the creek, but we see no natives. We are now approaching the spring country again.

Thursday, 16th June, The Neale. Started at 11.15. Still following the creek, which continues to spread widely over the plain. At five miles I observed some white patches of ground on the south-west side of Mount Dutton, resembling a batch of springs. I changed my course and steered for them, crossing the Neale at two miles and three-quarters. On the south-west side of the Neale the country is rather stony, and for about a mile from it the feed is not very good, in consequence of its being subject to inundation, but beyond that the feed is beautiful. At three miles and a half made the white patches, and found them to be springs covering a large extent of country, but not so active as those already described. Leaving the springs at two miles, crossed the Neale at a place where it becomes narrower and the channel much deeper, with long sheets of salt and brackish water. I shall now leave the creek. In the time of a flood an immense body of water must come down it. At the widest part, where it spreads itself out in the plain, the drift stuff is from fourteen to fifteen feet up in the trees. Camped at 4 p.m.

Friday, 17th June, The Neale. Discovered another large quantity of water supplied by springs. This country is a wonderful place for them. There is an immense quantity of water running now.

Saturday, 18th June, The Neale. Started early in the morning to examine the country. Found large quantities of quartz, samples of which I brought with me. Still well watered, but without any timber.

Sunday, 19th June, The Neale. Water in abundance, with large quantities of quartz. The course the quartz seems to take is from the south-west to the north-east. The plain we examined to-day is a large basin, surrounded by the hills from Mount Younghusband and Mount Kingston, with the creek running through the centre. To-morrow I shall have a look along the north-east side of Mount Kingston, for I see the quartz apparently goes through the range and breaks out again on the north-east side, which is very white.

Monday, 20th June, Mount Kingston. Started at 8 o'clock a.m. to examine the quartz on the east side of Mount Kingston. Crossed the creek, and at three miles struck a quartz reef. The Freeling Springs still continue, but seem inclined to run more to the eastward. Changed my course to a peak in a low range which has a white appearance. At eight miles reached the peak; the quartz ceases altogether, and the country is stony from here. I can see the line of the Neale running eastward; it spreads out over the plain. It was my intention to follow it until it reached the lake, but I find the ground too stony for me to do so. Being reduced to my last set of shoes, and some of them pretty well worn out, I am obliged to retreat. Changed my course at seven miles across the bed of the creek, three miles broad, with a number of brackish water holes in it, some very salt. At this point the trees cease. I can see nothing of the lake. Camped on a gum creek without water. The latter part of our course was over a very barren and rotten plain, surrounded by cliffs of gypsum, quite destitute of vegetation. It has evidently been the bed of a small lake at some time. There is no salt about it.

Tuesday, 21st June, Dry Gum Creek. At 7.40 started on the same course as last night, and after various changes of bearings arrived at the hill, whither I had sent Muller, and where he found two springs. Instead of two, they are numerous all round the hill; some are without water on the surface, and others have plenty. It is a perfect bed of springs. A little more east they are stronger, surrounded with green reeds and rushes.

Wednesday, 22nd June, Mount Younghusband. Started at 8.40. At three miles and a half came to a large bed of springs with reeds and rushes, water running and good, with numerous other small springs all round. They are a continuation of those we camped at last night, with an abundant supply of excellent water. At four miles crossed the salt creek coming from Hawker Springs. At eight miles crossed three salt and soda lagoons, surrounded by lime and gypsum mounds, in which are numerous springs up to the foot of the hills (ten miles and a half) and all round them. I have named these hills Parry Hills, after Samuel Parry, Esquire. It was my intention to have gone to the east from this, but the horses' shoes will not admit of it. To the south-east I observed three conical hills, for which I will now steer. At seven miles crossed a gum creek, in which are large water holes, where water had been lately, but there is now only mud. There must be water either up or down the creek, for there are numerous native tracks leading both ways. At ten miles crossed a large gum (stunted) creek with abundant springs of rather brackish water. At nineteen miles and a half camped on a broad creek, but no water. The country good.

Thursday, 23rd June, Dry Creek. Started at 8.30 on the same course for one of the conical hills. At three miles ascended it, and found it to be flat-topped. I can see nothing of any lake to the east. The view is interrupted by a flat-topped range. From this I changed my course, and at three miles and a half observed a peculiar-looking spot to the south-west, which had the appearance of springs. Changed my course for it, and at six miles came upon a hill of springs surrounded by a number of smaller ones, with an ample supply of first-rate water. The hill is covered with reeds and rushes; it is situated at the west side of a large plain, and is bounded by stony table land on the east side, which has an abrupt descent of about thirty feet into the plain. On the west side are a number of broken hills, and a small range composed of gypsum and lime, having the surface covered with fragments of quartz and ironstone, and a number of other pebbles. On the hill where the springs are we have found lava. There are numerous small creeks coming from the hill, and running in every direction. They seem to be all in confusion. The plain is about five miles wide. These I have named the Louden Springs.

Friday, 24th June, Louden Springs. I must remain here to-day, and put the last of the shoes upon some of the horses which are getting rather lame. I have been making them go without as long as I can.

Saturday, 25th June, Louden Springs. Started at 7.50. At 8.45 (three miles) crossed a gum creek, and at 12 o'clock (eleven miles) crossed the Douglas, but no water. The channel still broad and sandy.

Sunday, 26th June, The Douglas. Started at 8.25, on a bearing of 217 degrees. Crossed the lagoon, which was rather boggy in some places. It is now more than two miles broad, with a white crust on the top, composed of soda and salt, but mostly salt. It must be supplied by springs. At three miles crossed a salt creek, with salt water. It empties itself into the lagoon, and is the same that passes by the Strangway Springs. I can see nothing of any springs at this part of the creek. Steered upon the same course to intersect my outward tracks. Saw some natives walking along a valley. They did not observe us. I hailed them, and an old man came up to us. He was rather frightened, and trembled a good deal. He seemed to wonder and be pleased at my smoking a pipe of tobacco. I gave one to him and a piece of tobacco, but he did not know how to manage the cutting, filling, and lighting operations. I did these for him. In the first attempt he put the wrong end into his mouth, which he found rather hot, and quickly took it out. I then showed him the right end. He managed a whiff or two, but he did not fancy it. He seemed very much pleased with the pipe, which he kept. I then made him understand that I wanted water. He pointed the same course that I was steering. In a short time another made his appearance in the distance. By a little persuasion from the old fellow, he was induced to come up, and in a short time became very talkative, and very anxious to show us the water. In a few minutes a third made his appearance, and came up. He was the youngest—a stout, able-bodied fellow, about twenty-four years old. The others were much older, but were very powerful men, and all three in excellent condition. The women did not come up, but remained in the flat. I expected they were going to take us to some springs, and was disappointed when they showed us some rain water in a deep hole. They were quite surprised to see our horses drink it all. They would go no further with us, nor show us any more, and, in a short time after, left us. We struck our outward tracks, and steered for the Elizabeth Springs, where we arrived after dark.

Monday, 27th June, Elizabeth Springs. Gave the horses a half-day, and made the Mount Hamilton Springs in the afternoon.

Tuesday, 28th June, Hamilton Springs. Started for Chambers Creek to my first encampment. Arrived there in the afternoon. Distance, eighteen miles.

Wednesday, 29th June, Chambers Creek. Resting the horses and preparing for a trip down on the west side of Mount North-west, to see if I can find a road and water that way.

Friday, 1st July, Chambers Creek. Started at 8 a.m. on a bearing of 120 degrees. At twenty-four miles camped on a water hole in Gregory Creek, where it comes out of the hills. There are three remarkable peaks north of the water, one in particular having a white face to the east, with a course of black stones on the summit, distant about one mile. The first five miles was over a well-grassed country, with stones on the surface, slightly undulating, with a number of good valleys, very broad, emptying themselves into Gregory Creek. At twenty-two miles crossed the main channel of the creek. It is divided into a number of courses, with some very deep holes in them. When they are filled, they must retain water for a great length of time. There are a great many native encampments all about the creek. The gums are dwarf.

Saturday, 2nd July, Gregory Creek. Started at 10.8. Course, 120 degrees. At three miles, opposite a long permanent water hole, with rushes growing round it. At seven miles, crossed the upper part of the Gregory; eight miles and a half, top of dividing range; thirteen miles, crossed a creek with rain water; fourteen miles, crossed another deep channel. Camped at twenty-three miles, within twelve miles of Termination Hill. The country for ten miles before we halted was very good.

Sunday, 3rd July. Rounded Termination Hill, and arrived at Mr. Glen's station.

JOURNAL OF MR. STUART'S THIRD EXPEDITION (IN THE VICINITY OF LAKE TORRENS) NOVEMBER, 1859, TO JANUARY, 1860

FRIDAY, 4th November, 1859. Started from Chambers Creek for the Emerald Spring. At ten miles crossed nine fresh horse-tracks going eastward; I supposed them to be those of His Excellency the Governor-in-Chief. I have not as yet seen his outward track. Arrived at the spring before sundown.

Saturday, 5th November, Emerald Spring. Started at 7.30 on a course of 340 degrees. At seven miles and a half changed to 38 degrees, for three miles to a high sand hill, from which I could see two salt lagoons, one to the south and the other to the north; examined them, but could find no springs. Next bearing, 18 degrees, to clear the lagoon, two miles and a half sandy, with salt bush and grass. Changed to our first bearing, 340 degrees, for six miles, and then to 350 degrees, for five miles, when we reached the top of a high hill, from which we could see the lake lying to the north of us about three miles distant. Changed to 315 degrees for three miles and a half to get a good view of the lake. This is a large bay; from north-east to north-west there is nothing visible but the dark, deep blue line of the horizon. To the north-north-east there is an island very much resembling Boston Island (Port Lincoln) in shape; to the east of it there is a point of land coming from the mainland. To the north-north-west are, apparently, two small islands. A short distance to the east of the horn of the bay there seems to be much white sand or salt for two or three miles from the beach towards the blue water (on this side of which there is a white line as if it were surf): this again appears at the shores of the island, and also at the horn of the bay. From the south shore to the island the distance is great; I should say about twenty-five miles, but it is very difficult to judge correctly. At three miles and a half camped at sundown, without water.

Sunday, 6th November, Lake Eyre. Got up before daybreak to get the first glimpse of the lake, to see if there is any land on the horizon, and, with a powerful telescope, can see none. It has the same appearance as I described last night. I watched it for some time after sunrise, and it still continued the same. After breakfast went to examine the shore: course north, two miles and a half; found it to be caked with salt, with ironstone and lime gravel. When flooded, at about fifty yards from the hard beach, the water will be about three feet deep. I tried to ride to the water, but found it too soft, so I dismounted and tried it on foot. At about a quarter of a mile I came upon a number of small fish, all dried and caked in salt; they seem to have been left on the receding of the waters, or driven on shore by a heavy storm; they were scattered over a surface of twelve yards in breadth all along the shore; very few, especially of the larger ones, were perfect. I succeeded in obtaining three as nearly perfect as possible; one measured eight inches by three, one six inches by two and a half, and another five inches by two. They resemble the bream. I should think this a sufficient proof of the depth of the water. I then proceeded towards the water, but the ground became soft, and the clay was so very tenacious and my feet so heavy, that it was with difficulty I could move them, and so I was obliged to return. The salt is about three inches thick, and underneath it is clay. I would have tried it in some other places, but as my horses were without water (and as I intend to visit this place again), I think it more prudent to search for water for them, and, if I cannot find any, to return to the camp. Started on a south course to examine the country for springs. At six miles found we were running parallel to sand ridges, and no chance of water. Changed to 160 degrees, crossed a number of sand ridges, but no water, except a little rain water that we found in a hole. Proceeded to the camp, and arrived there about sundown.

Monday, 7th November, Emerald Springs. Finding that the weevil is at work with my dried beef, I must remain to-day and put it to rights. Prepared a package with the fish, etc., to be left for Mr. Barker when he comes here, to be sent to town. There are fish in this spring about three inches long. We have also found a cold-water spring among the warm ones.

Tuesday, 8th November, Emerald Springs. Not being satisfied about one of the lagoons I saw yesterday, I have sent Kekwick and Muller to see if there are any springs, while I and the others proceed to the Beresford Springs; they are to overtake me. Arrived at the springs at 3 p.m. We could find no fresh water on our way, but plenty of salt and brackish in the creek which we first struck at six miles from the Emerald Springs. Sundown: the two men have not come up; they must have found something to detain them; they had only to do about eight miles more than I had. I expect they will arrive during the night.

Wednesday, 9th November, Beresford Springs. No signs of the two men; they must have stopped at some water during the night. It is very tiresome to be delayed in this way: what can they be about? At 12 noon they arrived; they had passed my tracks and gone on to Mount Hugh instead of coming on here. I will give their horses an hour's rest and go on to the Strangway Springs. The Paisley Ponds are dry, but there is salt and brackish water three miles lower down the creek. Started at 2 p.m., and at 5 p.m. arrived at the springs, which are about ten miles from the Beresford. They are upon a high hill about one hundred feet above the level of the plains; there are a great number of them, and abundance of water, but very much impregnated with salt and soda. My eyes are very bad.

Thursday, 10th November, Strangway Springs. Suffering very much from bad eyes and the effects of the water of these springs; cannot help it, but must go and examine the country to north-west and west. Sent Muller to the east in search of springs, with instructions to strike my former tracks and examine all the country between. Started at 7 a.m. with one man, on a course of 315 degrees, and at one mile crossed a salt creek with water; at three miles the sand hills commenced, crossing our course at right angles. At 2 p.m. struck a large lagoon (salt) about two miles broad and five miles long, running north-east and south-west, narrowing at the ends; distance, fourteen miles; tried to cross it but found it too boggy; rounded it on the south-west point, where we discovered a spring; no surface water, but soft, and the same all round for about two acres square, covered with grass reeds of a very dark colour and very thick, showing the presence of water underneath. Proceeded round the lagoon to a high hill, which seemed to have reeds upon the top of it; after a good deal of bogging and crossing the bends of the lagoon, we arrived at the hill, and found it to be very remarkable. Its colour is dark-green from the reeds and rushes and water-grass which cover it. It is upwards of one hundred feet high, the lower part red sand; but a little higher up is a course of limestone. On the top is a black soil, sand and clay, through and over which the water trickles, and then filters through the sand into the lagoon. Where the water is, on the top, it is upwards of one hundred feet long. Immense numbers of tracks of emus and wild dogs, also some native tracks, all fresh. On the north-west side there is one solitary gum-tree, and about half a mile in the same direction is another bed of reeds, and a spring with water in it. All the banks round the lagoon are of a spongy nature. I am very glad I have found this; it will be another day's stage with water nearer to the Spring of Hope. We can now make that in one day, if we can get an early start. By the discovery of springs on this trip, the road can now be travelled to the furthest water that I saw on my last trip from Adelaide, and not be a night without water for the horses. The country to the south and south-east of the last springs (which I have named the William Springs, after the youngest son of John Chambers, Esquire), is sand hills and valleys, rich in grass and other food for cattle. Thence I proceeded to hill bearing 10 degrees south of north, distant three miles, from the top of which I could see no rising ground to the westward, nothing but sand hills. Changed my course to south, to a white place under some stony hills; at ten miles reached it, and found it to be a salt creek, but no springs. The last ten miles were through hills not so high as those I crossed on my way out, but more broken, with plenty of feed. It is my intention to push for the Strangway Springs tonight, so as to get an early start in the morning. Arrived at 10 p.m., found that one of the horses had not been seen all day; something always does go wrong when I am away; I shall have to make a search for him in the morning. My eyes very bad from the effects of the glare of the sun on the sand hills, and the heat reflected from them, and that everlasting torment, the flies.

Friday, 11th November, Strangway Springs. My eyes so bad I cannot see; unable to go myself in search of the missing horse; despatched two of the men at daybreak to circuit the spring, and cut her tracks if she has left them. They have returned, but can see no tracks leaving the spring; she must be concealed among the reeds; sent three men to examine them. They found her at 1 p.m. Started at 2 p.m., and arrived at William Springs at sundown. Distance, fourteen miles. By keeping a little more to the east, the sand hills can nearly be avoided, and a good road over stony country, with good feed, can be had to this spring.

Saturday, 12th November, William Springs. Very unwell, unable to move to-day; I am almost blind and suffering greatly from the effects of the water at Strangway Springs. As I wished to examine round this spring, I remained here to-day; and, as I could not go myself, sent two of the men in different directions. At sundown they returned, and reported that there are no springs for ten miles distant from east-south-east to north. To the east about three miles there is another lagoon resembling this one, but not so large, and no springs; plenty of grass about a mile from the lagoon. Saw two natives at a distance, but could not get near them.

Sunday, 13th November, William Springs. I feel a little better to-day, but suffer very much from the eyes. I hope I shall be able to travel to-morrow, for it is misery to remain in camp in the hot weather. Latitude, 28 degrees 57 minutes 24 seconds. Variation, 4 degrees 47 minutes east.

Monday, 14th November, William Springs. Started on a course of 317 degrees for the Hope Springs, and arrived at 5 p.m. I kept to the west in order to see what the country was in that direction, in the hope of finding some more springs. At twenty-one miles crossed the Douglas, coming from north-north-west; the country from it to the north-west and north looked quite white with quartz, and showed signs of being auriferous. From the Douglas to north-west the feed was not quite so plentiful, salt bush with grass, the salt bush predominating; but as we approached the Spring of Hope it improved, and became good as we neared the creek. Distance, thirty miles.

Tuesday, 15th November, Spring of Hope. The spring is still good, yielding a plentiful supply of water. Sent one of the men to the east and south-east to examine some white patches of country that I saw on our journey up here, while I, with one man and two days' provisions, started south-west to a high and prominent hill in the range. At 11 a.m. arrived at the top, from which I had a good view of the country all round. It is a table-topped hill, standing on high table land, which is intersected with numerous small watercourses, flowing towards the Douglas on the south and west sides of the mount, which I have named Mount Anna. It is compound of ironstone, quartz, granite, and a chalky substance, also an immense quantity of conglomerate quartz and ironstone, which has the appearance of having been run together in a smelting works. There are also numerous courses of slate of different descriptions and colours; the quartz, which exists in white patches, predominates, and gives the country the appearance of numerous springs. These patches have deceived me two or three times to-day. At twenty miles the sand hills begin again; the country being rather poor, with a number of isolated hills, and also some white chalky cliffs of twenty feet high and upwards. No water nor appearance of any to the west for a considerable distance. Changed to the north-west to look at some more white country. I am again disappointed; it turns out to be quartz with low chalky cliffs, and a large quantity of igneous stone. Country the same, with salt bush and a little grass in places. I can see no inducement for me to go further, so I shall return to the camp. Arrived after dark. My eyes are still very bad, and I suffer dreadfully from them. To-day has been hot, and the reflection from the white quartz and the heated stones was almost insufferable: what a relief it was when the sun went down! Distance, forty-five miles.

Wednesday, 16th November, Spring of Hope. Still very ill, and unable to go out myself. Sent Muller to examine the creek nearer Mount Margaret for water; if he finds any near the mount, I shall move there, as it will be nearer, for building the cone of stones on the top of the mount, than Hawker Springs. Shod our horses, and built a small cone of stones on a reef of rocks that runs along the top of a hill about half a mile west-north-west from the spring, to which it will act as a land mark. Muller has returned, and reports having found water in the other creek, about five miles north-north-west from this; the water is in the centre of the creek, in three or four holes, some of which are brackish, but one of them is very good. A number of natives were camped about it, but took to flight the moment they saw him; he tried to induce them to come near him, but they would not; they appeared to be very much frightened, and climbed up the cliffs to get out of his way. Plenty of feed between the two waters; through the hills there is an abundance. I find the water discovered to-day (which I have named The George Creek, after G. Davenport, Esquire), will be of no advantage to me when building the cone of stones; I shall therefore move to the Hawker Springs to-morrow.

Thursday, 17th November, Spring of Hope. Arrived at the Hawker Springs at noon, and commenced the survey. Springs still good; some of them at this point will require to be opened. We have opened one, and the water is beautiful. Immense quantities of reeds and rushes. Built a cone of stones on the hill at the westernmost spring.

Friday, 18th November, Hawker Springs. Building a cone of stones on the top of Mount Margaret, and making other preparations for the survey. To-day very hot, wind south-east; a great deal of lightning to the south. Obtained bearings of the following points from the hill at Hawker Springs—namely, Mount Margaret, Mount Younghusband, hill at Parry Springs, Mount Charles, and Mount Stevenson.

Saturday, 19th November, Hawker Springs. Sent the party on to Fanny Springs, where I intend to lay down my base-line. Went with Kekwick to the top of Mount Margaret. This hill is composed of grey and red granite, quartz, and ironstone; on the lower hill is a blue and brown stratum. I then proceeded to examine the creeks running to the east; in following one of them down we came upon another spring of water, running and very good. The creek is bounded on both sides for about a mile by nearly perpendicular cliffs, which appeared to get much lower and broken to the west. It is situated about one mile north of Mount Margaret, and runs into the Hawker Springs valley. Could see no more higher up. Followed the creek down to the opening. Proceeded about half a mile, entered another gorge, and rode up it about three-quarters of a mile; came upon another spring, running also, water excellent. Numerous native camps in the creek. Country the same as in the other creek; cliffs slate and not so high, but more broken, with watercourses between them, through which cattle could find their way to the tops of the hills, where there appears to be plenty of grass; there is also an abundance at the mouth of the gorge and on the plains. This creek also runs into the valley of the Hawker Springs. Distance from Mount Margaret, two miles and a half, 8 degrees east of north. As it was getting towards sunset I found I must make for the camp, which was about twelve miles off. Arrived after dark. Springs still as good as when I first saw them. Very tired, having had a very long day of it.

Sunday, 20th November, Fanny Springs. Got up at daybreak, and went to the top of Mount Charles, on which I had ordered the men to build a cone of stones after their arrival here yesterday. On my return to the camp the men informed me that Smith had absconded during the night. He generally made a practice of sleeping some little distance from the others, when I did not see him lie down; I had checked him for it several times. It did not appear that he had gone to sleep, but waited an opportunity to steal away, taking with him the mare which he used to ride, and harness, etc., also some provisions. As I had started very early to walk to Mount Charles, his absence was not observed until some time after I had left, and being detained some hours on the top of the hill, in consequence of the atmosphere being so thick that I could not obtain my observations, it was 7 a.m. before I heard of his departure. That moment I sent Kekwick for my own horse (he being the swiftest), and ordered him to saddle, mount, pursue, overtake, and bring Smith back; but during the time he was preparing, I had time to think the matter over, and decided upon not following him, as it would only knock up my horse and detain me three or four days. Smith must have started about midnight, for I was up taking observations from 12.30 a.m. until daybreak, and neither saw nor heard any one during that time. I could ill afford to lose the time in pursuing him, situated as I was in the midst of my survey, and he being a lazy, insolent, good-for-nothing man, and, worse than all, an incorrigible liar, I could place no dependence upon him. We are better without him; he has been a very great annoyance and trouble to me from the beginning throughout the journey. What could have caused him to take such a step I am at a loss to imagine; he has had no cause to complain of bad treatment or anything of that sort; he never mentioned such a thing to the other men, nor was he heard to complain of anything. Such conduct on an expedition like ours deserves the severest punishment: there is no knowing what fatal consequences may follow such a cowardly action. Had he not stolen the mare, I should have cared little about his running away, but I am short of riding horses and have a great deal for them to do during the time I am surveying and examining the country. The vagabond went off just as the heavy work was beginning, and it was principally for that work that I engaged him. He put on a pair of new boots, leaving those he had been wearing, evidently intending to push the mare as far as she would go, expecting he would be pursued, and then leave her and walk the rest. I expect, when he reaches the settled districts, he will tell some abominable lie about the matter. If such conduct is not severely dealt with, no confidence can be placed in any man engaged in future expeditions.

Monday, 21st November, Fanny Springs. Kekwick and I commenced chaining the base-line from the top of Mount Charles, bearing 131 degrees. Distance chained, four miles thirty chains. I ordered H. Strong to come to me with two horses, which he did about 1.30 p.m.; we had finished the line, and were waiting for him. I had seen some country that looked very much like springs, to the north-east, a mile or so from the line; went to examine it, and found some splendid springs—one in particular is a very large fountain, about twenty yards in diameter, quite circular and apparently very deep, from which there is running a large stream of water of the very finest description; it is one of the largest reservoirs I have yet seen, three times the size of the one at the Hamilton Springs, with abundance of water for any amount of cattle; the water is running a mile below it.

Tuesday, 22nd November, Fanny Springs. Engaged chaining the base-line to north-west. Saw some more springs a mile or two to the east; too tired to examine them to-day. It is dreadfully hot. Returned to the camp at sundown.

Wednesday, 23rd November, Fanny Springs. Finished the remaining part of base-line. The line is ten miles and forty chains long, crossing the top of Mount Charles.

Thursday, 24th November, Fanny Springs. Fixing the angles of runs. Found another batch of springs close to north-west boundary of large run, covering four or five acres of ground, with an immense quantity of reeds; they are not so active as the others. The ground round about is very soft, and the water is most excellent. After fixing the north-east corner, I proceeded to examine the country beyond the boundaries of the runs in search of springs. Having gone several miles north, I saw the appearance of a lagoon north-east, for which I started, but on my arrival found no springs round it. Still continued on the same course for a considerable distance further to a high sand hill, from which we could see the Neale winding through a broad valley. One part of the creek being much greener than the other, I went to examine it, and found the green appearance to be caused by fresh gum-trees, young saplings, rushes, and other fresh-water plants and bushes. The creek spreads over the plain in numerous channels, four miles wide, but the main channel has only gum-trees, with a chain of water holes, some salt, some brackish. By scratching on the bank where the rushes were growing we got some beautiful water in the gravel, a few inches below the surface. There was plenty of feed, and the wild currant, or rather grape, grew in great abundance, and was very superior to any I had tasted before. There were two kinds; one grew upon a dark-green bush, and had a tart and saltish taste, the other grew upon a bush of a much lighter colour, the fruit round and plump and much superior to the former; in taste it very much resembled some species of dark grape, only a little more acid. From this I went in a north-east direction to a mound I had seen on my former journey, and found it to be hot springs with a large stream of warm water flowing from them nearly as large as the Emerald Springs, and, as it seemed to me, warmer. It was a very hot day, and I had been riding fast. It was as much as I could bear to keep my hand in the spring for a few minutes, six inches below the surface. I put in a staff about four feet long, but could find no bottom—nothing but very soft mud; the staff came up quite hot. It is a very remarkable hill. From the west side it would be taken for a very high sand hill with scrub growing on it—in fact it is so. The springs are not seen until the top is reached. From them all the east side is covered with green reeds to the base of the hill. The hot springs are near the top, and cold ones on one side to the south; some at the bottom and some half-way up. There is a large lagoon to the east, which I will examine when I move the party up to this, for I have no time to-day. Returned towards the camp and fixed the north-west corner of the second run; I am obliged to drive pickets into the ground to show them. I would have built cones of stones, but could get none large enough to do it with. Arrived at the camp very late; fourteen hours on horseback.

Friday, 25th November, Fanny Springs. Started shortly after sunrise to mark the other two corners of the two runs. On approaching the south-west angle of the second run (Parry Spring run), I discovered three other springs close to the boundary of the first run. Two of them are outside, and one inside, or rather on the boundary. The latter is a large spring, having seven streams of water coming from it, one large, the others smaller. The other two have abundance of water, covered with reeds. Proceeded and marked the other corners, but, having no stones, was obliged to put down pickets. Returned to camp, keeping outside the south boundary in search of springs, but found none. Crossed over table land, salt bush and grass, with stones on the surface. Arrived at the camp a little before sundown.

Saturday, 26th November, Fanny Springs. Started for Parry Springs. In the evening commenced putting up a cone of stones on the northernmost hill. The day was excessively hot. One great thing here is that the nights are very cool, so that we are obliged to have a good fire on all night. We have had one or two warm nights since I have been out this time. I suppose the reason must be that a large body of water exists in the lake not far distant from us, the wind coming from north-east. From north-west to south-south-east the winds are generally cool. It is so cold in the morning that the men are wearing their top-coats; the day does not get hot until the sun is a considerable height.

Sunday, 27th November, Parry Springs. Cold wind this morning from the east. In the afternoon the sky became overcast, the clouds coming from the south-east.

Monday, 28th November, Parry Springs. Building a cone of stones on the northernmost of the hills, fixing the south-east corner of run Number 2, and moving to the hot springs. Arrived at sundown. Saw a number of holes where the natives had been digging for water. Cleaned out one, and found water at two feet from the surface, above the water in the creek. It is very good. On examining this spring, I find there is a great deal more water coming from it than from the Emerald Springs. The hot springs are on the top of the sand hill, and the cold ones at the foot. There are large quantities of the wild grape growing here, both red and white. They are very good indeed, and, if cultivated, would, I think, become a very nice fruit.

Tuesday, 29th November, Primrose springs. Surveying run. Sent Muller to the north to a distant range, and Strong to the north-east to look for springs. Towards evening both returned without being successful. They passed over plenty of good feeding country, but the range is high and stony, with very little grass, only salt bush. It is a continuation of Hanson range, all table land.

Wednesday, 30th November, Primrose springs. Surveying, etc. North-east corner of run Number 2 is about two miles west of the Neale. I scratched a few inches deep from the surface in the gravel, and found very good water. The wild grape is in abundance here, and grows as large as the cultivated one. I have obtained some choice seeds.

Thursday, 1st December, Primrose springs. At daybreak started with Kekwick to find the lake on an easterly course, keeping to south of east, to avoid a soft lagoon. Travelled over a fair salt-bush and grass country, with stones on the surface. In places the grass is abundant, though dry. At seven miles the sand hills commenced; they are low, with broad valleys between, covered with stone.

On the sand hills there was plenty of grass, and numerous native and emu tracks going towards the Neale, which is to the south of us. At fourteen miles struck a gum creek with salt water. Searched for springs, but could find none with fresh-water. Continued on a course east over sand hills and stony plain, and at twenty miles crossed the Neale. It is very broad, with numerous channels. In the main one there was plenty of water, but it was very brackish. We scratched a hole on the bank about two feet from the salt water, and found plenty of good water at six inches from the surface, of which our horses drank very readily. This seems to be the mode in which the natives obtain good water in a dry season like this. The emus and other birds also adopt the same plan. An immense quantity of water must come down this creek at times. The drift stuff was upwards of thirteen feet high in the gum-trees. A number of native tracks all about the creek, quite fresh, but we could not see any one. After giving our horses as much water as they would drink, we crossed the creek, which now runs north, and proceeded, still on our easterly course, over stony plains for four miles, then over sand hills, which continued to the lake, which we struck at thirty-five miles. The atmosphere is so thick, it is impossible to say what it is like to-night. Camped without water under a high sand hill, so that I may have a good view of the lake in the morning. I like not the appearance of it to-night; I am afraid we are going to lose it.

Friday, 2nd December, Lake Torrens. Got up at the first peep of day and ascended the sand hill. I fear my conjecture of last night is too true. I can see a small dark line of low land all round the horizon. The line of blue water is very small. So ends Lake Torrens! Started on a course of 30 degrees west of north to where the Neale empties itself into the lake. At seven miles struck it; found plenty of water, but very salt, with pelicans and other water-birds upon it. Traversed the creek to the south-west in search of water for the horses. At five miles came upon a number of water-bushes growing on the banks of a large brackish water hole. Scraped a hole about two feet from the bad water, and got good water six inches from the surface for ourselves and horses. Gave them an hour's rest and started on a west course for the camp, where we arrived at 9.30 p.m. The country was similar to that on our outward route; feed more abundant. At sundown we crossed the broad channel of a creek, with moisture in the centre. Having neither time nor light to examine it to-night, I must do so to-morrow, as I think there must be springs to supply the moisture.

Saturday, 3rd December, Primrose Springs. Sent Kekwick to examine the creek we crossed last night. I cannot go myself, for my eyes are so very bad I can scarcely see anything. This is the first time I have had such a long continuance of this complaint. I am trying every remedy I can imagine, but each seems to have very little or no effect. At sundown Kekwick returned, and reported having found the springs which supply the creek, but they are salter than the sea, or the strongest brine that ever was made. He brought in a fine sample of crystal of salt, which he got from under the water, attached to the branch of a bush which had blown into it. The creek is the upper part of the first gum creek crossed yesterday, and flows into the Neale, which accounts for the water being so salt at the mouth of it. No fresh-water springs to be seen round about.

Sunday, 4th December, Primrose Springs. Examining the Neale for fresh-water springs. The water holes are abundant, but all more or less brackish; plenty of rushes on the banks, where fresh water can be had by scratching a little below the surface. I have not the least doubt but there will be plenty of fresh water on the surface for a long time after the creek comes down and sweeps all the soda and salt into the lake. It is the rapid evaporation that causes it to be so brackish, and I should think the consumption by stock would make a great improvement in it; there would not be so much of it exposed to the sun, and the evaporation would be much less. After considering the matter of having seen the northern boundary of Lake Torrens, I am inclined to think I have been in error. What I have taken for the lake may have been a large lagoon, which receives the waters of the Neale before going into the large lake: I must examine it again. After my surveys are completed, I shall move my party down the creek to where we found the good water, and from there see what it really is. I cannot bring my mind to think it is the northern boundary of the lake.

Monday, 5th December, Primrose Springs. Moved the party down to the South Parry Springs. My eyes are still very bad.

Tuesday, 6th December, South Parry Springs. Shortly after daybreak started for Louden Springs, taking different courses, in search of more springs, but can find none. Examined the George Creek, where the small run is to be laid off; found some good water by scratching in the creek, where there are plenty of rushes. A little before sundown we arrived at the springs. I did not observe before that the higher springs on the top of the hill are warm, but not nearly so hot as the others; the lower ones are cold. Some other party has been here; we have seen their fresh tracks and the place where they have camped; they seem to have been wandering about a good deal before they found these springs.

Wednesday, 7th December, Louden Springs. Went to the top of Mount Stevenson, built a cone of stones, and obtained bearings to fix it. No appearance of any springs to the east of this, nor of the lake.

Thursday, 8th December, Louden Springs. Surveying and building trigonometrical station on a light-coloured hill to the south of this. My eyes very bad; can scarcely see; can do nothing.

Friday, 9th December, Louden Springs. Nearly blind; dreadful pain; can do nothing to-day; no sleep last night.

Saturday, 10th December, Louden Springs. All yesterday the wind was hot and strong from west and north-west; heavy clouds from south and south-west. In the evening the wind changed to south. This morning still the same; heavy clouds from same direction. My eyes are a little better, so that I shall be able to do something. The sky being overcast I shall put up some of the corners of this run.

Sunday, 11th December, Louden Springs. Still cloudy, but no rain.

Monday, 12th December, Louden Springs. Still very cloudy; wind south; heavy clouds to north-west; no rain. Finishing the east boundary of Number 3 run. Can find no more springs in or about this run. At sundown still very cloudy, but no rain.

Tuesday, 13th December, Louden Springs. Started at 7.15 a.m. to find the lake on an east course. The horses being a long distance off, it was late before they came up. At nine miles crossed the gum creek running north, spread out in a broad valley into numerous courses rich in food for cattle. At twelve miles sand hills commenced, and continued to the shores of the lake, with broad stony plains between, and plenty of grass. At twenty miles crossed the Douglas, running north through sand hills in a broad valley divided into numerous courses, with dwarf gum-trees, mallee, tea-tree, and numerous other bushes; the bed sandy, and no water. At thirty-five miles struck the lake where the Douglas joins it. The country travelled over to-day has been stony plain (undulating), and low sand hills, with abundance of feed, but no water. There is some water at the mouth of the Douglas, but it is salter than the sea. The water in the lake seems to be a long distance off, but the mirage is so very strong that I can form no opinion of it to-night. This seems also a bay I have got into. There is a point of land to the south bearing 25 degrees east of south, and the other bearing 25 degrees east of north. Searched about for water, but could find none. Camped in the creek without any. The country at this part is very low, and nearly on a level with the lake. The only sand hill I shall be able to get a view from is not above thirty feet high. At sundown I got on the top of the sand hill, but could see nothing distinctly; must wait until morning. This creek seems to be very little frequented by natives; can see very few tracks and no worleys.

Wednesday, 14th December, Lake Torrens. At the first dawn of day I got to the top of the hill, and remained there some time after sunrise. To the south-east there is the appearance of a point of land, which I suppose to be the island which I saw when I first struck the lake. There is the appearance of water between. A little more to the eastward I can see nothing but horizon. To the east there is again the appearance of very low distant land—a mere dark line when seen through a powerful telescope. To the north of that there is nothing visible but the horizon, with a blue and white streak between. To the north-north-east beyond the point, a little low land is to be seen running out from the point, with water in the far distance. Rode down to the beach to see what that was composed of; found it to be sand, mud and gravel; firm ground next the shore. Tried a little distance with the horses, but found it too soft to proceed with them. I then dismounted, and tried it on foot, but could only get about two miles; it became so soft, that I was sinking to the ankles, and the clay was so very tenacious that it completely tired me before I got back to the horses. The quantity of salt was not so great here as at the first place I examined. What I thought was a point of land bearing north-north-east turns out to be an island, which I can see from here. The point of the bay is north from where I took the bearings. Between the island and the point I can see nothing but horizon; too low to see any water. Traced the creek up for seven miles in search of water or springs, but could see none, nor any indications. Had breakfast, and started on a course of 20 degrees north of west in search of water or springs. Crossed the Davenport and ascended a low range, but still could not see any indications of water; the country similar to that passed over yesterday. Changed my bearing towards the camp, and arrived there a little before sundown. The horses were very thirsty, and drank an awful quantity of water, but being hot it will do them no harm. It is remarkable that to east of the hot springs I can find no others. This is the third time I have tried it, and been unsuccessful. I am almost afraid that the next time I try the lake I shall not find the north boundary of it. Where can all this water drain to? It is a mystery.

Thursday, 15th December, Louden Springs. Surveyed run Number 4, and sent Kekwick to correct observations from Mount Stevenson.

Friday, 16th December, Louden Springs. Finished Number 4 run. To-day we have discovered a large fresh-water hole in a creek joining the George and coming from the south-west. The water seems to be permanent; it is half a mile long and seems to be deep. On the banks a number of natives have been encamped; round about their fires were large quantities of the shells of the fresh-water mussel, the fish from which they had been eating: I should think this a very good proof of the water being permanent. After finishing the survey I followed the creek up for a number of miles in search of more water, but could find none. It spread into a number of courses over a large plain, on which there was splendid feed.

Saturday, 17th December, Louden Springs. Started for the springs under Mount Margaret to finish the western boundary of Number 1 run. Arrived towards sundown. Found the creek occupied by natives, who, as soon as they caught sight of us, bolted to the hill and got upon the top of a high cliff, and there remained for some time, having a good view of us. I did everything in my power to induce them to come down to us, but they would not, and beckoned us to be off back the road we came. At night they had fires round us, but at some distance off.

Sunday, 18th December, Mount Margaret. About 9 a.m. the natives made their appearance on the hill, and made signs for us to be off; they were eight in number. I found that we had camped close to a large quantity of acacia seed that they had been preparing when we arrived, but had no time to carry it away before we were on them. One old fellow was very talkative. I went towards them to try and make friends with them, but they all took to the hills. By signs I induced the old fellow to stop, and in a short time got him to come a little nearer. When I came to the steep bank of the creek he made signs for me to come no further. I showed him I had no arms with me, and wished him to come up. I could understand him so far that he wished us to go away, that they might get their seed. I thought it as well not to aggravate them, but to show them that we came as friends; and as I had completed all I had to do here, I moved the camp towards the Freeling Springs, at which they seemed very glad, and made signs for us to come back at sundown. They seemed to be a larger race than those down below; the men are tall and muscular, the females are low in stature and thin. I examined the Mount Margaret range in going along; there are a number of gum creeks coming from the north side which flow into the Neale. We searched them up and down, but could find no water. The number of channels that join them in the range is so great that it would take weeks to examine them minutely for water. We camped in one of them without water, although the country promises well for it.

Monday, 19th December, Gum Creek. Started on a north-west course to examine the country between this and the Mount Younghusband range. We could see no springs until we reached the Blyth, in which there is water, but a little brackish; it will do well for cattle. Rode through the middle of the range, and came upon some horse-tracks, not very old; saw where the party had camped, and a cairn of stones they had erected on the top of one of the hills. Followed their tracks some distance down the gully; they seemed to be going to the Burrow Springs; they appear, however, to have gone back again. Left the tracks, and proceeded to the Freeling Springs. Arrived there in the afternoon. No one has been here since I was, as far as I can see. The country we have passed over yesterday and to-day has been really splendid for feed. The springs continue the same, running in a strong stream and of the finest quality.

Tuesday, 20th December, Freeling Springs. Sent Kekwick and one of the men to examine the goldfield, and to select a place for sinking to-morrow morning. My eyes were so bad that I was unable to go. They returned in the afternoon, bringing with them samples from the quartz reefs, in which there was the appearance of gold. Kekwick said he had not seen such good quartz since he left the diggings in Victoria. There was every indication of gold, and I determined to give the place a good trial before leaving it.

Wednesday, 21st December, Freeling Springs. Commenced digging, but found the rocks too near. Surface indications were very slight here, but I found another place which seemed to promise better, so began sinking there, and at four feet came upon some large boulders, round which was very good-looking stuff for washing; took some of it to camp and washed it. No gold, but good indications; a quantity of black sand and emery, also other good signs. I shall continue the hole, and see what is in the bottom. Thunderstorm this afternoon; south-west hot wind.

Thursday, 22nd December, Freeling Springs. Occupied in sinking, but made little progress in consequence of the stones being so large, and the want of proper tools, crowbar, etc. Washed some more stuff from round about the boulders; the produce same as yesterday; no gold.

Friday, December 23rd, Freeling Springs. Found that we could do nothing with the stones with the tools we have. Examined the country round about, and found another place, which will be commenced to-morrow. Examined a quartz reef which had every indication of gold. I regretted that I had not another man, so that I might be able to examine the country for some distance round. It is necessary to have two men at the camp, which cannot be moved to where we are sinking, as there is no water within two miles. It would not be safe to leave the camp with one man only, and two digging, which is all our strength. Heavy thunderstorm from the south-west, but very little rain. The wind blew my tent in two. At sundown it passed over and cleared up, which I regretted to see, as I expected heavy rains at this season, to enable me to make for the north or north-west.

Saturday, 24th December, Freeling Springs. Sank upwards of six feet through gravel, shingle, stones, and quartz. Wind south-west. Heavy clouds; wind hot.

Sunday, 25th December, Freeling Springs. Wind south; heavy clouds, but no rain; towards evening changed to south-east. Cool.

Monday, 26th December, Freeling Springs. Got to the bottom of the hole; washed the stuff, but no gold. Commenced another hole by the side of the quartz reef, which looks well. In the morning the wind was from the north; at 10 a.m. it suddenly changed to south, and blew a perfect hurricane during the whole day, with heavy clouds; but no rain has fallen.

Tuesday, 27th December, Freeling Springs. The storm continued during the night, until about 3 o'clock this morning, when a few drops of rain fell, but not enough to be of any service to me. Bottomed the hole by the side of the quartz reef: no gold, and I think we shall not be able to sink any more; our tools are getting worn out. For the rest of the day examined the quartz reef, in which there is every appearance of gold; I shall stop the search for it and proceed to the north-east to-morrow, for I think some rain has fallen in that direction, which will enable me to examine the country and see if the lake still continues.

Wednesday, 28th December, Freeling Springs. At 7 a.m. started with Kekwick on a north-east course. At seven miles crossed the Neale, spread over a large grassy plain four miles broad, and ascended a low ridge of table-topped hills, stony, with salt-bush and grassed. Crossed another creek, at twenty miles, with myall and stunted gums running over a plain in numerous courses. Plenty of grass but no water. After crossing it, ascended a high peak, which I supposed to be the top of the Hanson range, but found another long table-topped hill, higher, about three miles distant. Ascended that, but could see nothing but more table-topped ranges in the distance. This hill is thirty-five miles from Freeling Springs. Searched for water, and after some time found a little water in one of the creeks, where we camped, it being after sundown. The country from the last creek is not so good, and very stony, so much so that it has lamed my horse, and nearly worn his shoes through at the tips. The horses have drunk all the water, and left none for the morning.

Thursday, 29th December, Hanson Range. Started at 6 a.m. on the same course for another part of the range. At six miles crossed a grassy creek of several channels, with myall and gum, but no water, running to north-east, nearly along our line. At seventeen miles struck the same creek again where it is joined with several others coming from the west-north-west and north. They are spread over a large broad plain covered with grass. Searched for water, but could not find any. Crossed the plains and creeks to a white hill on a north course, and at three miles reached the top; it was a low chalky cliff on the banks of the creek. Changed our course to the first hill I had taken. At seven miles and a half reached the top, which I found very stony. To the north can be seen the points of three other table-topped hills; to the north-east is a large stony plain about ten miles broad, beyond which are high sand hills, and beyond them again, in the far distance, is the luminous appearance of water. Not being on the highest part of the range I proceeded two miles to the south-east to get a better view. From here we could see the creek, winding in a south-east direction, until it reached the lake, which seemed to be about twenty-five miles off. We could not distinctly see it, the mirage and sand hills obscuring our view. My horse having lost both his fore shoes and there being no prospect of water further on, I was reluctantly obliged to return to the camp. We had seen a little rain water on the plain, about seven miles back, at which we decided to camp to-night. Arrived there a little before sundown. My horse very lame, scarcely able to walk along the stones. I am disappointed that there is not more rain water; there seems only to have been a slight shower.

Friday, 30th December, Hanson Range. The horses having strayed some distance, we did not get a start till half-past seven on a course of 323 degrees, to a white hill, to see whether there are any springs on the other side; at one mile and a half reached it, but no springs. Changed our course to a very prominent hill (which I have named Mount Arthur) bearing 275 degrees, and after crossing two small myall creeks and a stony plain with salt bush and grass, at ten miles we struck a large myall and gum creek, coming from the north-west, with some very deep channels. We went some miles up it, but could find no water, the courses for the water being too sandy and gravelly to retain it. At twenty-four miles from the last hill arrived at the summit of Mount Arthur. Changed course to 195 degrees. At ten miles struck another myall and gum creek of the same description as the others, coming from the range; no water. Camped. My horse is nearly done up; I am almost afraid he will not be able to reach the camp to-morrow.

Saturday, 31st December, Hanson Range. Started shortly after daybreak for the camp. At fifteen miles struck another myall and gum creek running into the Neale, and at twenty miles came upon the Neale, which is here three miles broad. Here we saw some recent native tracks and places where fires had been. Arrived at the camp at sundown; horses quite done up. I am sorry that I have been unable to make the lake on this journey; I could have done it, but should most likely have had to leave my horse; he never could have done it. I should then have been obliged to walk the distance back, with all the water dried up. Had I seen the least indication of water on ahead, I should have gone.

Sunday, 1st January, 1860, Freeling Springs. In the afternoon it became cloudy. Wind north. No rain.

Monday, 2nd January, Freeling Springs. Having observed a hill on Saturday that seemed to me a spring, where the Neale comes through the range, I sent Kekwick to examine it, my eyes being too bad. Sent Muller to examine some more quartz reefs in which I think gold exists. Towards sundown he returned with two good specimens, in which I am almost sure there is gold. The reef is twelve feet wide. Shortly after, Kekwick returned and reported springs and two large water holes, and numerous smaller ones, with abundance of permanent water, although slightly brackish. I shall move up and fix their position as soon as I am satisfied with the search for gold.

Tuesday, 3rd January, Freeling Springs. Sent Kekwick and Muller to get some more specimens of quartz. They returned with some in which there were very good indications of gold. It was useless for us to try any more, our tools being of no use. The reefs would require to be blasted. I am afraid there will be no surfacing here. I have done all that lies in my power to get at the gold; but without proper tools we can do nothing, so I shall be obliged to give it up, and start to-morrow for the Neale, to where I sent Kekwick yesterday.

Wednesday, 4th January, Freeling Springs. Started at 8 a.m., and arrived in about thirteen miles. The large water hole is upwards of a mile long, with fully forty yards of water: in width, from bank to bank, it is seventy yards, and upwards of fifteen feet deep; there are large mussel shells on the banks, and plenty of good feed. All round to the south there are low sand hills covered with grass. To the east, in some places, it is stony, with salt-bush, and many broad well-grassed valleys coming from the Mount Kingston range. About a quarter of a mile to the west of the large hole there is a course of springs coming from the Kingston Hills and sand hills, and emptying themselves into the creek. The water is delicious, and plentiful, and, if opened, these springs will yield an ample supply for all purposes. To the west are hills, with the creek coming through them, with water all the way up to where I crossed it in my return last trip. To the north are stony undulating rises, with salt-bush and grass.

Thursday, 5th January, The Neale. Examining the country round to the north and round Mount Harvey. It is poor and stony. On the eastern and northern sides it becomes bad at three miles from the creek. The country in the other directions is good, and will make a first-rate run. This, in connexion with the Mildred and McEllister Springs, will feed any number of cattle.

Friday, January 6th, The Neale. As my rations are now drawing to a close (for we started with provisions only for three months, and have been out now for three months and more), I must sound a retreat to get another supply at Chambers Creek. It was my intention to have sent two men down for them, but I am sorry to say that I have lost confidence in all except Kekwick. I cannot trust them to be sent far, nor dare I leave them with our equipment and horses while Kekwick and I go for the provisions. Situated as I am with them, I must take all the horses down; and if I can get men to replace them at Chambers Creek, I will send them about their business. They have been a constant source of annoyance to me from the very beginning of my journey. The man that I had out with me on my last journey has been the worst of the two. They seem to have made up their minds to do as little as possible, and that in the most slovenly and lazy manner imaginable. They appear to take no interest in the success of the expedition. I have talked to them until I am completely wearied out; indeed, I am surprised that I have endured it so long. Many a one would have discharged them, and sent them back walking to Adelaide; in fact, I had almost made up my mind to do so from here, and to run the chance of getting others at Mr. Barker's. Although they have behaved so badly, and so richly deserve to be punished (for they have taken advantage of me when I could get no others to supply their places), I could not find in my heart to do it. Kekwick is everything I could wish a man to be. He is active, pushing, and persevering. At any time, and at any moment, he is always ready, and takes a pleasure in doing all that lies in his power to forward the expedition. Would that the two others were like him! I should then have no trouble at all. Started at 7 a.m. on my return on a south-east course, and camped at a small spring on the east side of Mount Younghusband. Distance, twenty miles.

Saturday, 7th January, Mount Younghusband. Started at 7 a.m. for the Milne Springs, where I shall remain for a day or two to get all the horses fresh shod, and leave what things I do not require, intending to get them on my return. Arrived there at 11 o'clock. Found the water much the same as it was when I first saw it.

Sunday, 8th January, Milne Springs. Severe attack of lumbago. Sun hot; but cool breeze from south-east.

Monday, 9th January, Milne Springs. Unable to ride, so I was obliged to send Kekwick and one of the men to the westward. This was a great disappointment to me, as I should like to have seen the country myself to have connected it with my farthest north-west point on my first journey. The other man was shoeing the horses. Sun hot. Cool breeze from south-east. Very cold night and morning.

Tuesday, 10th January, Milne Springs. Latitude, 28 degrees 15 minutes 45 seconds. Shoeing horses. Flies a great trouble; can do nothing for them. If they are allowed to remain a moment on the eye, it swells up immediately, and is very painful. Kekwick and the other man returned at 9 o'clock p.m. They report having found two springs, one about nine miles west, and the other about thirty miles, in a large spring country, which they had not time to examine well. Although I am so unwell, I must start to-morrow and see what it is. Judging from their description, there must be something good; and I cannot leave without seeing it, although my provisions are nearly done.

Wednesday, 11th January, Milne Springs. Shortly after sunrise started with Kekwick on a west course for the larger spring country, leaving the near one until our return. At eleven miles and a half crossed the Blyth, coming from the south. At twenty-eight miles reached the spring country. Changed to 150 degrees, and at two miles camped at the spring. The springy place has the appearance of a large salt lagoon, three miles broad and upwards of eight miles long. At the south end of it is a creek with brackish water, and on its banks are the springs, the water from which is very good; they are not running.

Thursday, 12th January, West Springs. There are a number of natives at these springs. We have seen their smoke, and both old and recent tracks. Started on a south course. At four miles and a half came upon a creek, with reeds and brackish water, running a little to the west of north. Traced it down for upwards of a mile and a half. Saw that it ran into the swamp west of where we struck it. Could see no springs upon its banks. Returned to the place where we first struck it, and proceeded a mile on a course of 120 degrees to three large patches of very green reeds, which turned out to be eight feet high. Could find no surface water except what was brackish. The country was moist all round. Thence on the same bearing for two miles. Sent Kekwick to examine some places that looked like springs. They were in the middle of a large salt lagoon, having a crust of limestone, under which the water was, and if broken open, in many places where there was no sign of water, a beautiful supply could be obtained. Changed to 245 degrees, and, at about fifteen miles, changed to 90 degrees, through sand hills. We have seen many places where water can be obtained at a few inches below the surface. Camped at the spring. Feel very ill; can scarcely sit on my horse.

Friday, 13th January, West Springs. Being anxious to see the nature of the country between this and the Mount Margaret range, I started at 6.30 on a course of 110 degrees over occasional sand hills and stony places, with splendid feed. At ten miles and a half reached a stony rise, and changed my course to 76 degrees, for five miles, to a black hill composed of ironstone. Changed to 105 degrees, for one mile, to examine a white place coming down from the range, which had the appearance of springs, but found it to be composed of white quartz. Changed again to 50 degrees to a rough hill, which had also the appearance of springs. At two miles crossed the bed of the Blyth, which takes its rise in the range. No water in it, but loose sand and gravel. At seven miles reached the rough hill, after crossing three small tributaries; was disappointed in not finding water. Ascended the hill, from which we had a good view of the surrounding country, but see no indications of water. I must now make for the second spring found by my men three days ago. Course north, over stony hills and table land, in which I crossed my former tracks going to the Freeling Springs. Arrived at the spring at 7.30 p.m. All of us, men and horses, very tired.

Saturday, 14th January, Springs South of Mount Younghusband. Examined the spring, and found it to be a very good one; it is situated near the banks of the Blyth, on the same spongy ground that I discovered last time, and which was marked off as a run. Searched about, and found two more good springs. There was plenty of water in the creek, but the dry season had made it brackish. Discovered a spring in one of the creeks that runs east from Mount Margaret. The natives had cleared it out, and the water, which was very good, was about two feet from the surface. In the other two creeks we also found springs which only required opening. I then made for the camp, where I found everything all right.

Sunday, 15th January, Milne Springs. Preparing for a start to-morrow for Chambers Creek, by way of Louden Springs; I must endeavour to find some more springs, for I am not quite satisfied yet about that country. Very much annoyed by the misconduct of the two men I left behind at the camp; they have had the impertinence to open my plan-case, and have so damaged my principal plan with their hot moist hands, that I know not what to do with it. This is not the first time they have done it.

Monday, 16th January, Milne Springs. Started at 7.10 a.m. on a bearing of 138 degrees 30 minutes. At about twenty-two miles struck four other springs, beyond the Messrs. Levi's boundary; from one of them there is a strong stream of water flowing. They are almost completely hidden, and one cannot see them until almost on the top of them. I have taken bearings to fix them, and have named them Kekwick Springs. Five o'clock p.m. Arrived at Louden Springs. Distance, thirty-one miles.

Tuesday, 17th January, Louden Springs. Started shortly after daybreak, on a course of 110 degrees, over as fine a grass country as I have yet travelled over. At sixteen miles crossed the Douglas, running through sand hills covered with grass, but no water, nor any signs of springs. Proceeded in the same direction for eight miles, when we were stopped by a lagoon. Changed my course to south-south-west to a hill that had the appearance of water, but found beyond it another large dry lagoon, on the banks of which we saw the tracks of a single horse crossing the end of the lagoon, and steering for Lake Torrens; they seemed to be about two months old. Can they be the tracks of that infatuated man who left me on the 20th of November? In all probability he has lost my downward track and himself also. They are only about two miles to the east of mine. Camped without water on a sand hill.

Wednesday, 18th January, Sand Hill. Started shortly after daybreak on a south-south-east course, still in search of springs (crossing my outward track of last journey), at a place where I thought it most likely for them, but was unsuccessful. If I could have found one here, I should have gone direct to the Emerald Springs, but the horses would suffer very much if they were to be another night without water; the food is so dry, and the weather so hot, they cannot endure more than two days and one night without it. Changed my course to Strangway Spring. Arrived there at 2.30. Some of the horses very much done up. Camped, and gave them the rest of the day to recruit.

Thursday, 19th January, Strangway Springs. Started for the Beresford Springs. At nine miles and a half arrived there; and, at eight miles beyond, made the Hamilton Springs, where we camped for the night.

Friday, 20th January, Hamilton Springs. Started by way of the Emerald Springs, to see if Mr. Barker's party is there, or if any person had been there and got the parcel, and forwarded it to Mr. Chambers. Arrived at the springs, and found that some one had got it. Mr. B.'s party had gone. Went on to Chambers Creek, and found them there.

Saturday, 21st January, Chambers Creek. Here we found provisions awaiting us, as we expected; but the two men still exhibit a spirit of non-compliance, and refuse to proceed again to the north-west; they are bent upon leaving me and returning to Adelaide although they know that there are no men here to supply their places. They have demanded their wages and a discharge, which, under all the circumstances of the case, and considering how badly they have served me, I feel myself justified in withholding. I shall therefore be compelled to send Kekwick down as far as Mr. Chambers' station with my despatches, etc., and to procure other assistance. This will be a great loss of time and expense, which the wages these men have forfeited by not fulfilling their agreements will ill repay. Here we heard of the man Smith, who, it seems, left the mare, whether dead or alive we know not at present. He was lost for four days without water (according to his own account), and, after various adventures, and picking up sundry trifles from different travelling parties, who relieved him out of compassion, reached the settled districts in a most forlorn condition. Mr. Barker had left his station some three weeks before we arrived.

JOURNAL OF MR. STUART'S FOURTH EXPEDITION—FIXING THE CENTRE OF THE CONTINENT. FROM MARCH TO SEPTEMBER, 1860

FRIDAY, 2nd March, 1860, Chambers Creek. Left the creek for the north-west, with thirteen horses and two men. The grey horse being too weak to travel was left behind. Camped at Hamilton Springs.

Saturday, 3rd March, Mount Hamilton. Camped at the Beresford Springs, where it was evident that the natives, whose camp is a little way from this, had had a fight. There were the remains of a body of a very tall native lying on his back. The skull was broken in three or four places, the flesh nearly all devoured by the crows and native dogs, and both feet and hands were gone. There were three worleys on the rising ground, with waddies, boomerangs, spears, and a number of broken dishes scattered round them. The natives seemed to have run away and left them, or to have been driven away by a hostile tribe. Between two of the worleys we observed a handful of hair, apparently torn from the skull of the dead man, and a handful of emu feathers placed close together, the feathers to the north-west, the hair to the south-east. They were between two pieces of charred wood, which had been extinguished before the feathers and hair were placed there. It seemed to be a mark of some description.

Sunday, 4th March, Beresford Springs. Night and morning cold; day very hot. Wind south-west.

Monday, 5th March, Beresford Springs. Wind changed to the east during the night. Morning very cold. Arrived at the Strangway Springs. Day very hot. Wind variable.

Tuesday, 6th March, Strangway Springs. Very hot during the night. Made William Springs and camped. The day exceedingly hot, wind south-west, in which direction a heavy bank of clouds arose about noon; in the evening there was a great deal of lightning, and apparently much rain falling there, but none came down our way.

Wednesday, 7th March, William Springs. The night very hot and cloudy, with the wind from the west, but without rain. Started for Louden Spa,[13] the first few miles being over low sand rises and broad valleys of light sandy soil, with abundance of dry grass; by keeping a little more to the north-west the sand rises can be avoided. At seven miles we struck a swamp, but could see no springs. On approaching the Douglas the country becomes more stony, and continues so to the Spa, where we camped.

Thursday, 8th March, Louden Spa. Cold wind this morning from the south-east; the clouds are gone. Camped at Hawker Springs.

Friday, 9th March, Hawker Springs. Very cold last night. Wind from the south. During the day it changed to the south-east, and the sun was very hot. Camped at the Milne Springs, and found the articles we had left* there all right

[13] The Louden Springs of the two last expeditions.

(* See last expedition.); the natives had opened the place where we had put them, but had taken nothing.

Saturday, 10th March, Milne Springs. At half past 11 last night it began to rain, and continued doing so nearly all day. Wind south-east.

Sunday, 11th March, Milne Springs. About 10 o'clock last night we were flooded with water, although upon rising ground, and were obliged to move our camp to the top of a small hill. It rained all night and morning, but there are signs of a break in the clouds. During the day it has rained at intervals. The creek is coming down very rapidly, covering all the valley with a sheet of water.

Monday, 12th March, Milne Springs. A few heavy showers during the night, but now there seems a chance of a fine day, which will enable us to get our provisions dried again. The country is so boggy that I cannot proceed to-day, but if it continues fair I shall attempt it to-morrow morning. This rain is a great boon to me, as it will give me both feed and water for my horses, and if it has gone to the north-west it will save me a great deal of time looking for water.

Tuesday, 13th March, Milne Springs. Started for Freeling Springs. The country in some places is very soft, but the travelling is better than I expected. As we approached the Denison ranges the rain did not seem to have been so heavy, but when we came to the Peake, we found it running bank high, and very boggy. Impossible to cross it here, so I shall follow it up in a west-south-west direction. Camped at Freeling Springs.

Wednesday, 14th March, Freeling Springs. Started on a course a little to the south of west, to try and find a crossing-place. At two miles it turned a little to the north of west, but at ten miles it turned to the south-west, and was running very rapidly, about five miles an hour. I was obliged to stop at this point, as I could not cross the creek, the banks being so boggy. I have discovered another spring at eleven miles on the same bearing as the Freeling Springs, but I cannot get to it. From here it has the appearance of being very good; a hill covered with reeds at the top, the creek running round the east side of it. I shall endeavour to cross to-morrow and examine it.

Thursday, 15th March, The Peake. The creek being still impassable, I remained here another day. Yesterday the horse that was carrying my instruments broke away from the man who was leading him, burst the girths, and threw the saddlebags on the ground. The instruments were very much injured, in fact very nearly ruined; the sextant being put out of adjustment, has taken me all day to repair, and I am not sure now whether it is correct or not. It is a great misfortune. Wind north; clouds north-east.

Friday, 16th March, The Peake. Saddled and started to cross the Peake about three miles to the south-west, but had a fearful job in doing so, the banks being so boggy, and the current so strong. The horses could hardly keep on their feet, and most of them were up to their saddle-flaps, and some under water altogether. One poor old fellow we were obliged to leave in it, as he was unable to get out, and we were unable to help him, although we tried for hours. He is of very little use to me, for he has never recovered his trip to Moolloodoo and back. He has had nothing to carry since we started, and seemed to be improving every day. I wish now that I had left him at Chambers Creek along with the grey, but as he looked in better condition, I thought he would mend on the journey, and I intended him to bring the horses in every morning, when we got further out. We have been from 10 a.m. to 3.30 p.m. in getting across, including the time spent in trying to extricate Billy. I cannot proceed further to-day, and have therefore camped on the west side of the springs that we saw from the last encampment, which I named Kekwick Springs. There are six springs. The largest one will require to be opened; the reeds on it are very thick, and from ten to twelve feet in height. We tried again to get the horse on shore, but could not manage it; the more we try to extricate him, the worse he gets. I have left him; I do not think he will survive the night. It is now sundown, and raining heavily; the night looks very black and stormy. Wind from the south-west.

Saturday, 17th March, Kekwick Springs. About 8 o'clock last evening the wind changed to the north-west, and we had some very heavy rain, which lasted the greater part of the night. Early in the morning the wind changed again to the south-east, with occasional showers. At sunrise it looked very stormy. I must be off as soon as possible out of this boggy place. The old horse is still alive, but very weak. The water has lowered during the night. If no more rain falls to the south-west it will soon be dry, when he may have a chance of getting out. I cannot remain longer to assist him; it would only be putting the rest of my horses in danger. I would have remained here to-day to have dried my provisions, but the appearance of the weather will not allow me. They must take their chance. Started on a north-west course for the Neale. At fifteen miles struck it, and changed to the west to a creek coming south from the stony rises. The banks of the Neale are very boggy. The first four miles to-day were along the top of a sandy rise, with swampy flats on each side, with a number of reeds growing in them, also rushes and water-grass. At four miles was a strong rise, but before we arrived at it we had to cross one of the swamps, in which we encountered great difficulty. After many turnings and twistings, and being bogged up to the shoulders, we managed to get through all safe. It was fearfully hard work. For three miles, on the top of a stony rise, the country is poor (stones on the top of gypsum deposit), but after that it gradually improves, and towards the creek it becomes a good salt-bush country. Wind from the south-east; still very cloudy.

Sunday, 18th March, Neale River. Wind south-east; heavy clouds. I observed a bulbous plant growing in this creek resembling the Egyptian arum; it was just springing. I will endeavour to get some of the seed, if I can. I hoped we should have got our provisions dried to-day, but it was so showery we could not get it done. The creek is so boggy that we cannot cross it, and must follow it round to-morrow. A sad accident has happened to my plans. There was a small hole in the case that contains them, which I did not observe, and in crossing the Peake the water gained admittance and completely saturated them; it is a great misfortune. Sundown: still raining; wind same direction.

Monday, 19th March, Neale River. Rained during the night, and looks very stormy this morning. Followed the Neale round to where it goes through the gap in Hanson range; in places it was rather boggy, but good travelling in this wet weather—firmer than I expected. We had much difficulty in crossing some of the side creeks. Camped on the south side of the gap. Wind south-east; cloudy, with little rain.

Tuesday, 20th March, Neale River Gap, Hanson Range. Wind south-east; a few showers during the night. Still no chance of getting my provisions dried. It cleared off about noon, and became a fine day. Followed the Neale round, and camped on one of the side creeks coming from the south of west. Ground still soft. Wind south-east. Saw some smoke in the hills this morning, but no natives. Good country along both sides of the range on the west side of the Neale.

Wednesday, 21st March, Neale River. Beautiful sunshine. Shall remain here to-day, in order to dry my provisions. On examining them I find that a quantity of our dried meat is quite spoiled, which is a great loss—another wet day, and we should have lost the half of it.

Thursday, 22nd March, Side Creek of the Neale River. Wind south-west; clear sky. I intended to have gone north-west from this point, but, in attempting to cross the creek, we found it impassable. My horse got bogged at the first start, and we had some difficulty in getting him out. We were obliged to follow the creek westward for seven miles, where it passes between two high hills connected with the range. We managed at last, with great labour and difficulty, to get across without accident. At this place four creeks join the main one, and spread over a mile in breadth, with upwards of twenty boggy water-courses; water running. It has taken us five hours, from the time we started, to cross it. The principal creek comes from the south-west. I ascended the two hills to get a view of the surrounding country, and I could see the creek coming from a long way off in that direction. At this point the range seems broken or detached into numerous small ranges and isolated hills. I now changed my course to north-west, over table land of a light-brown colour, with stones on the surface; the vegetation was springing all over it and looking beautifully green. At six miles on this course camped on a myall creek. The work for the horses has been so very severe to-day that I have been induced to camp sooner than I intended. Wind south.

Friday, 23rd March, Myall Creek. Wind south. Started on the same course, north-west. At three miles crossed another tributary—gum and myall. The country, before we struck the creek, was good salt-bush country, with a plentiful supply of grass. The soil was of a light-brown colour, gypsum underneath, and stones on the surface, grass and herbs growing all round them. After crossing the creek, which was boggy, we again ascended a low table land of the same description. At ten miles came upon a few low sand rises, about a mile in breadth. We then struck a creek, another tributary, spread over a large plain, very boggy, with here and there patches of quicksand. We had great difficulty in getting over it, but at last succeeded without any mishap. We then entered a thick scrubby country of mulga and other shrubs; the soil now changed to a dark red, covered splendidly with grass. After the first mile the scrub became much thinner; ground slightly undulating. After crossing this good country, at twenty miles we struck a large creek running very rapidly at five miles per hour; breadth of water one hundred feet, with gum-trees on the bank. From bank to bank it was forty-four yards wide. This seemed to be only one of the courses. There were other gum-trees on the opposite side, and apparently other channels. Wind south. A few clouds from the north-west.

Saturday, 24th March, Large Gum-Tree Creek. Found it impossible to cross the Neale here; the banks were too boggy and steep. We therefore followed it round on a west course for three miles, and found that it came a little more from the north. Changed to 290 degrees, after trying in vain to cross the creek at this point. At about four or five miles south-south-west from this point there are two high peaks of a low range. The higher one I have named Mount Ben, and the range Head's Range; its general bearing is north-west to opposite this point; it turns then more to the west. I can see another spur further to the west, trending north-west. At four miles and a half after leaving we found a ford, and got the horses across all safe. I then changed to the north-west again, through a scrubby country—mulga, acacia, hakea, salt bush, and numerous others, with a plentiful supply of grass. The soil is of a red sandy nature, very loose, and does not retain water on the surface. We had great difficulty in getting through, many places being so very thick with dead mulga. We have seen no water since we left the creek. Distance, eighteen miles. I was obliged to camp without water for ourselves. As we crossed the Neale we saw fish in it of a good size, about eight inches long, from which I should say that the water is permanent. I shall have to run to the west to-morrow, for there is no appearance of this scrubby country terminating. I must have a whole day of it.

Sunday, 25th March, Mulga Scrub. I can see no termination, on this course, to this thick scrub. I can scarcely see one hundred yards before me. I shall therefore bear to the west, cut the Neale River, and see what sort of country is in that direction. At ten miles made it; the water still running, but not so rapidly. The gum-trees still existed in its bed, and there were large pools of water on the side courses. We had the same thick scrub to within a quarter of a mile of the creek, where we met a line of red sand hills covered with a spinifex. The range on the south-west side of the creek seemed to terminate here, and become low table land, apparently covered with a thick scrub, the creek coming more from the north. I did not like the appearance of the spinifex, an indication of desert to the westward. Camped on the creek. Wind north-west; heavy clouds from the same direction.

Monday, 26th March, The Neale River West. I am obliged to remain here to-day to repair damages done to the packs and bags, which have been torn all to pieces; it will take the whole of the day to put them in order. We have seen very few signs of natives visiting this part of the country. I shall go north to-morrow and try to get through this scrub. Wind south, sky overcast with heavy clouds; looks very like rain.

Tuesday, 27th March, West Neales. Rained very heavily during the night, and is still doing so, but less copiously. About noon it cleared up a little. I have sent Kekwick to get a notion of the country on the other side of the low range, while I endeavour to obtain an observation of the sun. The range is scrubby, composed of a light-coloured and dark-red conglomerate volcanic rock, easily broken. The view from it is not extensive. At a mile from the creek the sand ceases, and stony ground succeeds up to the range. Feed excellent south-west from the camp. To the eastward rugged hills, apparently with fine open grass and forest lands. Numerous rows of water holes visible. To the south-east, country more open. To the south-south-east and south still the same good country. From south to west the same; hills to the west from five to eight miles distant. View from another hill north-west two miles and a half. The hills on the west still continue towards the north-west, but become lower. Country scrubby, with occasional patches of open grass land. Creek coming in from north-north-west. From north-west to north-north-east mulga scrub. From north-north-east to east low range in the distance, like table land. Too cloudy to take an observation; occasional showers during the day. Wind south-south-east; still looking very black. Repairing my saddles; some of my horses are getting bad backs.

Wednesday, 28th March, West Neale River. Started on a north course to get through the mulga scrub. At ten miles could see the range to the north-east. The scrubby land now became sand hills; I could see no high ground on ahead, the scrub becoming thicker; it seemed to be a country similar to that I passed through on my south-east course (first journey), and I think is a continuation of it. I therefore changed my course to the north-east range, bearing 35 degrees. After five miles through the same description of country, mulga scrub with plenty of grass, we arrived at water, where three creeks join, one from the south-west, one west-north-west, and the other from about north-west. The water was still running in the one from the west-north-west with large long water holes; also water holes in the other two; gum-trees in the creek. I suppose this to be the Frew; excellent feed on the banks of the creek up to the range, which is stony. I ascended the table range in order to have a view of the country round. To this point the range comes from east-south-east, but here it takes a turn to the east of north, all flat-topped and stony, with mulga bushes on the top and sides; the rocks are of a light, flinty nature. At about six miles north the country seems to be open and stony. That country I shall steer for to-morrow. To the north-east is the range, but it seems to drop into low table land; distant about fifteen miles. To the north-west and west is the thick mulga, scrubby country. There are numerous tracks of natives in the different creeks, quite fresh, apparently made to-day. Wind south-east; clouds.

Thursday, 29th March, The Frew. Started on a north course. At one mile, after crossing a stony hill with mulga, we suddenly came upon the creek again; it winds round the hill. Here another branch joins it from the north, the other coming from the east of north. Along the base of the range there were very large water holes in both branches. The natives had evidently camped here last night; their fires were still alight; they seemed only just to have left. From the numerous fires I should think there had been a great number of natives here. All round about in every direction were numerous tracks. We also observed a number of winter habitations on the banks of the creek; also a large native grave, composed of sand, earth, wood, and stones. It was of a circular form, about four feet and a half high, and twenty to twenty-four yards in circumference. The mulga continued for about six miles; but at three miles we again crossed the north branch of the creek, coming now from the north-west. The mulga was not thick except on the top of the rises, where splendid grass was growing all through it. We now came upon the open stony country, with a few mulga creeks. There was a little salt bush, but an immense quantity of green grass, growing about a foot high, which gave to the country a beautiful appearance. It seemed to be the same all round as far as I could see. At fourteen miles we struck the other branch, where it joined, with splendid reaches of water, to the main one, which now came from the west of north, and continued to where our line cut the east branch. This seems to be the place where it takes its rise. Camped for the night. The whole of the country that we have travelled through to-day is the best for grass that I have ever gone through. I have nowhere seen its equal. From the number of natives, from there being winter and summer habitations, and from the native grave, I am led to conclude the water there is permanent. The gum-trees are large. I saw kangaroo-tracks.

Friday, 30th March, Small Branch of The Frew. Course north. At two miles and a half changed to 332 degrees to a distant hill, apparently a range of flat-topped hills. At sixteen miles crossed a large gum creek running to the south of east; it spreads out over a flat between rough hills of half a mile wide. The bed is very sandy; it will not retain water long. On the surface it very much resembles the Douglas, but is broader, and the gum-trees much larger. There were some rushes growing in its bed. I have named it the Ross. We then ascended the low range for which I had been steering. Four miles from the creek it is rough and stony, composed of igneous rock, with scrub, mulga, and plenty of grass quite to the top. To continue this course would lead me again into the mulga scrub, where I do not want to get if I can help it. It is far worse than guiding a vessel at sea; the compass requires to be constantly in hand. I again changed to the north, which appears to be open in the distance. I could see another range of flat-topped hills. After crossing over several small spurs coming from the range, and a number of small creeks, volcanic, and stony, we struck another large gum creek coming from the south of west, and running to the south-east. It was a fine creek. These courses of water spread over a grassy plain a mile wide; the water holes were long and deep, with numerous plants growing on their banks, indicating permanent water. The wild oats on the bank of the creek were four feet high. The country gone over to-day, although stony, was completely covered with grass and salt bush; it was even better than that passed yesterday. Some of the grass resembled the drake, some the wild wheat, and some rye—the same as discovered by Captain Sturt. There is a light shade over the horizon from south-east to north-west, indicating the presence of a lake in that direction. I have named it after my friend Mr. Stevenson. There are small fish in the holes of this creek, and mussel shells, also crabs about two inches by one inch and a half.

Saturday, 31st March, The Stevenson. I am obliged to remain here to-day; my horses require shoeing. The country cuts up the shoes very much.

Sunday, 1st April, The Stevenson. I find to-day that my right eye, from the long continuation of bad eyes, is now become useless to me for taking observations. I now see two suns instead of one, which has led me into an error of a few miles. I trust to goodness my other eye will not become the same; as long as it remains good, I can do. Wind east; cool. Heavy clouds.

Monday, 2nd April, The Stevenson. Started at 8 o'clock; course 355 degrees to distant hills. At six miles we struck a gum creek with water in it, but not permanent. At ten miles we crossed another, running between rugged hills; a little water coming from the west and running east-south-east through a mass of hills. At twelve miles crossed a valley a quarter of a mile broad, through which a gum creek runs, with an immense quantity of drift timber lying on its banks. At twenty miles arrived at the first part of the range, and at twenty-eight miles camped on a gum creek running east and coming from the south of west. The first three miles of to-day's journey were over good country; it then became rather scrubby, with numerous small creeks and valleys running to the east. Plenty of grass and salt-bush, with gravel, ironstone, and lime on the surface. At a mile before we made the rugged creek the ironstone became less, and a hard white stone took its place, and continued to the range, on which it is also found. Gypsum, chalk, ironstone, quartz, and other stones, are the chief materials of which it and the other hills are composed. There are also a few of a hard red sandstone. The range is broken, and running nearly east and west. The country round is slightly undulating; numerous small creeks running to the eastward, with a deal of grass and salt-bush. No water in this creek. Camped without. Wind east.

Tuesday, 3rd April, Gum Creek, South of Range. Ascended the hill at three miles from last night's camp. The country very rough, stony, and scrubby to the base. The view from it is very extensive. I have named it Mount Beddome, after S. Beddome, Esquire, of Adelaide. To the west is another broken range, about fifteen miles distant, of a dark-red colour, running nearly north and south. The country between is apparently open, with patches of scrub. A gum creek comes from the south-west and runs some distance to the north-east; it then turns to the east. In the distant west appears a dense scrub. On a bearing of 330 degrees there is a large isolated table hill, for which I shall shape my course, to see if I can get an entrance that way. To the north are a number of broken hills and peaks with scrub between; they are of every shape and size. To the east another flat-topped range; country between also scrubby; apparently open. Close to the range, distant about twenty miles saw hills in the far distance; to the east another flat-topped small range; between it and the other the creek seems to run. The highest point of it bears 80 degrees, and I have named it Mount Daniel, after Mr. Daniel Kekwick, of Adelaide. From east to south-east the country is open and grassy; low ranges in the distance. Saw some rain water, bearing 30 degrees, to which I will go, and give the horses a drink; they had none last night. Distance, two miles. Obtained an observation of the sun, 118 degrees 17 minutes 30 seconds. At six miles crossed the broad bed of a large gum creek; gravel; no water. At eight miles the red sand hills commence, covered with spinifex; and on the small flats mulga scrub, which continues to the base of the hill. Red loose sand; no water. Distance, twenty miles from Mount Beddome to this hill. The country good, until we get among the spinifex.

Wednesday, 4th April, Mount Humphries. At break of day ascended the mount, which is composed of a soft white coarse sandstone. On the top is a quantity of water-worn quartz, cemented into large masses. The view is much the same as from Mount Beddome, broken ranges all round the horizon, and apparently a dense scrub from south-west to west. It then becomes an open and grassy country, with alternate patches of scrub. I can see a gum creek about two miles distant; I can also see water in it, which the horses have not yet discovered. I shall therefore go in that direction, and give them a drink. To the north and eastward the country appears good. Went to the aforesaid water, to see if there is any that I can depend upon. On my return, wanting to correct my instrument, which met with an accident three or four days ago, by the girths getting under the horse's belly (he bolted and kicked it off), I sent Kekwick to examine the creek that I saw coming from the north. He says there is plenty of water to serve our purpose. The creek is very large, with the finest gum-trees we have yet seen, all sizes and heights. This seems to be a favourite place for the natives to camp, as there are eleven worleys in one encampment. We saw here a number of new parrots, the black cockatoo, and numerous other birds. The creek runs over a space of about two miles, coming from the west; the bed sandy. After leaving it, on a bearing of 329 degrees, for nine miles, we passed over a plain of as fine a country as any man would wish to see—a beautiful red soil covered with grass a foot high; after that it becomes a little sandy. At fifteen miles we got into some sand hills, but the feed was still most abundant. I have not passed through such splendid country since I have been in the colony. I only hope it may continue. The creek I have named the Finke, after William Finke, Esquire, of Adelaide, my sincere and tried friend, and one of the liberal supporters of the different explorations I have had the honour to lead. Wind south-east. Cloudy.

Thursday, April 5th, Good Country. Started on the same course to some hills, through sand hills and spinifex for ten miles. Halted for half an hour to obtain an observation of the sun, 117 degrees 6 minutes. Within the last mile or two we have passed a few patches of shea-oak, growing large, having a very rough and thick bark, nearly black. They have a dismal appearance. The spinifex now ceased, and grass began to take its place as we approached the hills. From the top of the hill the view is limited, except to the south-west, where, in the far distance, is a long range. The country between seems to be scrub, red sand hills, and spinifex. To the west the country is open, but at five miles is intercepted by the point of the range that I am about to cross. To the north-west and east is a mass of flat-topped hills, of every size and shape, running always to the east. Camped on the head of a small gum creek, among the hills, which are composed of the same description of stones as the others. This water hole is three feet deep, and will last a month or so. The native cucumber is growing here.

Friday, 6th April, Small Gum Creek in Range of Hills. Started on the same course, 330 degrees, to a remarkable hill, which has the appearance at this distance of a locomotive engine with its funnel. For three miles the country is very good, but after that high sand hills succeeded, covered with spinifex. At six miles we got to one of the largest gum creeks I have yet seen. It is much the same as the one we saw on the 4th, and the water in it is running. Great difficulty in crossing it, its bed being quicksand. We were nearly across, when I saw a black fellow among the bushes; I pulled up, and called to him. At first he seemed at a loss to know where the sound came from. As soon, however, as he saw the other horses coming up, he took to his heels, and was off like a shot, and we saw no more of him. As far as I can judge, the creek comes from the south-west, but the sand hills are so high, and the large black shea-oak so thick, that I cannot distinguish the creek very well. These trees look so much like gums in the distance; some of them are very large, as also are the gums in the creek. Numerous tracks of blacks all about. It is the upper part of the Finke, and at this point runs through high sand hills (red), covered with spinifex, which it is very difficult to get the horses through. We passed through a few patches of good grassy country. In the sand hills the oak is getting more plentiful. We were three-quarters of an hour in crossing the creek, and obtained an observation of the sun, 116 degrees 26 minutes 15 seconds. We then proceeded on the same course towards the remarkable pillar, through high, heavy sand hills, covered with spinifex, and, at twelve miles from last night's camp, arrived at it. It is a pillar of sandstone, standing on a hill upwards of one hundred feet high. From the base of the pillar to its top is about one hundred and fifty feet, quite perpendicular; and it is twenty feet wide by ten feet deep, with two small peaks on the top. I have named it Chambers Pillar, in honour of James Chambers, Esquire, who, with William Finke, Esquire, has been my great supporter in all my explorations. To the north and north-east of it are numerous remarkable hills, which have a very striking effect in the landscape; they resemble nothing so much as a number of old castles in ruins; they are standing in the midst of sand hills. Proceeded, still on the same course, through the sand rises, spinifex, and low sandstone hills, at the foot of which we saw some rain water, where I camped. To the south-west are some high hills, through which I think the Finke comes. I would follow it up, but the immense quantity of sand in its bed shows that it comes from a sandy country, which I wish to avoid if I can. Wind south-east. Heavy clouds; very like rain.

Saturday, 7th April, Rain Water under Sandstone Hills. Started on the same course 330 degrees, over low sand rises and spinifex, for six miles. It then became a plain of red soil, with mulga bushes, and for seven miles was as fine a grassed country as any one would wish to look at; it could be cut with a scythe. Dip of the country to the east, sand hills to the west; afterwards it became alternate sand hills and grassy plains, mulga, mallee, and black oak. From the top of one of the sand hills, I can see a range which our line will cut; I shall make to the foot of that to-night, and I expect I shall find a creek with water there. Proceeded through another long plain sloping towards the creek, and covered with grass. At about one mile from the creek we again met with sand hills and spinifex, which continued to it. Arrived and camped; found water. It is very broad, with a sandy bottom, which will not retain water long; beautiful grass on both banks. Wind east, and cool.

Sunday, 8th April, The Hugh Gum Creek. I have named this creek the Hugh, and the range James Range. It is scrubby on this side and is not flat-topped as all the others have been, which indicates a change of country. On the other side the bearing is nearly east and west. Examined the creek, but cannot find sufficient water to depend upon for any length of time; the gum-trees are large. Numerous parrots, black cockatoos, and other birds. Wind east; very cold during the night.

Monday, 9th April, The Hugh Gum Creek. Started for the highest point of the James range. At four miles arrived on the top, through a very thick scrub of mulga; the range is composed of soft red sandstone, long blocks of it lying on the side. To the east, apparently red sand hills, beyond which are seen the tops of other hills to the north-east. On the north-west the view is intercepted by a high, broken range, with two very remarkable bluffs about the centre. I shall direct my course to the east bluff, which is apparently the higher of the two. In the intermediate country are three lower ranges, between which are flats of green grass, and red sand hills. To the west are grassy flats next to the creek; beyond these are seen the tops of distant ranges and broken hills; at about six miles the Hugh seems to turn more to the north, towards a very rough range of red sandstone. We then descended into a grassy flat with a few gum-trees. We have had a very great difficulty in crossing the range, and now I am again stopped by another low range of the same description, which is nearly perpendicular—huge masses of red sandstone on its side, and in the valley a number of old native camps. After following the range three miles, we at last found out a place to cross it. Although this is not half the height of James range, we encountered far more difficulty; the scrub was very dense, a great quantity having withered and fallen down: we could scarcely get the horses to face it. Our course was also intercepted by deep, perpendicular ravines, which we were obliged to round after a great deal of trouble, having our saddlebags torn to pieces, and our skin and clothes in the same predicament. We arrived at the foot nearly naked, and got into open sandy rises and valleys, with mulga and plenty of grass, among which there is some spinifex growing. At sundown, after having gone about eight miles further, we made a large gum creek, in which we found some water; it is very broad, with a sand and gravel bottom. Camped, both men and horses being very tired.

Tuesday, 10th April, Gum Creek, Bend of the Hugh. I find our saddle-bags and harness are so much torn and broken that I cannot proceed until they are repaired. I am compelled with great reluctance to remain here to-day. This creek is running to the west. On ascending a sand hill this morning, I find that it is the Hugh (which seems to drain the sand hills) that we saw to the east from the top of James range. There is another branch between us and the high ranges. At about four miles west it seems to break through the rough range and join the

Hugh. A large number of native encampments here, and rushes are growing in and about the creek: there is plenty of water.

Wednesday, 11th April, Bend of the Hugh. Got the things put pretty well to rights, and started towards the high bluff. I find that my poor little mare, Polly, has got staked in the fetlock-joint, and is nearly dead lame; but I must proceed. At six miles and a half we again crossed the Hugh, and at another mile found it coming through the range, which is a double one. The south range is red sandstone, the next is hard white stone, and also red sandstone, with a few hills of ironstone; a well-grassed valley lying between. The two gorges are rocky, and in some places perpendicular, with some gum-trees growing on the sides. The cucumber plant thrives here in great quantities, and water is abundant. At twelve miles we got through both the gorges of the range, which I have named the Waterhouse Range, after the Honourable the Colonial Secretary. The country between last night's camp and the range is a red sandy soil, with a few sand hills, on which is growing the spinifex, but the valleys between are broad and beautifully grassed. At fifteen miles again crossed the Hugh, coming from the east, with splendid gum-trees of every size lining the banks. The pine was also met with here for the first time. There is a magnificent hole of water here, long and deep, with rushes growing round it. I think it is a spring; the water seems to come from below a large bed of conglomerate quartz. I should say it was permanent. Black cockatoos and other birds abound here, and there are numbers of native tracks all about. I hoped to-day to have gained the top of the bluff, which is still seven or eight miles off, and appears to be so very rough that I anticipate a deal of difficulty in crossing it. I am forced to halt at this bend of the creek, in consequence of the little mare becoming so lame that she is unable to proceed further to-day. Our hands are very bad from being torn by the scrub, and the flies are a perfect torment. Indications of scurvy are beginning to show themselves upon us. Wind west; cool night.

Thursday 12th April, The Hugh. Started for the bluff. At eight miles we again struck the creek coming from the west, and several other gum creeks coming from the range and joining it. We have now entered the lower hills of the range. Again have we travelled through a splendid country for grass, but as we approached the creek it became a little stony. At twelve miles we found a number of springs in the range. Here I obtained an observation of the sun. As we approached near the bluff, our route became very difficult; we could not get up the creek for precipices, and were obliged to turn in every direction. About two miles from where I obtained the observation, we arrived with great difficulty at the foot of the bluff; it has taken us all the afternoon. I expected to have gone to the top of it to-night, but it is too late. It will take half a day, it is so high and rough. We are camped at a good spring, where I have found a very remarkable palm-tree, with light-green fronds ten feet long, having small leaves a quarter of an inch in breadth, and about eight inches in length, and a quarter of an inch apart, growing from each side, and coming to a sharp point. They spread out like the top of the grass-tree, and the fruit has a large kernel about the size of an egg, with a hard shell; the inside has the taste of a cocoa-nut, but when roasted is like a potato. Here we have also the india-rubber tree, the cork-tree, and several new plants. This is the only real range that I have met with since leaving the Flinders range. I have named it the McDonnell Range, after his Excellency the Governor-in-Chief of South Australia, as a token of my gratitude for his kindness to me on many occasions. The east bluff I have named Brinkley Bluff, after Captain Brinkley, of Adelaide, and the west one I have named Hanson Bluff, after the Honourable R. Hanson, of Adelaide. The range is composed of gneiss rock and quartz.

Friday, 13th April, Brinkley Bluff, McDonnell Range. At sunrise I ascended the bluff, which is the most difficult hill I have ever climbed; it took me an hour and a half to reach the top. It is very high, and is composed principally of igneous rock, with a little ironstone, much the same as the ranges down the country. On reaching the top, I was disappointed; the view was not so good as I expected, in consequence of the morning being so very hazy. I have, however, been enabled to decide what course to take. To the south-west the Waterhouse and James ranges seem to join. At west-south-west they are hidden by one of the spurs of the McDonnell range. To the north-west the view is intercepted by another point of this range, on which is a high peak, which I have named Mount Hay, after the Honourable Alexander Hay, the Commissioner of Crown Lands. About five miles to the north are numerous small spurs, beyond which there is an extensive wooded or scrubby plain; and beyond that, in the far distance, is another range, broken by a high conical hill, bearing about west-north-west, to which I will go, after getting through the range. To the north-east is the end of another range coming from the south. On this, which I have named Strangway Range, after the Honourable the Attorney-General, is another high hill. Beyond is a luminous, hazy appearance, as if it proceeded from a large body of water. A little more to the east there are three high hills; the middle one, which I should think is upwards of thirty miles from us, is the highest, and is bluff at both ends; it seems to be connected with Strangway range. To the east is a complete mass of ranges, with the same luminous appearance behind them. I had a terrible job in getting down the bluff; one false step and I should have been dashed to pieces in the abyss below; I was thankful when I arrived safely at the foot. I find that I have taken the wrong creek to get through the bluff. The Hugh still comes in that way, but more to the westward. Started at 10 o'clock; the hills very bad to get over; wind easterly. Camped at sundown on the creek; there is an abundance of water, which apparently is permanent, from the number of rushes growing all about it. The feed is splendid. There are a number of fresh native tracks.

Saturday, 14th April, McDonnell Range. Started at 8 o'clock to follow the creek, as it seems to be the best way of getting through the other ranges; but, as it comes too much from the east, I must leave it, and get through at some of the low hills further down. This we at last contrived to do after a severe struggle. It has taken us the whole day to come about five miles. We are now camped, north of the bluff, at a gorge, in which there is a good spring of water; the creeks now run north from the range.

Sunday, 15th April, The North Gorge of McDonnell Range. I ascended the high hill on the east side of the gorge; the atmosphere being much clearer, I got a better view of the country. To the north-west, between the McDonnell range and the conical hill north-north-west, is a large plain, apparently scrub; no hills on the horizon, but a light shade in the far distance; the conical hill bears 340 degrees from this; it appears to be high. From the foot of this, for about five miles, is an open grassy country, with a few small patches of bushes. A number of gum creeks come from the ranges, and seem to empty themselves in the plains. The country in the ranges is as fine a pastoral hill-country as a man would wish to possess; grass to the top of the hills, and abundance of water through the whole of the ranges. I forgot to mention that the nut we found on the south side of the range is not fit to eat; it caused both men to vomit violently. I ate one, but it had no bad effect on me.

Monday, 16th April, The North Gorge of McDonnell Range. Started at 9 o'clock to cross the scrub for the distant high peak. For five miles the plain was open and well grassed: afterwards it became thick, with mulga bushes and other scrubs. At twenty miles we again encountered the spinifex, which continued until we camped after dark. Distance, thirty miles. Met with no creek or watercourse after leaving the McDonnell ranges.

Tuesday, 17th April, In the Scrub. Got an early start, and continued through the scrub and spinifex on the same course, 340 degrees. At three miles passed a small stony hill, about two miles to the west of our course. At eighteen miles saw to the west two prominent bluff hills, and two or three small ones, about ten miles distant from us. At thirty-two miles crossed a strong rise. There are three reap-hook hills about three miles west, their steep side facing the south. At sundown reached the hills. At two miles passed a small sandy gum creek, the only watercourse we have seen between the two ranges. Followed the range to the north-west till after dark, hoping to find a gum creek coming from the range, but without success; nothing but rocky and sandy watercourses. Camped. The poor horses again without water; I trust that I shall find some for them in the morning; if not, I shall have to return to the McDonnell range. Very little rain seems to have fallen here; the grass is all dried up. The spinifex continues until within a mile of the range. The small gum creek that we passed is running south-west into the scrub.

Wednesday, 18th April, Under the High Peak, Mount Freeling. At daybreak sent Kekwick in search of water, while I ascended the high mount to see if any could be seen from that place. To my great delight I beheld a little in a creek on the other side of the range, bearing 113 degrees, about a mile and a half. I find this is not quite the highest point of the range; there is another hill, still higher, about fifteen miles further to the north-north-west. About two miles off I can see a gum creek looking very green, coming from the range in the direction in which I have sent Kekwick, where I hope he will find water. The country from west to north-east is a mass of hills and broken ranges; to the south-west high broken ranges. To the north-north-east is another hill, with a plain of scrub between. To the south-east scrub, with tops of hills in the far distance. Brinkley Bluff bears 166 degrees and Mount Hay 186 degrees. Returned to the camp, and find to my great satisfaction that Kekwick has discovered some water in the creek about two miles off. I am very glad of it, for I am sure that some of my horses would not have stood the journey back without it. I must not leave this range without endeavouring to find a permanent water, as no rain seems to have fallen to the north of us; everything is so dry, one would think it was the middle of summer. The sun is also very hot, but the nights and mornings are cool. Wind east. Old tracks and native camps about. The range is composed of the same description of rocks as the McDonnell ranges, with rather more quartz than mica. We here found new shrubs and flowers, also a small brown pigeon with a crest. I have built a small cone of stones on the peak, and named it Mount Freeling, after the Honourable Colonel Freeling, Surveyor-General. The range I have called the Reynolds, after the Honourable Thomas Reynolds, the Treasurer.

Thursday, 19th April, Mount Hugh. The horses separated during the night, and were not found until after one o'clock. Moved to the east side of the mount to where I had seen the water from the top. We found plenty of water in the gum creek which is the head of the one we crossed on Tuesday night, just before making the range. We were obliged to come a long way round before we could get to it, the hills being all rough sharp rocks, impassable for horses; abundance of grass with a little spinifex on the hills. At this camp I have marked a tree "J. M.D. S."; the cone of stones on the top of the mount bears 293 degrees. Ten miles distant in a branch creek about half a mile to the north of this is more water; and a little higher up, in a ledge of rocks, is a splendid reservoir of water, thirty yards in diameter and about one hundred yards in circumference. We could not get to the middle to try the depth, but where we tried it it was twelve feet deep. A few yards higher up is another ledge of rocks, behind which is a second reservoir, but smaller, having a drainage into the former one. Native tracks about. Wind north. I have named this Anna's Reservoir, after Mr. James Chambers' youngest daughter.

Friday, 20th April, East Side of Mount Hugh. Started to the south-east to find a crossing place over the range; this was not an easy matter, from the roughness of the hills; at last, however, we got over it. On the other side we found a large gum creek with water in it, running to the north-east. Camped. The range is well grassed, with gum creeks coming from it, and a little mulga scrub. Here we have discovered a new tree, whose dark-green leaf has the shape of two wide prongs; the seed or bean, of which I have obtained a few, is of a red colour; the foliage is very thick. The stem of the largest we have seen is about eighteen inches in diameter. The wood is soft; when in the state of a bush it has thorns on it like a rose. Here we have also obtained some seed of the vegetable we have been using; we have found this vegetable most useful; it can be eaten as a salad, boiled as a vegetable, or cooked as a fruit. We have also some other seeds of new flowers. The bearing from this to the cone of stones on Hugh Mount, 233 degrees 45 minutes.

Saturday, 21st April, Gum Creek, East Side of Mount Freeling. Started at half-past seven across the scrub to another high hill. For seven miles the scrub is open, and the land beautifully grassed. At twelve miles from the camp we crossed another gum creek, coming from the range; as far as I could see it ran to the north-east. After seven miles the scrub became much thicker. We had great difficulty in getting through, from the quantity of dead timber, which has torn our saddle-bags and clothes to pieces. There are a number of gum-trees, and the new tree that was found on Captain Sturt's expedition, 1844, but mulga predominates. At fourteen miles we struck a large gum plain, but after a short time again entered the scrub. At about twenty-two miles met another arm of the gum plains, with large granite rocks nearly level with the surface. We found rain water in the holes of these rocks. At thirty-two miles crossed the sandy bed of a large gum creek divided into a number of channels; too dark to see any water. Four miles further on, camped on a small gum creek with a little rain water; the creeks are running to the north-east. The soil is of a red sandy colour: the grass most abundant throughout the whole day's journey. Occasionally we met with a few hundred yards of spinifex. Wind south-east. Native tracks quite fresh in the scrub and plain; we also passed several old worleys.

Sunday, 22nd April, Small Gum Creek, under Mount Stuart, Centre of Australia. To-day I find from my observations of the sun, 111 degrees 00 minutes 30 seconds, that I am now camped in the centre of Australia. I have marked a tree and planted the British flag there. There is a high mount about two miles and a half to the north-north-east. I wish it had been in the centre; but on it to-morrow I will raise a cone of stones, and plant the flag there, and name it Central Mount Stuart. We have been in search of permanent water to-day, but cannot find any. I hope from the top of Central Mount Stuart to find something good to the north-west. Wind south. Examined a large creek; can find no surface water, but got some by scratching in the sand. It is a large creek divided into many channels, but they are all filled with sand; splendid grass all round this camp.

Monday, 23rd April, Centre. Took Kekwick and the flag, and went to the top of the mount, but found it to be much higher and more difficult of ascent than I anticipated. After a deal of labour, slips, and knocks, we at last arrived on the top. It is quite as high as Mount Serle, if not higher. The view to the north is over a large plain of gums, mulga, and spinifex, with watercourses running through it. The large gum creek that we crossed winds round this hill in a north-east direction; at about ten miles it is joined by another. After joining they take a course more north, and I lost sight of them in the far-distant plain. To the north-north-east is the termination of the hills; to the north-east, east and south-east are broken ranges, and to the north-north-west the ranges on the west side of the plain terminate. To the north-west are broken ranges; and to the west is a very high peak, between which and this place to the south-west are a number of isolated hills. Built a large cone of stones, in the centre of which I placed a pole with the British flag nailed to it. Near the top of the cone I placed a small bottle, in which there is a slip of paper, with our signatures to it, stating by whom it was raised. We then gave three hearty cheers for the flag, the emblem of civil and religious liberty, and may it be a sign to the natives that the dawn of liberty, civilization, and Christianity is about to break upon them. We can see no water from the top. Descended, but did not reach the camp till after dark. This water still continues, which makes me think there must certainly be more higher up. I have named the range John Range, after my friend and well-wisher, John Chambers, Esquire, brother to James Chambers, Esquire, one of the promoters of this expedition.

Tuesday, 24th April, Central Mount Stuart. Sent Kekwick in search of water, and to examine a hill that has the appearance of having a cone of stones upon it; meanwhile I made up my plan, and Ben mended the saddlebags, which were in a sad mess from coming through the scrub. Kekwick returned in the afternoon, having found water higher up the creek. He has also found a new rose of a beautiful description, having thorns on its branches, and a seed-vessel resembling a gherkin. It has a sweet, strong perfume; the leaves are white, but as the flower is withered, I am unable to describe it. The native orange-tree abounds here. Mount Stuart is composed of hard red sandstone, covered with spinifex, and a little scrub on the top. The white ant abounds in the scrubs, and we even found some of their habitations near the top of Mount Stuart.

Wednesday, 25th April, Central Mount Stuart. There is a remarkable hill about two miles to the west, having another small hill at the north end in the shape of a bottle; this I have named Mount Esther, at the request of the maker of the flag. Started at 9 o'clock, on a course a little north of west, to the high peak that I saw from the top of Mount Stuart, which bears 272 degrees. My reason for going west is that I do not like the appearance of the country to the north for finding water; it seems to be sandy. From the peak I expect to find another stratum to take me up to the north-north-west. Around the mount and on the west side, the country is well grassed, and red sandy soil; no stones. To the north and south of our line are several isolated hills, composed principally of granite. At ten miles there is a quartz reef on the north side of the south hills. At twelve miles struck a gum creek coming from the south and running to the north; it has three channels. We found a little rain water in one, and camped, to enable us to finish the mending of the saddle-bags. Wind east; very cold morning and night. The large creek that flowed round Mount Stuart is named the Hanson, after the Honourable R. Hanson, of Adelaide.

Thursday, 26th April, Gum Creek on West Course. Started at a quarter past 8 o'clock on the same course for the high peak. At two miles crossed some low granite and quartz hills; and at four miles crossed a gum creek running to the north with sand and gravel beds. No water. The country then became difficult to get through, in consequence of the number of dead mulga bushes. At ten miles the grass ceased, and spinifex took its place, and continued to the banks of the next gum creek, which we crossed at twenty-two miles; the bed sandy, and divided into a number of channels, coming from the south-east, and running a little to the east of north, but no water in them. Native tracks in its bed. On the west side of the creek the grass again begins, and continues to the hills, where we arrived at five minutes to 7. Camped without water. There seems to have been very little rain here—the grass and everything else is quite dry. Distance, thirty-eight miles.

Friday, 27th April, East Side of Mount Denison. Sent Kekwick to the south-west to a remarkable hill which I hope may yield some water, with orders to return immediately if he should find any nearer, so that we might get some for the horses. I waited till past 12, but he did not return, so I started, intending to go to the top of the mount. On getting to the north-east side of the ranges, I liked the appearance of the country for water, and seeing that the top of the mount was still some distance off, and that it would make it too late to return, I set to work myself to look for water. After an hour's search I was successful, finding some rain water in a gum creek coming from the hills. The natives must have been there quite recently, as their fires were still warm; and, as I had left the camp and provisions with only one man, I hurried back, had the horses saddled and packed, and brought them down to the water, leaving a note for Kekwick to follow in a west-north-west direction to a gum creek about three miles distant. Kekwick's search was also successful; he found permanent water under the high peak to which I sent him, and which I have named Mount Leichardt, in memory of that unfortunate explorer, whose fate is still a mystery. I have seen no trace of his having passed to the westward. Kekwick describes the water he has found as abundant and beautifully clear, springing out of conglomerate rock much resembling marble; its length is upwards of a quarter of a mile, falling into natural basins in the solid rock, some six feet in depth and of considerable capacity. The country round the base of the range is covered with the most luxuriant grass and vegetation. Mount Leichardt and the range are composed, at their base, of a soft conglomerate rock in immense irregular masses, heaped one on the other; the higher part where the spring appears is of the same conglomerate, but broad and solid, having smooth faces, which makes the ascent very difficult.

Saturday, 28th April, Gum Creek under Mount Denison. As soon as the horses were caught I started for the top of the mount. I left my horse in a small rocky gum creek which I thought would lead me to the foot of the mount. At about a quarter of a mile from the mouth of the gorge, I came upon some water in a rocky hole, followed it up, and, two hundred yards further, was stopped by a perpendicular precipice with water trickling over it into a large reservoir. I had now to take to the hills, which were very rough, and after a deal of difficulty I arrived, as I thought, at the top, but to my disappointment I had to go down a fearfully steep gully. At it I went, and again I arrived, as I fancied, at the top, but here again was another gully to cross, and a rise still higher. I have at last arrived at the summit, after a deal of labour and many scratches. This is certainly the highest mountain I have yet ascended; it has taken me full three hours to get to the summit. The view is extensive, but not encouraging. Central Mount Stuart bears 95 degrees. Mount Leichardt, 155 degrees 30 minutes. To the south, broken ranges with wooded plains before them, and in the far distance, scarcely visible, appears to be a very high mountain, a long, long way off. To the south-west the same description of range. About thirty miles to the west is a high mount with open country, and patches of woodland in the foreground. At the north-west there appears to be an immense open plain with patches of wood. To the north is another plain becoming more wooded to the north-east. As this is the highest mountain that I have seen in Central Australia, I have taken the liberty of naming it Mount Denison, after his Excellency Sir William Denison, K.C.B., Governor-General. The next range (bearing 334 degrees), being the last of the highest ones north, I have named Mount Barkly, after his Excellency Sir Henry Barkly, Governor-in-Chief of Victoria. When on the second highest point of this mount, I saw a native smoke rise up in the creek below, a short distance from where I had tied my horse. This naturally made me very anxious for his safety, and when I descended I was rejoiced to find him safe. The natives have been in the creek and on the mount: their tracks, which are quite fresh, lead me to conclude that they have been running. The descent was difficult, but I discovered a shorter route, and it has taken me two hours to come down. Arrived at the camp at 4.30, and found all right. I intended to have built a large cone of stones on the summit; but, when I arrived there, I was too much exhausted to do so. I have, however, erected a small one, placing a little paper below one of the stones, to show that a white man has been there. I have also marked a tree "J. M.D. S." on the creek where we are now camped. Mount Denison bears from here 249 degrees.

Sunday, 29th April, Gum Creek under Mount Denison. Latitude, 21 degrees 48 minutes. Variation, 3 degrees 20 minutes east. Mount Denison and the surrounding hills are composed of a hard reddish-brown sandstone. About one hundred yards from the summit is a course of conglomerate, composed of stones from half an inch to four inches in diameter, having the appearance of being rounded at a former period by water. From the foot to the top of this course is about ten feet, and the breadth on the top is about twelve feet. There is red sandstone on the summit, with three or four pines growing. The mount and adjoining hill are covered with spinifex, but the plain is grassed. The wind has now changed to the west, and it is much hotter.

Monday, 30th April, Under Mount Denison. The wind changed again to the south-east during the night, and is much colder. Started on a course, 315 degrees, across the plain towards Mount Barkly. The highest point of the mount is eighteen miles distant from our camp on the creek. We had to round the west side of it, finding no water until we came upon a little in the gorge coming from the highest point. It was dark before we arrived, so that we could not take the horses up to-night. Wind south-east, blowing a hurricane, and very cold.

Tuesday, 1st May, North-west Side of Mount Barkly. On examining the water, I find it is only a drainage from the rocks, and there is not more than two gallons for each horse. I ascended the hill, but could see nothing more than I had seen from Mount Denison. The base is composed of a hard red sandstone, the top of quartz rock. I do not like the appearance of the country before us. Started on a course of 335 degrees, and at six miles and a half came upon a large gum creek divided into numerous channels: searched it carefully, without finding any surface water; but I discovered a native well about four feet deep, in the east channel, close to a small hill of rocks. Cleared it out, and watered the horses with a quart pot, which took us long after dark—each horse drinking about ten gallons, and some of them more. Natives have been here lately, and from the tracks they seem to be numerous. We also observed the rose-coloured cockatoo. I have named this creek The Fisher, after Sir James Hurtle Fisher; it runs a little east of north.

Wednesday, 2nd May, The Fisher. We did not start until 11 o'clock in consequence of it taking a long time to water the horses. We steered for some hills that I had seen from the top of the last two mounts. At thirteen miles arrived at the hills, but found them low, and no appearance of water. Changed my course west 35 degrees north to some higher hills. At 6.30 camped in the scrub without water. The country from Mount Denison to this is a light-red sandy soil, covered with spinifex, with very little grass, and is nearly a dead level. In some places it is scrubby, having a number of gum-trees, and the new tree of Captain Sturt growing all over it. From a distance it has the appearance of a good country, and is very deceiving; you constantly think you are coming upon a gum creek. Wind south-east; very cold at night and morning.

Thursday, 3rd May, Spinifex and Gum Plains. Started on the same course, west 35 degrees north, and at four miles reached the top of the hills, which are low and composed of dark-red sandstone and quartz. The bearings to Mount Denison, 146 degrees; Mount Barkly, 142 degrees; to another hill west-north-west, 302 degrees, distant about ten miles, which I have named Mount Turnbull, after the late Gavin Turnbull, Esquire, Surgeon in the Indian Army. The morning is very hazy, and I cannot see distinctly; besides, my eyes are again very bad. The appearance of the country all round is that of having gum creeks everywhere. To the north there are some more low hills. A short distance off, on a bearing of 328 degrees, there appears to be a gum creek with something white as if it were water, so I shall change my course. At 3.50 camped, some of my horses being nearly done up from want of water, and having nothing to eat but spinifex. I have now come eighteen miles, and the plain has the same appearance now as when I first started—spinifex and gum-trees, with a little scrub occasionally. We are expecting every moment to come upon a gum creek, but hope is disappointed. I have not so much as seen a water-course since I left the Fisher, and how far this country may continue it is impossible to tell. I intended to have turned back sooner, but I was expecting every moment to meet with a creek. It is very alluring, and apt to lead the traveller into serious mistakes. I wish I had turned back earlier, for I am almost afraid that I have allowed myself to come too far. I am doubtful if all my horses will be able to get back to water. In rainy weather this country will not retain the water on the surface, and we have not so much as seen a clay-pan of the smallest dimensions. The gum-trees on this plain have a smooth white bark, and the leaves are some light-green and some dark. Most of the trees seem very healthy; there are very few dead ones about. To-morrow morning I must unwillingly retreat to water for my horses. There is no chance of getting to the north-west in this direction, unless this plain soon terminates. From what I could see there is little hope of its doing so for a long distance.

Friday, 4th May, Gum and Spinifex Plains. At times this country is visited by blacks, but it must be seldom, as since we left the Fisher we have only seen the track of one, who seems to have come from the east, and to have returned in that direction. The spinifex in many places has been burnt, and the track of the native was peculiar—not broad and flat, as they generally are, but long and narrow, with a deep hollow in the foot, and the large toe projecting a good deal; the other in some respects more like the print of a white man than of a native. Had I crossed it the day before, I would have followed it. My horses are now suffering too much from the want of water to allow me to do so. If I did, and were not to find water to-night, I should lose the whole of the horses and our own lives into the bargain. I must now retreat to Mount Denison, which I do with great reluctance; it is losing so much time, and my provisions are limited. Started back at 7.10 a.m., and at thirty miles came upon a native well, with a little grass round it; the bottom was moist. Unsaddled, and turned the horses out. Commenced clearing out the well the best way we could, with a quart pot and a small tin dish, having unfortunately lost our shovel in crossing the McDonnell ranges. We had great difficulty in keeping the horses out while we cleared it. To our great disappointment we found the water coming in very slowly. We can only manage, in an hour and a half, to get about six gallons, which must be the allowance for each horse, and it will take us till to-morrow morning to water them all. One of us is required to be constantly with them to keep them back, and that he can hardly do; some of them will get away from him do all he can. Kekwick's horse was nearly done up before we reached this place; also one of the others. Those nearest to the cart-breed give in first.

Saturday, 5th May, Native Well. Got all the horses watered by 11 o'clock a.m., and could only get about five gallons for each horse, although we were employed the whole of the night, and got no sleep. Started for the Fisher, and arrived at the native well at sundown. Were obliged to tie the horses up, to keep them from getting into it. We could scarcely get some of them as far as this, as they are quite done up. What was still worse, we found the native well had fallen in since we left. It cannot be helped: we must take things as they come. Commenced immediately to cut a number of stakes, rushes, and grass, to keep the sand back, and by 3 o'clock in the morning we got them all watered, and very thankful we were to do so. It has been, and is still, bitter cold throughout the night and morning, the wind still coming from the south-east. We had a pot of tea, although we could ill afford it, and lay down and got a little sleep, completely tired and worn out with hard work and want of rest.

Sunday, 6th May, The Fisher. Got up at daybreak and went to the well, but found that the rascals of horses had been there before us, and trodden in one side of the well. They had as much water last night and morning as they could drink, and the quantity that some of them drank was enormous. I had no idea that a horse could hold so much, yet still they want more. I shall remain here two days, put down more stakes, clear out the well, and give them as much as they will drink. During this trying time I have been very much pleased with the conduct of Kekwick and Ben; they have exerted themselves to the utmost, and everything has been done with the greatest alacrity and cheerfulness. Although they have only had two hours' sleep during the last two nights, there has not been a single word of dissatisfaction from either of them, which is highly gratifying to me. It is, indeed, a great pleasure to have men that will do their work without grumbling. Watered the horses as they came in. They do not now drink a fourth part of what they did at first.

Monday, 7th May, The Fisher. Had a good night's rest, and felt recovered from the past fatigue. Started for the creek on the east side of Mount Denison, to the water at which we camped before, keeping to the north side of Mount Barkly in search of water, but could find none. Arrived at the creek after dark. Kekwick's horse is entirely done up; he had to get off and lead him for two miles. Another of the horses is nearly as bad, but he managed to get to the creek. We found the water greatly reduced, but still enough for us.

Tuesday, 8th May, Creek East of Mount Denison. I must remain here two days to allow the horses to recover. I am afraid if we have such another journey, I shall have to leave some of them behind. I do not know what is the cause of their giving in so soon; I have had horses that have suffered three times as much privation, and yet have held out. The light ones are all right; it is the heavy ones, of the cart-horse breed, that feel it most. I had been keeping them up on purpose for an occasion like this, and they all looked in first-rate condition, but the work of the past week has made a great alteration in some of them. I suppose the young grass is not yet strong enough for them. It is very vexing to be thus disappointed and delayed. To think that they should fail me at the very moment when I expected them to do their best, and after all the trouble and loss of time I have incurred in giving them short journeys! However, I cannot improve it by complaining, and must rest contented and hope for the best. Wind south-east. Storm brewing.

Wednesday, 9th May, Creek East of Mount Denison. Resting horses and putting our things in order. Wind blowing very strong from the south-east; it has continued nearly in the same quarter since March.

Thursday, 10th May, Creek East of Mount Denison. I find that I must give the horses another day; they have not yet recovered, and I expect we shall have some more hard work for them. We have not quite finished mending.

Friday, 11th May, Creek East of Mount Denison. Ben was taken very ill during the night, and is still so bad that I am obliged to remain here another day. Afternoon: Ben feels much better, so I shall start to-morrow.

Saturday, 12th May, Creek East of Mount Denison. Ben is better, and the horses look as if they can stand a little more hardship. Started at 8.20 on a bearing of 28 degrees east of north, to see if I can get on in that direction. For fourteen miles our course was through mulga scrub and spinifex, in some places very thick. At twenty-seven miles camped without water. The country that we have passed over the last two days is apparently destitute of water, even in rainy weather. I do not think the ground would retain it a single day. Very little feed for the horses.

Sunday, 13th May, Scrub and Gum Flat. I do not like the appearance of the country. As I can see no hope of obtaining water on this course, I shall change to the east, in order to cut the large gum creek that I crossed on the 26th ultimo, and, if I find water in it, to follow it out to wherever it goes. At three miles cut a small gum creek: searched for water both up and down, but could find none, nor any appearance of it. Still keeping my east course, we then passed through a very thick mulga scrub, and at ten miles struck a low range of hills, composed of quartz, with a conical peak, which I ascended. The prospect from this is very extensive, but disheartening, apparently the same sort of scrubby country that I have endeavoured to break through to the north-west. The view to the north is dismal; there are a few isolated hills, seemingly the termination of John range, and of the same formation as this that I am now on. To east-south-east there appears to be a creek, to which I shall now go. At three miles I reached what I had supposed to be a creek, but it is a small narrow gum flat which receives the drainage from this low range. We found a hole where there had been water, but it was all gone. I have named the peak Mount Rennie, after Major Rennie of the Indian army. In this small flat we shot a new macaw, which I shall carry with me, and preserve the skin, if we get to water to-night. The front part of the neck and underneath the wings is of a beautiful crimson hue, the back is of a light lead colour, the tail square, the beak smaller than a cockatoo's, and the crest the same as a macaw's. After leaving this flat, we passed through some scrub, and came upon another of the same description. Here I narrowly escaped being killed. My attention being engaged looking for water, my horse took fright at a wallaby, and rushed into some scrub, which pulled me from the saddle, my foot and the staff that I carry for placing my compass on catching in the stirrup-iron. Finding that he was dragging me, he commenced kicking at a fearful rate; he struck me on the shoulder joint, knocked my hat off, and grazed my forehead. I soon got clear, but found the kick on my shoulder very painful. Mounted again, and at seven miles we came upon some more low hills with another prominent peak of a dark-red sandstone. This I have named Mount Peake, after E.J. Peake, Esquire, of Adelaide. I now find that the gum creek which I crossed between Central Mount Stuart and Mount Denison runs out and forms the gum plains we have just passed. No hope of water. I must now bear in for the centre to get it. Passed through a very thick, nasty mulga scrub for five miles, and camped again without water under some low stony hills. I feel the effects of my accident very much.

Monday, 14th May, Stony Hills, Mulga Scrub. Feel very stiff and ill. Started at daylight, and passed through three belts of thick mulga scrub, between which there were low stony hills. At three miles passed a small gum creek, emptying itself into the scrub. At seventeen miles passed another, doing the same; at twenty miles another, and at twenty-four miles a third, under the hills north-west of Central Mount Stuart. This has a very remarkable hill at the north-west, in the shape of a large bottle with a long neck. We have had the greatest difficulty in getting all our horses to the water; three of them are very bad; two have been down a dozen times during the journey to-day. On approaching the range, we passed through some large patches of kangaroo grass, growing very thickly, and reaching to my shoulder when in the saddle.

Tuesday, 15th May, Centre. The horses look very bad to-day; I shall therefore give them three or four days' rest. It is very vexing, but it cannot be helped. The water here will last about ten days. I shall cause another search for more to be made; I myself am too unwell to assist. Yesterday I rode in the greatest pain from the effects of my fall, and it was with great difficulty that I was able to sit in the saddle until we reached here. Scurvy also has taken a very serious hold of me; my hands are a complete mass of sores that will not heal, but, when I remain for two or three days in some place where I can get them well washed, they are much better; if not, they are worse than ever, and I am rendered nearly helpless. My mouth and gums are now so bad that I am obliged to eat flour and water boiled. The pains in my limbs and muscles are almost insufferable. Kekwick is also suffering from bad hands, but, as yet, has no other symptoms. I really hope and trust that it will not be the cause of my having to turn back. I suffered dreadfully during the past night. This afternoon the wind has changed to the west—the first time since March; a few clouds are coming up in that direction.

Wednesday, 16th May, Centre. I despatched Kekwick at daybreak in search of permanent water, with orders to devote the whole of two days to that purpose. I must now do everything that is in my power to break this barrier that prevents me from getting to the north. If I could only get one hundred and twenty miles from this, I think there would be a chance of reaching the coast. I wish the horses could endure the want of water a day or two longer, but I fear they cannot; this last journey has tried them to their utmost. Two of them look very wretched to-day, and will with difficulty get over it; one I scarcely think will do so. I should not have been afraid to have risked two more days with five of them. If they had been all like these five, I should have tried to the north-west a degree and back again without water. I have been suffering dreadfully during the past three weeks from pains in the muscles, caused by the scurvy, but the last two nights they have been most excruciating. Violent pains darted at intervals through my whole body. My powers of endurance were so severely tested, that, last night, I almost wished that death would come and relieve me from my fearful torture. I am so very weak that I must with patience abide my time, and trust in the Almighty. This morning I feel a little easier; the medicines I brought with me are all bad, and have no effect. The wind still from the north-west, with a few light clouds. Towards sundown the wind has changed to the south-west; heavy clouds coming from the north-west.

Thursday, 17th May, Centre. Wind from the south; the heavy clouds continued until sunrise, and then cleared off. I fully expected some rain, but was disappointed. I have again had another dreadful night of suffering; I had, however, about two hours' sleep, which, as it was the only sleep I have had for the last three nights, was a great boon. This morning I observe that the muscles of my limbs are changing from yellow-green to black; my mouth is getting worse, and it is with difficulty that I can swallow anything. I am determined not to give in; I shall move about as long as I am able. I only wish the horses had been all right, and then I should not have stayed here so long. Kekwick returned at 3 o'clock, and reported having found water in the Hanson, about fifteen miles from Central Mount Stuart, but only a small supply. Beyond that the creek divides into two, one running north and the other east, but he could see no more water further down. He also saw two natives, armed with long spears, about three hundred yards off; they did not observe him, and he thought it most prudent not to show himself, but to remain behind a thick bush until they were gone. In this instance I regret his caution, for I am anxious to see or hear what is the appearance of the Central natives. Wind variable, with heavy clouds from north-west.

Friday, 18th May, Centre. I have again had a very bad night, and feel unable to move to-day. Wind the same.

Saturday, 19th May, Centre. I had a few hours' sleep last night, which has been of great benefit to me. I shall attempt to move down to the water in the Hanson. Arrived there about 1.30 completely done up from the motion of the horse. The water is a few inches below the surface in the sand. East side of Mount Stuart bearing 250 degrees, about ten miles distant. I do not think the water is permanent.

Sunday, 20th May, The Hanson. Another dreadful night for me. Wind and clouds still coming from the north-west, but no rain.

Monday, 21st May, The Hanson. Unable to move; very ill indeed. When shall I get relief from this dreadful state?

Tuesday, 22nd May, The Hanson. I got a little sleep last night, and feel a great deal easier this morning, and shall try my horse back again. I shall now steer north-east to a range of hills that I saw from the top of Central Mount Stuart, and hope from these to obtain an entrance to the north-west or north-east. I also hope to cut the creek that carries off the surplus water from all the creeks which I have passed since March. It must go somewhere, for it is difficult to believe that those numerous bodies of water can be consumed by evaporation. Started on a bearing of 48 degrees, crossed the Hanson, running a little on our right; at six miles crossed it again, running more to the north for two miles further. We crossed four more of its courses, all running in the same direction. The most easterly one is spread over a large salt-creek valley, and forms a lagoon at the foot of some sand ridges, the highest of which is ten miles and a half from our last camp. On the east side of it there is a large lagoon, five miles long by one mile and a half broad, in which water has lately been, but it is now dry. We then proceeded through a little scrub, with splendid grass, and at twelve miles cut a small gum creek, coming from the range. We saw a number of birds about, and there were tracks of natives, quite fresh, in the creek. Sent Kekwick down it to see if there were water, while I went up and examined it. This is the large gum plain that we met with the day we made the Centre; it is completely covered with grass. Kekwick ran the creek out. At about two miles he observed a little water in the creek, where the natives had been digging. He also came upon two of them, and two little children. They did not observe him until he was within fifty yards, when they stood for a few minutes paralysed with astonishment; then, snatching up the children, ran off as quickly as their legs could carry them. They did not utter a sound, although he called to them. He remarked that they had no hair on their heads, or it was as short as if it had been burned off close. I wish I had seen them; I should have overtaken them and seen if it were a fact that the hair was burnt. It is reported in Adelaide that there are natives in the interior without hair on their bodies. At fourteen miles we again struck the creek, and found plenty of water in it. It winds all over the plain in every direction. Camped for the night very much done up. I could hardly sit in my saddle for this short distance. Wind north-west.

Wednesday, 23rd May, Gum Creek, East Range, the Stirling. The wind has changed again to the south-east. I have named this creek the Stirling, after the Honourable Edward Stirling, M.L.C. Followed it into the range on the same course towards a bluff, where I think I shall find an easy crossing. At one mile from the camp the hills commenced on the south-east side of the creek, but on the north-west side they commenced three miles further back. There was abundance of water in the creek for thirteen miles; at ten miles there was another large branch with water coming from the south-east. At fourteen miles ascended the bluff and obtained the following bearings: South side of the creek, to a high part of the range about two miles off (which I have named Mount Gwynne, after his Honour, Justice Gwynne), 186 degrees. North side of the creek, to another hill about two miles and a half off (which I have named Mount Mann, in memory of the late Commissioner of Insolvency), 249 degrees. Central Mount Stuart bears 131 degrees to the highest point. At the north-west termination of the next range, to which I shall now go, there are two very large hills, the north one, which is the highest, I have named Mount Strzelecki, after Count Strzelecki, bearing 358 degrees. I have named the high peak on the same range Mount Morphett, after the Honourable John Morphett, M.L.C. The view from this bluff is extensive, except to the west-north-west, which is hidden by this range just alluded to, which I have named Forster Range, after the Honourable Anthony Forster, M.L.C. From the south-west it has the appearance of a long continuous range, but, on entering it, it is much broken into irregular and rugged hills: on this side, the north-east, it consists of table-hills, with a number of rugged isolated ones on the north side. To the north-west there is another scrubby and gum-tree plain; to the north-north-west are some isolated low ranges; to the north are grassy plains and low ranges; to the east are several spurs from this range, which is composed of a very hard dark-red stone, mixed with small round quartz and ironstone, and in some places a hard flinty quartz. The range and hills are covered with spinifex, but the valleys are beautifully grassed. We descended, and at four miles struck a creek coming from the range, and running between two low ranges towards the north-east. At seven miles changed my course to north-east to camp in the creek, and endeavour to get water for the horses before encountering the scrubby plains to-morrow morning. At five miles came upon a low range, but no creek; it must have gone further to the eastward. It being now quite dark, we camped under the ranges. Since I changed my course I have come through a patch of mulga and other scrubs with plenty of grass, but no watercourses. Wind south-east;

heavy clouds from the north-west; lightning in the south and west.

Thursday, 24th May, Range of Low Hills. This morning I feel very ill from climbing the bluff yesterday; I had no sleep during the night, the pains being so very violent. About 9 o'clock we had a heavy shower of rain, and a little more during the night. Very late before the horses were found, and the atmosphere very thick, with the prospect of rain for the rest of the day. This and my being so ill have decided me to remain here until to-morrow, there being sufficient rain water for the horses. A few more light showers during the afternoon and evening. Wind still the same; heavy clouds from the north-west.

Friday, 25th May, Range of Low Hills. I feel better this morning. The clouds have all gone during the night, and it is now quite clear. Started for Mount Strzelecki, passing through some very thick mulga scrub, with a few gum-trees and plenty of grass. At twenty-one miles came upon a small gum creek, where we gave the horses water, filled our own canteens, and proceeded to the foot of the mount and camped. At a mile from its base the spinifex begins again. Wind south-east. Very cold.

Saturday, 26th May, Mount Strzelecki. Ascended the mount, and built a cone of stones. To the east are hills connected with this range, which I have named Crawford Range, after —— Crawford, Esquire, of Adelaide. To the east-north-east is a large wooded undulating plain, with another range in the extreme distance. To the north-east the distant range continues with the same plain between. At a bearing of 55 degrees is a large lagoon, in which there appears to be a little water. To the north-north-east the plain appears to be rather more scrubby, and with a few sand hills. To the north the point of the distant range is lost sight of by some high scrubby land. To the west there are a few low hills, from fifteen to twenty-five miles distant. This range is composed of a hard flinty quartz, partly of a blue colour, with a little ironstone. We can find no permanent water in this range, but, from the two or three native tracks, quite fresh, which we have passed, I think there must be some about. Descended, and proceeded round the range to the lagoon, the range being too rough to cross. There is not enough water to be a drink for the horses. Camped. Very heavy clouds from the north-west. The mount is about four miles distant. At sundown there was a beautiful rain for an hour. It is very strange, the clouds come from the north-west, and the wind from the south-east. The rain seems to be coming against the wind.

Sunday, 27th May, Lagoon North-east of Mount Strzelecki. We had a few heavy showers during the night, but it seems as if the rain would now clear off. I hope not, for there is only about two inches of water in the lagoon. I am again suffering much pain from the exertion it cost me to climb Mount Strzelecki, and from assisting in building the cone of stones; but if I did not put my hands to almost everything that is required, I should never get on. My party is too small. It is killing work.

Monday, 28th May, Lagoon North-east of Mount Strzelecki. We could not get a start till 9.15, the horses having strayed to a distant bank for shelter from the wind, which was piercingly cold. I had, in the first instance, to go three miles north-north-west, in order to clear the low stony range that runs on to the east side of the lagoon. I then changed to 22 degrees to the far-distant range. For the first three miles our course was through a very thick mulga scrub, with plenty of grass, and occasionally a little spinifex; it then changed to a slightly undulating country of a reddish soil, with gum and cork-trees, and numerous low sandy plains, much resembling the gum and spinifex plains to the west, where I was twice beaten back. It certainly is a desert country. Camped without water on a little patch of grass. Distance to-day, twenty-eight miles. Wind south-east. Very cold all day.

Tuesday, 29th May, Scrub, Spinifex and Gum-Trees. Started at 8 on the same course for the range, which is still distant, through the same description of country. At seven miles we came upon a plain of long grass, which seems to have been flooded. It is about two miles broad. Between this and the first hill of the range we passed four more of the same description. Distance to the first hill, fourteen miles. In another mile we struck a small creek; searched for water, but could find none, although birds were numerous; thence through another mulga scrub, and after crossing a number of rough stony hills, we arrived at the top of the range, which I have named Davenport Range, after the Honourable Samuel Davenport, M.L.C. It is composed of hard red sandstone, with courses of quartz. I find this is not the range for which I am bound. Although this one is high, the other is still higher, and, I should think, is still forty or fifty miles distant. The day is thick, and I cannot see distinctly. Between these ranges is a large plain, more open than those we have come over. To the north the range appears to terminate; to the west of north, in the far distance, just visible, are two high hills, the northernmost of which is conical. To the east and south-east is the plain and range; to the west, continuation of the same plain that we have come over in the last two days' journey. Although we had some heavy showers at the lagoon, we have not passed a single water-course, except the one we crossed a few miles before we made this range, nor did we see a drop of surface water: it seems to be all absorbed the moment that it falls. Descended the north-north-east side of the range, and at a mile and a half found some rain water in a creek, coming from the range. Camped. Wind south-east. Distance, twenty miles.

Wednesday, 30th May, The Davenport Range. I find this water will not last more than three days. I have determined to remain here to-day, and have sent Kekwick in search of more water. As I am now a little better, I must get my plan brought up. It has got in arrear, in consequence of my hands being so bad with the scurvy. My limbs are much easier, yet the riding is still very painful; my mouth also is much better, so that I am led to hope that the disease will soon leave me. Native tracks about here, and when I was on the top of the range I saw smoke in the scrub a few miles to the north-west. Sundown: I am quite surprised that Kekwick has not returned, as my instructions to him were not to go above five or six miles, and then to return whether he found water or not. I am very much afraid that something has happened to him.

Thursday, 31st May, The Davenport Range. Kekwick has not returned. I begin to feel very uneasy about him. I must be off and follow up his tracks. Sent Ben for the horses. He was a long time in finding them, as is generally the case when one wants a thing in a hurry. 9.30: Kekwick has arrived before the horses; he overshot his mark last night, and got beyond the camp. I am very glad he is all safe. He informs me that he came upon plenty of water a few miles from here, which compensates for the anxiety he caused me during the night. His reason for not returning as I had directed was that he crossed a gum creek which had so promising an appearance, that he was induced to follow it to the plains, where he found an abundance of water. While he was riding he was taken very ill, and was unable to come on for some time, which made it so late that he could not see to reach the camp. He is unable to proceed to-day, which is vexing, for I wish to get on as quickly as possible.

Friday, 1st June, The Davenport Range. The horses having strayed, we did not get a start till late. Our course was 22 degrees, and at two miles we struck a small gum creek coming from the range and running west-north-west. At three miles and a half we crossed a larger one coming from, and running in, the same direction. Then commenced again the same sort of country that we passed through the other day. At eight miles struck a splendid large gum creek or river, having long and deep reaches of water with fish four or five inches in length; it is running through the plain as far as I can see, which is only a short distance, the ground being low and level. Its course at this place is to the west-north-west; it is very broad, and in some places the banks are perpendicular, and are well grassed and covered with fine gum-trees, mulga and other bushes. From bank to bank its width is about ten chains. This is the finest creek for water that we have passed since leaving Chambers Creek. The day being far advanced, I shall camp here, and get to the range to-morrow. I am very much inclined to follow this creek and see where it empties itself; but I expect to find a large one close to the range, or on the other side. I wish also to get on the top to see what the country on ahead is like. The fact of fish being in this creek leads me to think that it does not empty itself into the gum plains, like others lately passed, but that it must flow either into the sea on the north-west coast, or into a lake. I have named it the Bonney Creek, after Charles Bonney, Esquire, late Commissioner of Crown Lands for South Australia.

Saturday, 2nd June, The Bonney Creek. Started at 8.20 on the same course, 22 degrees, for the range, through a country of alternate spinifex and grass with a little mulga scrub. At seven miles we struck another large gum creek with every appearance of water, but I had no time to look for it, being anxious to make the range to-night, and endeavour to find water either on this side or on the other. The creek is large, and resembles the last. I have named it the McLaren, after John McLaren, Esquire, late Deputy Surveyor-General of South Australia. At seventeen miles, after passing through a well-grassed country with a little scrub, we reached the top of the first range, which is composed of a hard white granite-looking rock, with courses of quartz running through it. I have three or four spurs to cross yet before I make the main range. So far as I can see, McLaren Creek is running much in the same direction as the Bonney. Started from the top of the range and had a very difficult job in crossing the spurs. About sundown arrived all safe on a gum flat, between the ranges, and attempted to get upon what appears to be the highest range, but getting up the horses deterred us. We then sought for water among the numerous gum creeks which cover the plain, and at dark found some, and camped. There is a good supply of water, but I do not think it is permanent; it will last, however, for a month or six weeks. I have named these ranges the Murchison, after Sir Roderick Murchison, President of the Royal Geographical Society, London. Wind varying.

Sunday, 3rd June, Murchison Ranges. I feel very unwell this morning, from the rough ride yesterday. It was my intention to have walked to the top of the range to-day, but I am not able to do so. The small plain between the ranges is a bed of soft white sandstone, through which the different creeks have cut deep courses; the stones on the surface (igneous principally), are composed of iron, quartz, dark black and blue stone, also a bright red one, all run together and twisted into every sort of nick, as also with the limestone, and many other sorts which I do not know. This plain is covered with a most hard spinifex, very difficult to get the horses to face. In another creek, about one mile south-west from the camp, is a large water hole which will last six months; it is ten yards long by twenty yards wide.

Monday, 4th June, Murchison Ranges. Started on a course of 330 degrees to round this spur of the ranges, and at four miles and a half changed to 15 degrees to the high point of the range, and at three miles arrived on the top. I have named it Mount Figg. The view from this is extensive. The course of this range from the south to this point is 25 degrees; it then makes a turn to the north-north-west, in which direction the country appears more open, with some patches of thick scrub, and high ranges in the distance. From north-west to west it appears to be gum plain, with open patches of grass, and a number of creeks running into it from the range. I shall change my course to a high peak on the north-west point of the range, which bears from this 340 degrees 30 minutes. This range is volcanic here, and is of the same formation as I have already given. Started from the top of the mount at 12 o'clock. Went for eight miles along the side of the range, and met with a small gum creek running on our course; followed it up for three miles without finding water; it then took a more westerly course, so I left it to pursue my route. After leaving the mount, the range is composed of red sandstone with a little quartz. We have occasionally met with a little limestone gravel. Camped at 6 o'clock, without water.

Tuesday, 5th June, Gum-Tree Plain. Started on the same course at 7 o'clock for the high peak, through the same sort of country as yesterday. No watercourse. At fifteen miles ascended the peak, which I have named Mount Samuel, after my brother. The top is a mass of nearly pure ironstone. It attracted the compass 160 degrees. From north to west are broken ranges and isolated hills of a volcanic character, in all sorts of shapes. The isolated hills seem to be the termination of these ranges, which run nearly north and south. I have named them the McDouall Ranges, after Colonel McDouall, of the 2nd Life Guards,

Logan, Wigtownshire. I then changed my course to the north-north-east in search of water, there being no appearance of any to the north-north-west. After travelling five miles over small grassy, scrubby plains, between isolated hills and gum-trees, I could not find a water-course, so I changed to the east, to try if I could see anything from a high hill, which I ascended, and discovered a gum creek coming from the range on the east side. Followed it down, and, one mile and a half from the top, found a splendid hole of water in the rock, very deep, and permanent. The creek is very rocky, and its course here is north-east into the plain. Wind south-east. Clouds from the north-west.

Wednesday, 6th June, Gum Creek, North-east Side of the McDouall Ranges. There being nothing but spinifex on the ranges and creeks, the horses had been travelling nearly all night in search of food, and had gone a long way before they were overtaken. This morning saddled and got a start by 11 o'clock on a course of 340 degrees, crossing numerous creeks and stout spinifex, through which we had great difficulty in driving the horses. At five miles struck a gum creek in which we found water. The banks have excellent feed upon them, and in abundance, so, for the sake of the horses, I have determined to remain here to-day. This creek, which I have named Tennant Creek, after John Tennant, Esquire, of Port Lincoln, runs east. In searching for the horses this morning Ben found three or four more large water holes in the adjoining creek, a little south-east from this. Before we reached this, we crossed some marks very much resembling old horse-tracks.

Thursday, 7th June, Tennant Creek, McDouall Ranges. Started at 7.20. Course, 340 degrees. At three miles passed through an immense number of huge granite rocks piled together and scattered about in every direction, with a few small water-courses running amongst them to the eastward. We then encountered a rather thick scrub, and occasionally crossed a few low quartz rises coming from the McDouall ranges. At fourteen miles ascended the highest of them, which I have named Mount Woodcock, after the Venerable the Archdeacon of Adelaide. To the north-west and north is another range, about ten miles distant, which seems to continue a long way. I will change my course to 315 degrees, which will take me to the highest point. At two miles on this course came upon a gum creek running to the north-east, which I named Bishop Creek; followed it for one mile and a half, and found water, which will last a month or six weeks, and an immense number of birds. This is a camping-place of the natives, who seem to have been here very lately. We watered the horses and proceeded towards the range. At about two miles passed a low rugged ironstone range, peculiar in having a large square mass of ironstone standing by itself about the centre. I have named it Mount Sinclair, after James Sinclair, of Port Lincoln. Passed through a thick scrub, among which we saw a very handsome bush that was new to us, having a blue-green leaf ten inches long by six inches broad. We looked for some seed, but could not find any. At five miles crossed a grassy gum plain, where a creek empties itself. The same scrub continues to the range, which we reached at twelve miles from the water. It is not very high, but rough and steep, and we had great difficulty in getting to the top, but after many twistings and turnings and scramblings, we arrived there all right, and found it to be table land. At fourteen miles camped without water. The range is composed of ironstone, granite, quartz and red sandstone, running north of west and south of east. I have named it Short Range, after the Right Reverend the Lord Bishop of Adelaide.

Friday, 8th June, Short Range. Started at 8 o'clock on the same course, 315 degrees, to some very distant rising grounds. Short range seems to run nearly parallel to our course, as also does another distant range to the north, which I have named Sturt Range, after Captain Sturt. The table land continued about two miles, and then there was a gradual descent to the plains, and we entered a thick scrub with spinifex and gums. At eighteen miles came upon a beautiful plain of grass, having large gum-trees, and a new description of tree, the foliage of which is a dark-green and rather round, and the bark rough and of a dark colour. Here also was the cork-tree, and numerous other shrubs. This grassy plain continued for thirty-one miles, until we camped, but the last part is not so good. When I struck this plain, I was in great hopes of finding a large creek of water, but have been disappointed; we have not crossed a single water-course in thirty-one miles. Camped at sundown. No water. Wind south-east.

Saturday, 9th June, Grassy Plain. There is some rising ground a few miles further on, to which I shall go in search of a creek; I might be able to see something from it. If I do not find water I shall have to retreat to Bishop Creek, as the horses have now been two nights without water. Started at 7 o'clock, same course, 315 degrees, through scrub and a light sandy soil. At four miles got to the rise, which is a scrubby sand-hill. From this I can see nothing, the scrub being so thick; it is of a nasty, tough, wiry description, and has torn our hands and saddle-bags to pieces. I got up a tree to look over the top of this scrub, which is about twelve feet high, and I could see our course for a long distance; it appears to be the same terrible scrub, with no sign of any creeks. It is very vexing to get thus far, and have to turn back, when perhaps another day's journey would bring me to a better country. I shall now try a south course, and cut the grassy plains to the westward, in the hope of finding water; if so, I shall be able to make two days' journey to the north-west. Started on a south course for fourteen miles, through scrub and small grassy plains alternately, but we could find neither creek nor water. I now regret that I attempted the south course, which makes the distance from the water so much greater. Wind still south-east; heavy clouds coming from the north-west, I trust it will rain before morning.

Sunday, 10th June, Grassy Plains. Started at sunrise, and at two miles again got into the scrub. Three of the horses we can scarcely get along; they are very much done up. At 11 o'clock, one horse gave in altogether. We cannot get him up; we have tried everything in our power to do something for him. The other horses have been carrying his load, and he has had nothing to carry for this last hour and a half; all our efforts are in vain, and I am obliged, although with great reluctance, to leave him to his fate. Had this occurred nearer the water, I should have put an end to his existence and taken part of him to eat, for we are now very short of provisions, and the other horses have quite enough to carry without sharing his load; I wish I had left him sooner. At 12 o'clock, I find I shall lose some more of them, if they do not get water to-night, and it will be to-morrow before I can reach Bishop Creek. I shall now go to Short range and try to find some. The little bay mare Polly has become nearly mad, running about among the other horses, and kicking them as she passes; even the men do not escape from her heels. At five miles made the range. There are no large creeks coming from this side—nothing but small ones which empty themselves into the plain; sand up to the foot of the hills. Before we reach the range another of the horses is done up; he has only been carrying about 30 pounds in consequence of his back having been bad for the last three weeks. We lightened all the weak horses two days since. We shall now try if he will go without anything on his back. We are now amongst the granite ridges, and hope we shall find water on this side. The horse has given in before we can get to the other side. We must leave him for the sake of the others. Too much time has already been lost in endeavouring to get them on. Reached the other side and searched the different creeks, but cannot find any water. Crossed a spur of the range running south, and can see a nice-looking creek with gum-trees. Our hopes and spirits are again revived; the sight of it has even invigorated the horses, and they are hurrying on towards it. Traversed it down, but, to our great disappointment, find that it loses itself in a grassy plain. It is now dark, so I must remain here for the night. The sky is quite overcast, and I trust that Providence will send us rain before morning. An accident has happened to the water we were carrying; it was all lost yesterday. If it clears during the night, so that I can see the stars to guide me, I shall move on.

Monday, 11th June, Short Range. During the night there were a few drops of rain, which again raised our hopes, and about 4 o'clock it looked as if we were to have a deluge, but, alas! it only rained for about two minutes, and as much fell as would wet a pocket-handkerchief. Saddled and started through the range, my poor little mare looking very bad this morning; I have taken everything off her, so that she may hold out until we get to water, and I have been obliged to leave as many things at this camp as I could possibly do without. The mare lies down every few yards, I am therefore compelled to leave her for the sake of the others. From the number of birds about here, I think there must be water near; I hope she may find it, although I am afraid she is too far gone even to try it. At 1 o'clock, at the foot of Mount Woodcock, the horses' spirits revived at sight of their old track. I shall now be able to get all the rest of them safe to water, although there is one still doubtful. My own black mare shows a few symptoms of madness, but still keeps on, and does her work well. About an hour before sundown arrived at the water without any more losses, for which I sincerely thank the Almighty. We have had a terrible job to keep the horses from drinking too much water, but, as they have now eaten a few mouthfuls of grass, I have allowed them to drink as much as they thought proper. The natives have been here since we left.

Tuesday, 12th June, Bishop Creek. Resting: the horses look very bad; they remained by the water all night.

Wednesday, 13th June, Bishop Creek. The horses still look very bad this morning; they have again stayed by the water nearly all night; they had been one hundred and one hours without a drop, and have accomplished a journey of one hundred and twelve miles; they will require a week to recover; one of them is very lame from a kick the little mare gave him in her madness. Thus ends my last attempt, at present, to make the Victoria River; three times have I tried it, and have been forced to retreat. About 11 o'clock I heard the voice of a native; looked round and could see two in the scrub, about a quarter of a mile distant. I beckoned to them to approach, but they kept making signs which I could not understand. I then moved towards them, but the moment they saw me move, they ran off immediately. About a quarter of an hour afterwards they again made their appearance on the top of the quartz reef, opposite our camp, and two others showed themselves in about the same place as the two first did. Thinking this was the only water, I made signs to the two on the reef to go to the water; but they still continued talking and making signs which I could not understand; it seemed as if they wished us to go away, which I was determined not to do. They then made a number of furious frantic gestures, shaking their spears, and twirling them round their heads, etc. etc., I suppose bidding us defiance. I should think the youngest was about twenty-five years of age. He placed a very long spear into the instrument they throw them with, and, after a few more gestures, descended from the reef, and gradually came a little nearer. I made signs of encouragement for him to come on, at the same time moving towards him. At last we arrived on the banks of the creek, he on one side, and I on the other. He had a long spear, a womera, and two instruments like the boomerang, but more the shape of a scimitar, with a very sharp edge, having a thick place at the end, roughly carved, for the hand. The gestures he was making were now signs of hostility, and he came fully prepared for war. I then broke a branch of green leaves from a bush, and held it up towards him, inviting him to come across to me. As he did not seem to fancy that, I crossed to where he was, and got within two yards of him. He thought I was quite near enough, and would not have me any nearer, for he kept moving back as I approached. I wished to get close up to him, but he would not have it; we then stood still, and I tried to make him understand, by signs, that all we wanted was water for two or three days. At last he seemed to understand, nodded his head, pointed to the water, then to our camp, and held up his five fingers. I then endeavoured to learn from him if there was water to the north or north-east, but I could make nothing of him. He viewed me very steadily for a long time, began talking, and seeing that I did not understand him, he made the sign that natives generally do of wanting something to eat, and pointed towards me. Whether he meant to ask if I was hungry, or to suggest that I should make a very good supper for him, I do not know, but I bowed my head as if I understood him perfectly. We then separated, I keeping a careful watch upon him all the time I was crossing the creek. Before I left him the other one joined. The first was a tall, powerful, well-made fellow, upwards of six feet; his hair was very long, and he had a red-coloured net tied round his head, with the ends of his hair lying on his shoulders. I observed nothing else that was peculiar about them. They had neither skins nor anything round their bodies, but were quite naked. They then took their departure. A short time afterwards I saw them joined by five others. We have seen nothing more of them to-day, and I hope they will not trouble us any more, but let me get my horses rested in peace. Wind south, all the clouds gone; nights and mornings very cold. Occupied during the day in shoeing horses, and repairing and making saddle-bags.

Thursday, 14th June, Bishop Creek. On examining the water holes, I find there are small crab fish in them, which leads me to think this water is permanent. This morning we again hear the voices of the natives up the creek to the west. There must be plenty more water up there, as most of the birds go in that direction to drink, passing by this water. The natives have not come near us to-day, but we have seen the smoke of their fires. Shoeing horses, repairing and making saddle-bags, which were torn all to pieces by the scrub.

Friday, 15th June, Bishop Creek. Resting horses, and getting our equipment in order for another trial, as I think the horses will be ready to start on Monday morning. No more of the natives but their smoke is still visible. Wind south; day hot, night cool.

Saturday, 16th June, Bishop Creek. The horses are still drinking an immense quantity of water; they are at it five and six times a day; they must have suffered dreadfully. The grass here is as dry as if it were the middle of summer, instead of winter. I hope we may soon have rain, which would be a great blessing to me.

Sunday, 17th June, Bishop Creek. The horses still pay frequent visits to the water. We have found more about a mile up the creek, and there seems to be plenty further up in the hills; I cannot examine it just now, in consequence of the natives being about. It would not do for me to leave, as the party is so small, nor do I like sending one of them, for he might be taken by surprise and cut off, which would ruin me altogether, being able to do scarcely anything myself. Although I am much better, I am still very weak; the pains in my limbs are not so constant. I attribute the relief to eating a number of native cucumbers which are in quantities on this creek. The horse that was kicked by the mare is still very lame. Wind south-east.

Monday, 18th June, Bishop Creek. Started at 9.30 on a bearing of 18 degrees, through a plain of alternate grass, scrub, and spinifex, and at five miles passed a number of isolated hills close together, composed of large masses of ironstone, quartz, and a hard brown rock, very irregular, and all sorts of shapes; the stones seem as if they had undergone the action of fire. We then proceeded through some very bad spinifex, dark-coloured, long, hard and dry; we could scarcely get the horses to face it. We then came upon a grassy plain, and at ten miles struck a gum creek coming from the west of north-west, and running (at this place) east-north-east; followed it and found an abundance of water in long deep holes, with shells of the crab fish lying on the banks. The water is upwards of a mile in length; the creek then spreads out over a grassy plain with scrub and gum-trees, and is joined by the other creeks coming from the McDouall range. I thought it advisable to camp here for the rest of the day, as a further journey would be a risk for the horse that is lame, and I do not wish to lose any more; as it is, I am afraid he will not be able to cross Short range, which I hope to do in a few hours. Natives about. Splendid grass on this plain, and on the banks of the creek, which I have named Phillips Creek, after John Phillips, Esquire, J.P., of Kanyaka. Wind variable.

Tuesday, 19th June, Phillips Creek. Started at 8 o'clock on the same bearing, 18 degrees. We first passed through a well-grassed plain with a little scrub, then again through hard spinifex to the range. At one mile crossed another gum creek with water in it, coming from Short range. At four miles reached the top of the spur of the range; and at seven miles, the top of the range. About two miles to the east, the range seems to terminate in a gum plain, a spur from the McDouall range running on the other side of the plain, and crossing our line a few miles further on. Short range here is composed of quartz, ironstone, and red granite, with a little limestone. Descended into the plain, and at ten miles came upon another gum creek, spreading over a grassy plain, but could find no water. At thirteen miles came upon some dry swamps with a number of birds about them. At fourteen miles reached the top of the next range. From this the appearance of the country, on this course, is evidently very scrubby. On a bearing of 55 degrees, in the far distance, is the termination of another range. I do not like facing the scrub again so soon after my late loss, and with my horses not yet recovered. I shall return to the swamps and look for water. If I find any, I shall start in the morning for the end of the distant range. My lame horse is unable to do more to-day; crossing the range has been very hard upon him. Returned to the swamps and found a fine pond of water. Camped. The water is derived from the creek that we passed in the middle of the day. I have named these ponds after Kekwick, in token of the zeal and activity he has displayed during the expedition.

Wednesday, 20th June, Kekwick Ponds. Saddled at sunrise, and proceeded to the top of the low range, from which I turned back yesterday, and changed my course to 56 degrees to the northernmost point of the distant hills, through a plain of alternate grass and spinifex. At 3 o'clock struck the William Creek again, with splendid grass on its banks. It ran nearly our course for about three miles, and then turned to the east. We then entered the same sort of scrub as that in which I lost my horses; this continued until we reached the hills, which we did in about eighteen miles. From this we can see a range to the south-south-east. About ten miles off there is a large lake, with red sand hills on the east side. I cannot see the extent of it, the hills that I am now on being so low; they are composed of granite, and run north and south. To the north and north-east is another lake, about the same distance, to which I shall go on a course of 32 degrees 30 minutes. On the north side of this one there are also sand hills with scrub. For two miles after leaving the hills we passed through a soft, sandy, scrubby country and spinifex. It then became harder, with grass and spinifex alternately. At four miles from the hills we camped without water. My horses have not recovered from their last trial, and seem to be very tired to-night, although to-day's journey was not a long one, but it has been very hot, and the scrub thick and difficult to get through.

Thursday, 21st June, Scrub. The horses having gone back on the track, we did not get a start until 8.30—course, 32 degrees 30 minutes to a high hill on the other side of the lake, passing through a thick scrub of cork-tree and gums, with spinifex and grass. At seven miles came upon what I thought was the lake, but it turns out to be a large plain of rich alluvial soil covered with dry grass, which gave it the appearance of a lake. It was three miles across to the top of the hill; no water-course through, nor any water to be seen. The hills on the north side are composed of ironstone and granite, and, from the distance, looked very much like sand hills. From the top of the hill I can see the plain extending a little to the west of north, but I cannot see far for the mirage. To the north-north-east is another plain of the same description, but much smaller, about a mile and a half broad, and nearly circular. To the north-east is another very extensive one; its dimensions I cannot see. I seem to have got into the land of grassy plains and low stony hills. I wish my horses had had water last night or yesterday. They seem to be very much in want of it. I must devote the rest of this day to a search for it. I shall now direct my course for the south part of the plain that I have just crossed; it seems to be the lowest part, and the flight of the birds is directed that way. Searched all round, but can find no water; so I must return to Kekwick Ponds. The day is extremely hot, and my horses cannot stand two more nights without water. Would that they had more endurance! It is dreadful to have to turn back almost at the threshold of success. I cannot be far from the dip of the country to the Gulf. Returned by another course to where I camped last night, but still no water. I would fain try the plain to the south, but I dare not risk the loss of more horses. Proceeded to the low range that I crossed yesterday; examined round it, but cannot find any water. Camped. Two of the horses very much done up. I must go back through that nasty scrub again.

Friday, 22nd June, Under the West Low Range. Started at sunrise for the ponds, and at 1.30 arrived; the horses being very much exhausted. I am glad I did not remain another night without water; three of them are completely done up, and it has been with difficulty that we have got them here. Wind south-west.

Saturday, 23rd June, Kekwick Ponds. Resting horses. About 1 o'clock we were visited by two natives, who presented us with four opossums and a number of small birds and parrots. They were much frightened at first, but after a short time became very bold, and, coming to our camp, wanted to steal everything they could lay their fingers on. I caught one concealing the rasp that is used in shoeing the horses under the netting he had round his waist, and was obliged to take it from him by force. The canteens they seemed determined to have, and it was with difficulty we could get them from them. They wished to pry into everything, until I lost all patience and ordered them off. In about half an hour two other young men approached the camp. Thinking they might be in want of water, and afraid to come to it on account of the horses, I sent Ben with a tin dishful, which they drank. They were very young men, and too much frightened to come any nearer. About an hour before sundown, one of the first that had come, returned, bringing with him three others, two of whom were young, tall, powerful, well made, and good-looking, and as fine specimens of the native as I have yet seen. On their heads they had a neatly-fitting hat or helmet close to the brow, and rising straight up to a rounded peak, three or four inches above the head and gradually becoming narrower towards the back part. The outside was net-work; the inside was composed of feathers very tightly bound together with cord until it was as hard as a piece of wood; it may be used as a protection from the sun, or as armour for the battle-field. One of them had a great many scars upon him, and seemed to be a leading man. Only two had helmets on, the others had pieces of netting bound round their foreheads. One was an old man, and seemed to be the father of these two fine young men. He was very talkative, but I could make nothing of him. I have endeavoured, by signs, to get information from him as to where the next water is, but we cannot understand each other. After some time, and having conferred with his two sons, he turned round, and surprised me by giving me one of the Masonic signs. I looked at him steadily; he repeated it, and so did his two sons. I then returned it, which seemed to please them much, the old man patting me on the shoulder and stroking down my beard. They then took their departure, making friendly signs until they were out of sight. We enjoyed a good supper from the opossums, which we have not had for many a day. The men are complaining of weakness from the want of sufficient nourishment. I find the quantity of rations is not enough; five pounds of flour per week is too little for many weeks together. It may do very well for a month or so, but when it comes to the length of time we have been out, we all feel it very much; and the dried meat that I brought with me being very young, it has not half the strength in it that old meat has.

Sunday, 24th June, Kekwick Ponds. Our black friends have not made their appearance to-day.

Monday, 25th June, Kekwick Ponds. Started again on a bearing of 345 degrees to some very distant hills, to see if I can get into the face of the country to the Gulf of Carpentaria. At two miles crossed a large gum creek (with long beds of concrete ironstone), which I have named Hayward Creek, after Frederick Hayward, Esquire. The banks are beautifully grassed, and extend for four miles on the north side. At fourteen miles struck a gum creek with large sheets of water in which were plenty of ducks, native companions, black shags, cranes, and other birds. Camped here for the remainder of the day. The course of the creek at this point is to the north of east, and coming from the north of west, apparently from the range, which is distant about ten miles. It very much resembles Chambers Creek. The ponds (in which we found some small fish) are about eighty yards broad, and about three quarters of a mile long, having large masses of concrete ironstone at both ends, separating the one pond from the other; large gum-trees being in the ponds. Wind north-west. Very hot.

Tuesday, 26th June, Large Gum Creek, with Sheets of Water. I have resolved to follow this creek down to-day, and, if the water continues, to follow it out. Started on a course 77 degrees, and at six miles crossed the creek, which is running a little more to the north. There are long sheets of water all the way down to this, the banks in some places being steep, with the lower part formed of concrete, and the upper red sandy soil, which gives me a bad opinion of it for water, if the concrete ceases. Here we saw some blacks; they would not come near us, but walked off as fast as they could. From the top of the rise we saw where they were camped, on the banks of a large sheet of water; we passed on without taking any more notice of them, and at nine miles, not seeing any appearance of the creek, I changed my course to 25 degrees. At three quarters of a mile cut it again, but without water in it; it is much narrower and deeper, having sandy banks and bed. Changed again to 77 degrees, the creek frequently crossing our course, and at fifteen miles saw there was no hope of obtaining water. The country is becoming more sandy, and is thickly covered with spinifex and scrub. We crossed down to the banks of the creek; no rising ground visible. I must keep closer to the hills, and, as the day has been very hot, I shall return and camp at nine miles from our last camp, if there is water; if not, I shall have to camp a short way above where we saw the natives this morning. I do not wish to get too near them, or to annoy them in any way. We could find no water below where they were camped; I therefore pushed on to get above them before dark. At half-past one o'clock, about three miles from the creek, we saw where they had been examining our tracks, and as we approached the creek their tracks became very numerous on ours. When we arrived on the top of the rise, where we had previously seen their camp and fires, we could now see nothing of them, neither smoke, fires, nor anything else: it was then nearly dark. I concluded they had left in consequence of having seen us pass in the morning, as natives in general do. I was moving on to the place where we crossed the creek in the morning, when suddenly from behind some scrub which we had just entered, up started three tall powerful fellows fully armed, having a number of boomerangs, waddies, and spears. Their distance from us was about two hundred yards. It being so nearly dark, and the scrub we were then in placing us at a disadvantage, I wished to pass without taking any notice of them, but such was not their intention, for they continued to approach us, calling out and making all sorts of gestures apparently of defiance. I then faced them, making every sign of friendship I could think of. They seemed to be in a great fury, moving their boomerangs above their head, bawling at the top of their voices, and performing some sort of a dance. They were now joined by more of their tribe, so that in a few minutes their numbers had increased to upwards of thirty; every bush seemed to produce a man. Putting the horses on towards the creek, and placing ourselves between them and the natives, I told my men to get their guns ready, for I could see they were determined upon mischief. They paid no regard to all the signs of friendship I kept constantly making, but were still gradually approaching nearer and nearer to us. I felt very unwilling to fire upon them, and still continued making signs of peace and friendship, but all to no purpose. Their leader, an old man, who was in advance, made signs with his boomerang, which we took as a signal for us to be off. They were, however, intended as tokens of defiance, for I had no sooner turned my horse's head to comply with what I thought were their wishes, than we received a shower of boomerangs, accompanied by a fearful yell; they then set fire to the grass, and commenced jumping, dancing, yelling, and throwing their arms into all sorts of postures, like so many fiends. In addition to the thirty that already confronted us, I could now see many others getting up from behind the bushes. Still I felt unwilling to fire upon them, and tried again to make them understand that we wished to do them no harm. Having now approached within about forty yards of us, they made another charge, and threw their boomerangs, which came whistling and whizzing past our ears, one of them striking my horse. I then gave orders to fire, which stayed their mad career for a little. Our pack-horses, which were on before us, took fright when they heard the firing and fearful yelling, and made off for the creek. Seeing some of the blacks running from bush to bush, with the intention of cutting us off from our horses, while those in front were still yelling, throwing their boomerangs, and coming nearer to us, we gave them another reception, and I sent Ben after the horses to drive them on to a more favourable place, while Kekwick and I remained to cover our rear. We soon got in advance of those who were endeavouring to cut us off, but they still kept following, though beyond the reach of our guns, the fearful yelling still continuing from more numerous voices, and fires springing up in every direction. It being now quite dark, with the country scrubby, and our enemies bold and daring, we could be easily surrounded and destroyed by such determined fellows as they have shown themselves to be. Seeing there is no hope with such fearful odds (ten to one at least) against us, and knowing all the disadvantages under which we labour, I very unwillingly make up my mind to push on to our last night's camp. We have done so, and now I have had a little time to consider the matter over I do not think it prudent to remain here to-night; I shall therefore continue on until I reach the open grassy plain or gum creek. They are still following us up; I only wish that I had four more men, for my party is so small that we can only fall back and act on the defensive. If I were to stand and fight them (which I wish I could) our horses must remain unprotected, and we, in all probability, should be cut off from them. Our enemies seem to be aiming at that, and to prevent our advance up the creek; by this time they have found out their mistake, as we did not go a step out of our course for them. Arrived at Hayward Creek at 11 o'clock at night.

Wednesday, 27th June, Hayward Creek. This morning we see signal fires all around us. It was my intention last night to have gone this morning to Kekwick Ponds to water the horses, then to give them the day to rest, and proceed to-morrow back again to the large creek, and go on to the distant hills that I was steering for on the 25th instant, but, after considering the matter over the whole night, I have most reluctantly come to the determination to abandon the attempt to make the Gulf of Carpentaria. Situated as I now am, it would be most imprudent. In the first place my party is far too small to cope with such wily, determined natives as those we have just encountered. If they had been Europeans they could not better have arranged and carried out their plan of attack. They had evidently observed us passing in the morning, had examined our tracks to see which way we had gone, and knew we could get no water down the creek, but must retrace our steps to obtain it above them; they therefore lay in wait for our return. Their charge was in double column, open order, and we had to take steady aim, to make an impression. With such as these for enemies in our rear, and, most probably, far worse in advance, it would be destruction to all my party for me to attempt to go on. All the information of the interior that I have already obtained would be lost. Moreover, we have only half rations for six months, four of which are gone, and I have been economizing as much as I possibly could in case of our having to be out a longer time, so that my men now complain of great weakness, and are unable to perform what they have to do. Again, only two showers of rain have fallen since March, and I am afraid of the waters drying up to the south, and there is no appearance of rain at present. The days are now become very hot again, and the feed for the horses as dry as if it were the middle of summer. The poor animals are very much reduced in condition, so much so that I am afraid of their being longer than one night without water. Finally, my health is so bad, that I am hardly able to sit in the saddle. After taking all those things into consideration, I think it would be madness and folly to attempt more. If my own life were the only sacrifice, I would willingly risk it to accomplish my purpose; but it seems that I am destined to be disappointed; man proposes, but the Almighty disposes, and his will must be obeyed. Seeing the signal fires around, and dreading lest our black friends at Kekwick Ponds might have been playing a double part with us, in spite of their Masonic signs, I gave them a wide berth, and steered for Bishop Creek. Arrived there in the afternoon, and found that the creek had not been visited by natives since we left. These natives do not deposit their dead bodies in the ground, but place them in the trees, and, judging from the number of these corpses which we have passed between this and the large creek, where they made their attack upon us, they must be very numerous. These natives have quite a different cast of features from those in the south; they have neither the broad flat nose and large mouth, nor the projecting eyebrows, but have more of the Malay; they are tall, muscular, well-made men, and I think they must have seen or encountered white men before.

Thursday, 28th June, Bishop Creek. Camped at the rocky water hole north-east side of the McDouall range.

Friday, 29th June, Anderson Creek. Crossed the McDouall ranges and camped on a gum creek on the north-east side of the Murchison ranges, which I have named Gilbert Creek, after Thomas Gilbert, Esquire, late Colonial Storekeeper.

Saturday, 30th June, Gilbert Creek. Crossed the Murchison ranges, and the large gum creek coming from them, and running west-north-west, which I have named Baker Creek, after the Honourable John Baker, M.L.C. I did not examine it, but should think from its appearance that there is water in it; besides, I can distinguish the smoke of a native encampment. Proceeded to the creek where we camped before, but found all the water gone, except a little moisture in the bottom of the holes. I was rather surprised at this, for I thought it would have lasted three months at least. Went to another creek, where there was a large hole of water in conglomerate rock; this we found also to be very much reduced; when we last saw it, its depth was four feet, and now it is only eighteen inches. Camped.

Sunday, 1st July, Murchison Ranges. My horses very tired, and three of them are nearly done up.

Monday, 2nd July, Murchison Ranges. Proceeded to the Bonney Creek to get feed for the horses, there being very little besides spinifex under the ranges. Smoke of native encampments on and about the creek; I must be very careful.

Tuesday, 3rd July, The Bonney Creek. We have not seen any more of the natives yet. I shall rest the horses to-day, there being plenty of feed, which they very much want. Being so very few of us, I am obliged to turn them out with the saddles on; so that, if we are attacked again, one can put the packs on, while I and the other defend him. The water in this hole is very much reduced, but I think it will not fail altogether, in consequence of the small fish being in it. From the diminution of the water in this creek since I left it, a month ago, I am inclined to think that I shall have a very hard push to get back; my horses being so weak from the hardships they have undergone, that they are now unable to do as much as they did before. I fear that I shall not get any water between this and Forster's range, a distance of upwards of eighty miles, so I shall rest them here for a week, if the natives will be quiet; if not, I must run the risk of losing more of them. To-day, I had made up my mind to follow out this creek, to see if the waters continue, and if it would take me to the north of the spinifex and gum-tree plain which I had to turn back from on my north-west course from Mount Denison, and if rain falls to try again for the Victoria River. I am, however, disappointed, for, on weighing the rations, I find I am terribly short, which I did not expect, and which cuts off all hope of my attaining that point. My troubles and vexations seem to come upon me all at once. Had I but a stronger party, and six months' rations, I think I should be able to accomplish something before my return. I have done my best, and can do no more. My eyesight is now so bad that I cannot depend upon my observations, which will be a great loss to me; and the scurvy has returned with greater severity. Before I start on my return, if everything goes right, I shall run down this creek a short distance. It may, at some future time, turn out to be the road to the Victoria River, or one of its tributaries. Wind south and south-west.

Wednesday, 4th July, The Bonney Creek. The water in this hole has been diminishing very rapidly since we were here; it is falling at the rate of six inches per day, which is a poor look-out for us on our homeward course. I have not a day to spare now, as the weather is becoming very hot, and will dry it up much faster. I must push back as soon as my horses are rested and able to undergo the eighty miles without water. I must give up the examination of this creek, for every day now is of the utmost importance, and I must not give the horses one mile more than I can help. Oh! that rain would fall before I leave this. It would indeed be an inestimable blessing. Wind from all points. At sundown a few clouds have made their appearance.

Thursday, 5th July, The Bonney Creek. During the night it became very cloudy from the west, and this morning still continues. My hopes are again raised. If it should rain, I shall try for the Victoria River again, even though I should be without rations for my return; I could kill one of the horses and dry his flesh, and that would take me back. Still very cloudy, and every sign of rain. I am making preparations for another trial. At sundown there are still heavy black clouds coming from the west, which have raised our hopes of success to the highest point, and I ardently trust they will be realized. No natives have come near us, yet they are still about.

Friday, 6th July, The Bonney Creek. A sad, sad disappointment; all our most sanguine hopes are again gone, for, during the night, the clouds broke up and have all vanished; it is very vexing. I shall rest the horses till Monday, and then, ill and dispirited, commence my homeward journey. I dare not venture into a new route, for, want of water, and the low condition of my horses, compel me to keep my former track. Last night about 10 o'clock, I observed the comet for the first time, above the west horizon; it set at 7 o'clock 20 degrees north of west. At sundown it has become overcast with heavy clouds, and my hopes are again raised; I trust we may get it now. Midnight: still cloudy, and every appearance of rain. Wind changeable.

Saturday, 7th July, The Bonney Creek. Alas! all the clouds are again gone; our hopes were only raised to be dashed down with greater disappointment. The wind has returned to its old quarter, south-east. Natives still about, but they do not come near us. I shall now prepare for my return on Monday morning; it is very disheartening.

Sunday, 8th July, The Bonney Creek. The weather has every appearance of being dry for some time to come, not a cloud to be seen; the wind south-east, and very cold night and morning. All hope of making the coast is now gone. On weighing our rations to-day, I find that we are again short since we halted here. The man Ben has been making it a regular practice to steal them since he has been with me. I have caught him several times doing so, and all the threats and warnings of the consequences have had no effect upon him. They deter him for a day or two, and then he is as bad as ever. I have been in the habit of reducing our allowance to make up for the loss, which has been very hard upon Kekwick and myself; he has helped himself to about double his allowance during the journey.

Monday, 9th July, The Bonney Creek. Started for the Davenport range, where we camped before; the water is all dried up. Ascended the range, and changed my bearing to Mount Morphett, 196 degrees, in the Crawford range, in the hope of finding water there. At four miles struck the creek that I have before crossed nearer to the range, found water, and camped to give my horses every chance. I have named this creek Barker Creek, after Mr. Chambers' brother-in-law. I do not think this water is permanent, but, from the number of birds that are passing up the creek, I think there must be permanent water higher up. This range seems to yield a deal of water on both sides. Native graves about.

Tuesday, 10th July, Barker Creek. Started at 6.30 on a bearing of 196 degrees towards Mount Strzelecki. At six miles crossed a gum creek, coming from the range, and running to the west, on my former track. I crossed it where it lost itself on the plain. The country is well grassed, with a little spinifex occasionally, from the range to this point. At twelve miles it became scrubby and sandy with a little grass, spinifex predominating, which continued to where we camped. Wind, south-east.

Wednesday, 11th July, Scrub North-north-east of Mount Strzelecki. One of the horses having parted from the others, and gone a long distance off in search of water, it was 9 o'clock before we could get a start. At seven miles arrived at a lagoon north-east of Mount Strzelecki. Found a little water and feed for the horses. Camped to give them the benefit of it. Wind, south-south-east. Cold.

Thursday, 12th July, Lagoon North-east of Mount Strzelecki. Made an early start, crossing the range, on a south course. Very rough and difficult. Could see no water. To the south-east of Mount Morphett there is the appearance of a creek, and on the south-west there are also the signs of a watered country, which is more hilly. Proceeded on through the thick dead mulga scrub, to the north side of Forster range, where we camped at dark without water. The country passed over to-day is splendidly grassed, especially as we approached the range. There is also a little spinifex, but not much. Distance to-day, thirty-two miles.

Friday, 13th July, North Side of Forster Range. Started early, proceeding to the gum creek coming from the north side of Forster's range, where we found a little water, numerous fresh tracks of natives, and a great number of birds. I have named this the Barrow Creek, after J.H. Barrow, Esquire, M.P. Crossed the range to the Stirling Creek, which we followed down, and found an abundant supply of water. The upper part of it is now dry, and it is difficult to say whether it is permanent or not; but, to judge from the number of native tracks and encampments, and the many birds, I should think it is. The wood-duck is also on some of the pools. At dark we can hear the natives down the creek.

Saturday, 14th July, Stirling Creek. I shall give the horses a rest to-day and to-morrow, for I do not expect to get water before we reach the reservoir in the Reynolds range. I am afraid it will be all gone in the Hanson and at the Centre.

Sunday, 15th July, Stirling Creek. Resting horses, etc., etc.

Monday, 16th July, Stirling Creek. The natives were prowling about during the night, and startled three horses, which separated from the others, went off at full gallop, and were not recovered till noon, about four miles off. Too late to start to-day, for which I am very sorry, as every hour is now of the utmost value to us, in consequence of the evaporation of the water. Not the slightest appearance of any rain yet. Wind, south.

Tuesday, 17th July, Stirling Creek. Proceeded to the Hanson. Shortly after we started, we were followed by the natives, shouting as they came along, but keeping at a respectful distance. They followed us through the scrub for about two hours, but when we came to the open ground at the lagoons they went off. I intended to have halted and spoken to them there, thinking it would not be safe to do so in the scrub. They were tall, powerful-looking fellows, and had their arms with them. We then went on to the Hanson, crossing numerous fresh native tracks. On nearing the water, we saw five blacks, who took fright and went off at full speed. There were many more in the distance; in fact, they seemed to be very numerous about here. The country all round was covered with their tracks. Found water still there, but had to clear the sand away a little to give the horses a drink. Thinking that it would not be safe to camp in the neighbourhood of so many natives, I went on to the Central Creek, and in going through some scrub, we again disturbed some more, but could only see children, one a little fellow about seven years old, who was cleaning some grass seeds in a worley, with a child who could just walk. The moment he saw us he jumped up, and, seizing his father's spear, took the child by the hand and walked off out of our way. It was quite pleasing to see the bold spirit of the little fellow. On nearing Central Mount Stuart we saw two men, who made off into the scrub. Arrived at the creek after dark, but the water is all gone. On examining the hole where the water was, we discovered a small native well, with a very little water, too little to be of any service to me. To-morrow morning I must push on through the scrub to Anna's Reservoir. My horses are still very weak, and I do not think they will be able to do it in a day. Wind variable.

Wednesday, 18th July, Centre. Starting early, we crossed the Hanson, and got through the scrub to the gum plains, where we camped at sundown, the horses not being able to do the whole journey in one day. The creeks empty themselves into the plains, but there is no water. Still, from the number of birds that are about, I think there must be water not far away, but I have no time to search for it. If I do not find water in the gum creek (which is doubtful) the horses will have another long day's journey. They are suffering much from the dryness of the feed, three of them being infected with worms. Wind, south-east.

Thursday, 19th July, Gum Plains. Made our way through the remainder of the scrub, and arrived in the afternoon at the gum creek, where we found a little water, and clearing away the sand, obtained enough for our horses. There will be enough for them to-day and to-morrow morning. I shall therefore stop here for the rest of the day. There are some heavy clouds coming up from the west and south-west, which I hope will give us rain. Wind still from the south-east. The natives have been upon our old tracks through the whole of the scrub in great numbers, and there are many traces of them about this creek, some of which are quite fresh. The drying up of the water round about has compelled them to collect round this and other creeks which are permanent.

Friday, 20th July, Gum Creek North-east of Mount Freeling. Crossed the Reynolds range to Anna's Reservoir, which is still full of water. I may now say that this is permanent. The water we camped at is gone, but there is still a little down the creek. We could not get enough for the horses this morning in the creek we have left. Judging from the number of native tracks that we have crossed this morning, there must be permanent water on the north side of the range, which is composed of immense blocks of granite, apparently on the top of mica slate, with occasional courses of quartz and ironstone. To the north-east of where we camped last night, about three miles distant, is the point of the range, on which there is a very remarkable high peak, composed of ironstone, with a number of very rough rounded ironstone hills. I have named this Mount Freeling. Here I found indications of copper, the only place I have seen it in all this journey. The natives do not seem to have frequented this reservoir much of late, as there were no fresh tracks within two miles of it. In the creek close by, there were some very old worleys. No rain; clouds all gone. Wind, still south-east.

Saturday, 21st July, Anna's Reservoir, Reynolds Range. I shall remain here till Monday morning to rest the horses, for they need it much; they all have sore backs. A small pimple made its appearance under the saddle, and has gradually spread into a large sore, which we cannot heal up; it makes them very weak. The clouds have again made their appearance from the north-west, and the wind has also changed to that quarter. I hope we shall now get some rain, so that I can make short journeys for my horses, to enable them to gather strength. Two long journeys on successive days without water would reduce them again to the same state of weakness as they were in at the Bonney Creek. For the last fourteen days we have been getting a quantity of the native cucumber and other vegetables, which have done me a great deal of good; the pains in my limbs and back are much relieved, and I trust will soon go away altogether if these vegetables hold out. We boil and eat the cucumbers with a little sugar, and in this way they are very good, and resemble the gooseberry; we have obtained from one plant upwards of two gallons of them, averaging from one to two inches in length, and an inch in breadth.

Sunday, 22nd July, Anna's Reservoir. On examining the creek near the reservoir, we have found some more large and deep water holes. I have named this Wicksteed Creek. The clouds are again heavy, and have every appearance of rain; they and the wind both come from the north-west.

Monday, 23rd July, Anna's Reservoir. No rain has fallen; again all the clouds are gone. Started early for the spring in the North gorge, McDonnell range, which we noticed on April 14th. Camped at dark in the thick scrub and spinifex. No feed for the horses, so we had to tie them up during the night. Wind, south-east again.

Tuesday, 24th July, Dense Scrub and Spinifex. Started through the remainder of the scrub to the gorge, where we arrived at 7 o'clock, after twelve hours' journey. Camped outside, and drove the horses up to the spring. There is still the same supply of water; it is an excellent spring, and might be of great importance to future exploration. I have named it Hamilton Spring. Wind, variable.

Wednesday, 25th July, Hamilton Spring, McDonnell range. Resting the horses. Yesterday afternoon we passed a great number of fresh tracks of natives apparently going to Hamilton Peak, which leads me to think there must be permanent water there. The peak is very high—quite as high as Mount Arden, but there is another part of the spur higher than it, to which I have given the name of Mount Hugh; further to the west-north-west is a mount, still higher, which I have named Mount Hay. Wind, north-east. It has been very hot to-day.

Thursday, 26th July, Hamilton Spring, McDonnell Range. Started across the ranges to Brinkley Bluff, and camped on the east side. There is still plenty of water in the Hugh, although greatly reduced. The natives have been following our former tracks in great numbers; some of their foot-prints are very large. There is a great quantity of marble in this creek.

Friday, 27th July, Brinkley Bluff, McDonnell Range. Started down the Hugh, and camped on the south side of Brinkley Bluff, finding plenty of water all the way, in holes of various sizes, with reeds and rushes growing round them, with plenty of feed on the banks. Wind, variable.

Saturday, 28th July, The Hugh, South Side of Brinkley Bluff, McDonnell Range. Proceeded towards the Waterhouse range, and stopped at my former camp of the 11th April. The spring still gives out an abundance of water; we have also found another good spring on the south side of the creek, which is here very broad, nearly two hundred yards wide, with a good feeding country all round, and a small strip of salt-bush on the banks. Splendid gum-trees in the creek. Wind, east; sun, hot.

Sunday, 29th July, The Hugh, between McDonnell and Waterhouse Ranges. Wind variable; some clouds coming from the south-west.

Monday, 30th July, The Hugh, between McDonnell and Waterhouse Ranges. Proceeded towards the range; at four miles crossed the creek, and half a mile further entered the ranges. We made our former camp of April 9th on the creek, but no water, so followed it down to the westward, and after clearing a hole, found sufficient for our wants in the sand. Camped. Very unwell. Wind, south-east. Not a drop of rain has fallen since we were here before.

Tuesday, 31st July, Between the Waterhouse and James Ranges. Started on a course of 220 degrees, following down the creek through James range, instead of crossing it. I am afraid there will be no water at our camp on the south side. I have a chance of getting some in the range. At two miles met with a good water hole, under a sandstone hill. At seven miles the creek enters the range; the bed is broad, sandy, and gravelled. At twelve miles we found some water, and camped, as I am too unwell to continue in the saddle any longer. Cleared a hole, and obtained water sufficient for our purpose. Wind, south-east.

Wednesday, 1st August, In James Range, on the Hugh. Followed the creek through the remainder of the range, and found water in four different places. I have not the least doubt that there is plenty, but the creek is so broad, and divided into so many courses, that it would require four men at least to examine it well. On arriving at our camp of the 7th April, we found all the water gone. Scratched in the sand, and found a little moisture, but no water; after a fruitless search of an hour, I was going back to the last water that I had seen, six miles distant, when two emus came into the creek, and made for a large gum-tree in the middle. On going to it, I found a fine hole of water round its roots. Camped. Wind the same.

Thursday, 2nd August, The Hugh, South Side of James Range. Went down the south side of the creek, through good grassy country. At fourteen miles in a side creek we found a native well about four feet deep. We camped here, as there is little prospect of finding any more water in the Hugh, which is become broad and sandy. As to surface water, my men have neither the strength nor the appliances for digging. There is plenty of water under this sand, but having only a small tin dish, the labour is too great. My men have now lost all their former energy and activity, and move about as if they were a hundred years old; it is sad to see them; our horses, too, suffer very much from their sore backs. On the south side of the creek are some isolated hills, chiefly composed of limestone, ironstone, quartz, and granite. This morning there was ice on the water left in the tin dish, and also in the canteens, an eighth of an inch thick. It was very cold.

Friday, 3rd August, James Range. I find the water in the well is nearly all gone this morning. It would take us nine hours to water the horses here, so slowly does it come in; I must therefore go back to our last camp. I shall follow the creek round, for there might be a chance of getting some nearer. Saddled, and proceeded up the creek, and at four miles found a little under the limestone rocks coming from a small side creek; gave the horses a drink turning back, and made for the Finke on a course of 160 degrees. Crossing a few stony hills and small plains, at ten miles, we ascended a broken table range, which I have named Warwick Range; it is composed of hard grey limestone and ironstone. We then proceeded through a well-grassed country, with mulga bushes, and at twenty miles camped under a redstone hill, not being able to get any further. No water.

Saturday, 4th August, Small Hill between the Hugh and the Finke. The horses strayed a long way in the night, so that I did not get them till after 11 o'clock this morning, and could not start until noon. Passed over a country of much the same description as yesterday, crossing three stony hills running nearly east and west, and at nine miles camped, without water, in a fine grassy country, which, as the grass is green, will be quite a treat for the horses. About six miles north of Chambers Pillar. Wind, south-east.

Sunday, 5th August, North of Chambers Pillar. At sunrise heavy clouds came up from the south-east, bringing with them a very thick fog, through which I had great difficulty in steering my course; it cleared off about 10 o'clock. I expected rain, but none has fallen; it is now quite clear again. Arrived at the Finke at 12 o'clock, and was very much surprised to find so little water. I had no idea it would have gone away so soon. The bed is very broad and sandy, which is the cause of the rapid disappearance of the large quantity that I saw when I crossed before. This is a great disappointment, as it was my intention to run it down, in the hope that it would take me into South Australia. I shall go one day's journey down, and see what it is; if I can find no more water I must return to this, to rest my horses, and push for the Stevenson. I cannot remain here, for this water will only last a short time. My provisions will barely carry me down, and there is not the least appearance of rain. I am afraid my retreat is cut off. Wind, south-east. Clouds.

Monday, 6th August, The Finke. Thick fog again this morning. From the heavy clouds that have passed yesterday to the south of us, I think a shower of rain may have fallen there; I ought not to allow the chance of it to escape, as it is likely to be my only one until the equinox, and I have not provisions sufficient to remain until that time, so I must push the horses as far as they will go, and then we must walk the rest, which is a very black prospect, considering the weak state we all are in. Proceeded to the south-east, having camped on my former course at two clay-pans, where I think there is a chance of water, if a shower has fallen there. Started on our former course and arrived at the clay-pans without seeing a drop of water; neither is there any in them. Camped; the horses being very tired, from coming through so many sand hills.

Tuesday, 7th August, Clay-pans in Sand Hills. A light dew fell last night and this morning, which I am very glad of; it will be a good thing for the horses. Kekwick was unwell last night, but I cannot stop on his account. He must endure it the best way he can. If I find water at where I suppose the Finke joins the gum creek that runs a little north of Mount Humphries, I will remain there a day to give him rest. He is completely done up. I hope he will not get worse. I must push back as quickly as possible, and get him into the settled districts. At noon we made the Finke. Still the same white, sandy bed; but here it is about a quarter of a mile broad, and the east bank is composed of white sandstone, with a course of light slate on the top of it, then courses of limestone and other rocks, and, on the top of all, red sand hills. The gum-trees are not so large as they are further north. On first striking the creek we could find no water, but, by following it down for a short distance, we discovered a little, which will do for us. It is more than I expected, and I feel most thankful for it. Kekwick still very ill. Poor fellow, he is suffering very much. I dare not show him much pity, or I should have the other giving in altogether. I hope and trust he will soon get better again, and that to-morrow's rest may do him good. He has been a most valuable man to me. I place entire confidence in him. A better one I could not have got. I wish the other had been like him, and then neither he nor I should have suffered so much from hunger. Wind, south-east.

Wednesday, 8th August, The Finke. Resting Kekwick and shoeing horses. This water was going away very rapidly, so I rode down the creek for ten miles to see if there were any more, that I may risk following it down. After joining the West Creek it spreads itself over a broad valley, bounded on the north by sand hills and on the south by stony hills. Course, eastward. It is divided into numerous courses; very sandy, and immense quantities of drift wood about it. Some very large gum-trees piled high on the banks, and a great number of birds of every description; but I could find no water. It is so broad, with so many courses, that it would require half a dozen men to examine it well. If we were to stay searching for water here, and be unsuccessful, and the creeks on ahead were to be dried up, we should lose our horses and have to walk, which Kekwick could not do. I do not consider it would be right thus to risk his life. I shall therefore make for the Stevenson, where I am almost certain to find water. Wind, east.

Thursday, 9th August, The Finke. Started early on our former tracks, passing Mount Humphrey and Mount Beddome. Camped at our old place. I should think from the appearance of the country that the Finke takes a south-east course from where I left it yesterday. The hills run that way. Wind, south-south-east.

Friday, 10th August, South of Mount Beddome. Proceeded on our former course to the Stevenson, which we made a little before dark, and found water, but I am quite surprised to see so little of it left. The fine large holes are nearly dry. Wind, east.

Saturday, 11th August, The Stevenson. The horses having lost some shoes, I am forced to remain here to-day to put others on. There is more water a little further down the creek, at which I camped. No rain seems to have fallen since I was here before. The sun has been very hot to-day. Wind, east-south-east.

Sunday, 12th August, The Stevenson. I was too unwell to move yesterday, but, feeling a little better this morning, I rode down the creek. For three miles it takes a south-east course, then east-south-east through table land, with rocky and precipitous hills on each side. I then went on a south-east course for nine miles, through a splendidly-grassed country, with numerous small creeks running into the Stevenson. During my ride I found plenty of water, and splendid grass, up to the saddle-flaps, and quite green. Ducks and numerous other birds abound here; the water is quite alive with them. I regret that I have not provisions enough to enable me to follow this creek round its different bends. It is a splendid feeding country for cattle, and much resembles Chambers Creek. Wind, south-east.

Monday, 13th August, The Stevenson. Started on a course of 135 degrees to see if the Stevenson comes from the south; continued on the table land, from where I left it yesterday for sixteen miles from last night's camp, when we suddenly dropped into the bed of a large broad sandy gum-creek, coming from the west, which I find to be the Ross. There are many rushes about it; it runs in three or four courses, in all of which water can be obtained by scratching in the sand. There are plenty of birds. It is evidently raining to the east of this. Camped. My course takes me across the middle of a range, which I shall endeavour to cross to-morrow. There are two small springs, but they are brackish. Wind, south.

Tuesday, 14th August, The Ross. Started on the same course, 135 degrees, and again ascended the stony table land. Crossing thence, we met two small myall-creeks running north-east with birds upon them. At seven miles crossed another, and found a fine large deep water hole with ducks on it. We again ascended the table land, which continued to the range, and at sixteen miles gained the top, which is table land about a mile broad; the view is extensive to the east-north-east and north. We descended on a course of 175 degrees to search for water in the creek below. We crossed a number of myall-creeks, coming from the range, and running south-east; in many the water has just dried up. At six miles on the same course we found water and camped, the horses being tired by their rough journey. This water hole is not permanent although when full it is deep and large, and will last a considerable time. The Stevenson and Ross seem to take a north-east course. On a further examination of this creek I found a large hole of water about two hundred yards long and thirty broad, with birds upon it, and plants that grow round permanent water. I also found shells. This creek I have named Anderson Creek, after James Anderson, Esquire, of Port Lincoln, and the range Bagot Range, after the Honourable the Commissioner of Crown Lands.

Wednesday, 15th August, Anderson Creek. Started towards the south-east point of Bagot range, which I find to be five miles distant. The country between is undulating and stony, with plenty of grass. To the east, about thirty miles, is a high isolated hill, bearing 100 degrees. At six miles and a half crossed a myall and gum creek, in which, about a mile to the east, under a red bank, is a large water hole, seemingly permanent. At ten miles crossed the Frew, whose bed is sandy, and has many courses, the banks being covered with rushes. The rest of the day's journey was through mallee scrub and sand hills, in which we camped without water; the feed, however, is abundant, yet not so thick as when I crossed before.[14] Wind, south.

Thursday, 16th August, Mulga Scrub and Sand Hills. Started at 7 o'clock on a course of 170 degrees, and in four hours made the Neale, and camped, as there was still plenty of water.

Friday, 17th August, The Neale. Proceeded on a south-east course, and camped on a side branch of the Neale, with plenty of water in large holes. Wind, east.

Saturday, 18th August, Side Branch of the Neale. Proceeded towards the gap in Hanson range, and camped near one of the large water holes. It is very cloudy.

Sunday, 19th August, Gap in the Hanson Range. Still cloudy, and looks like rain, so we must push on to-day, in case the Peake River should come down and stop us, which would not suit the state of my provisions, as we have lost a quantity of flour by the scrub scoring the bags, and we have not enough to take us to Chambers Creek. At eight miles camped west-north-west of Freeling Springs, having given the horses a drink in crossing the Neale.

Monday, 20th August, Sand Hills West-north-west of Freeling Springs. It still threatens for rain. Proceeded to Kekwick Springs to see if the horse we had left in the Peake had got out. We found his bones; he does not seem to have made a struggle since we left him, as he is in the same position. From the number of tracks, the natives must have visited him. Proceeded to Freeling Springs and camped. There were a number of ducks and two swans on the large water hole. We shot one of the latter, which was a great treat to our half-starved party. Wind variable.

Tuesday, 21st August, Freeling Springs. Still cloudy, and we had a few drops of rain during the night; also distant thunder and lightning. Resting horses. Wind, north-east.

Wednesday, 22nd August, Freeling Springs. Proceeded through Denison range, and camped at the Milne Springs. Wind, north-east. Still cloudy, but no rain.

Thursday, 23rd August, Milne Springs. Went on and camped at Louden Spa. Wind variable.

Friday, 24th August, Louden Spa. Camped at the William Springs. Wind, north-west.

Saturday, 25th August, William Springs. Proceeded to the Strangway and

---

[14] See ante, March 28, 29, and 30.

Beresford Springs, and camped at Paisley Ponds. Wind, north-east.

Sunday, 26th August, Paisley Ponds. During the night thunder and lightning from the north-west, with a few drops of rain. Cloudy this morning; had a few showers on our journey to Hamilton Springs. Found Mr. Brodie camped there three miles south-east of Mount Hamilton. He received and treated us with the greatest kindness.

Mr. Stuart and his party remained at Hamilton Springs until 1st September, when they proceeded to Chambers Creek, where, having reached the settled districts, his journal ends.

JOURNAL OF MR. STUART'S FIFTH EXPEDITION, FROM NOVEMBER, 1860, TO SEPTEMBER, 1861

WHEN Mr. Stuart reached Adelaide, in October, 1860, on his return from his last expedition, bringing with him the intelligence that he had penetrated to the northward almost as far as the eighteenth degree of south latitude, and had only been forced to retreat by the hostility of the natives, the South Australian Parliament voted a sum of 2500 pounds for a larger, better-armed, and more perfectly organized party, of which he was to be the leader. The ill-fated Victorian expedition, under Burke and Wills, had already started from Melbourne, on the previous 20th of August, amid all the excitement of a popular ovation, but a messenger was instantly despatched by the Victorian Government to overtake them, in order to give them what information the South Australian Government allowed to be known. On the 29th of November Mr. Stuart was ready to start once more, and left Moolooloo with seven men and thirty horses, arriving at Mr. Glen's station on the 1st of December, and at Goolong Springs on the 4th. He was delayed at the latter place for several days, in consequence of the horses, and more especially the town horses, being unmanageable and unequal to their work. The party reached Welcome Springs on the 8th, and Finniss Springs on the 11th. The water at Finniss Springs seemed to have an injurious influence on the town horses, but those that had been with Mr. Stuart on his previous journeys were not so much affected. The following evening they arrived at Chambers Creek, where they remained until the end of the month.

During their stay at Chambers Creek they were occupied in killing and drying bullocks, mending saddles, weighing rations, shoeing horses, and generally preparing to start. Several of the horses, which had been knocked up and left behind on the way, had to be brought up; others became quite blind, one was lost, and one died. On the 31st of December four fresh horses arrived, which had been kindly sent up by Mr. Finke the moment he heard of the difficulty in which Mr. Stuart was placed. The party was also further increased, both by horses and men, so that when it left Chambers Creek, on the 1st of January, 1861, it numbered twelve men and forty-nine horses. The following is the list of those who started:—

John McDouall Stuart, Leader of the Expedition.
William Kekwick, Second in Command.
F. Thring, Third Officer.
— Ewart, Storekeeper.
— Sullivan, Shoeing Smith.
— Thompson, Saddler.
— Lawrence.
— Masters.
J. Woodforde.
— Wall.
E.E. Bayliffe.
J. Thomas.

Shortly after starting, the horses that Mr. Finke sent up went off at a gallop, taking with them one of the others; but, at about a mile, they were headed by Ewart, Wall, and Lawrence, and brought back covered with sweat. Not content with this gallop, in a short time afterwards they bolted again. This last one seemed to content them, for they went very quietly for the rest of the day; they had, however, lost a pick, which could not be found. The party arrived at Mr. Ferguson's station, at Hamilton Springs, that evening. Louden Spa was reached on the 8th of January. The next day Mr. Stuart writes:

"Wednesday, 9th January, Louden Spa. I am obliged to leave two horses. I thought that I should have been able to have got them down as far as Mr. Levi's station. There are three others that I must leave behind; they are now nearly useless to me, and cause more delay than I can afford. I shall reduce my party to ten individuals, in order to lighten the horses that I take with me. I shall take thirty weeks' provisions; the rest I shall leave there (Mr. Levi's station). The two men who are to return are to have a month's provisions to carry them down. They will be here two weeks, and if the horses have not recovered by that time, they will remain another week, when they will have one week's provisions to take them to Chambers Creek, where they will get enough to carry them to the mine."

Bayliffe and Thomas were the two men selected to return, and it may not be without interest to follow them back to the settled districts. They did not arrive at Melrose, Mount Remarkable, until the latter end of March. Thomas was suffering severely from rheumatism, and had to be conveyed in a cart for the last six miles of his journey from a place where he and his companion had camped for the purpose of recruiting themselves. They had been obliged to leave two of the horses at Mr. Mather's station, and two more had died on the road. The men arrived with one horse only, which they were using as a pack-horse.

But to return to the rest of the party, who reached Mr. Levi's station the same evening (January 9th) on which they parted from the two men. On Friday, January 11th, Mr. Stuart writes:

"I have now all put in order, and consider myself fairly started, with thirty weeks' provisions. Day extremely hot. An eclipse of the sun took place at noon. Although our poor little dog Toby is carried on one of the pack-horses, he is unable to bear this great heat. I fear he will not survive the day. Arrived at Milne Springs about 5 p.m. At sundown poor little Toby died, regretted by us all, for he had already become a great favourite."

On January 21st Mr. Stuart reached the Neale Creek, a little to the east of where he struck it before, but found that the large bodies of water had nearly all gone; by digging in the sand of the main channel, however, they obtained sufficient for their immediate wants. Exploring parties were despatched up and down the creek, and returned, reporting abundance of water eight miles above and five miles below where they were. They also brought back with them some fish, resembling the bream, which were very palatable when cooked. An attack of dysentery prevented Mr. Stuart from proceeding for a few days, and, during his stay, the natives, while studiously keeping themselves out of sight, set fire to the surrounding grass. On the 27th the expedition arrived at the Hamilton, after a heavy journey of thirty-five miles. "I observed," says Mr. Stuart, "a peculiar feature in one of the families of the mulga bushes; the branches seemed to be covered with hoar frost, but on closer examination it turned out to be a substance resembling honey in taste and thickness. It was transparent, and presented a very pretty appearance when the sun shone upon it, making the branches look as though they were hung with small diamonds."

The course now taken was through Bagot range to the Stevenson, where they arrived on February 1st. The next day they proceeded northward, and at eight miles came upon a large water hole, which was named Lindsay Creek, after J. Lindsay, Esquire, M.L.A. This water hole was one hundred and fifty yards long, thirty wide, and from eight to fifteen feet deep in the deepest parts. The native cucumber was growing upon its banks, and the feed was abundant. Here they met with immense numbers of brown pigeons, of the same description as those found by Captain Sturt in 1845. There were thousands of them; in fact, they flew by in such dense masses that, on two occasions, Woodforde killed thirteen with a single shot. The travellers pronounced them first-rate eating. Many natives, tall, powerful fellows, were seen, but they did not speak with them. After trying for water in the neighbourhood of Mount Daniel, they were compelled to return to Lindsay Creek, which they did not quit until February 9th, when they camped on another creek, which was named the Coglin, after P.B. Coglin, Esquire, M.L.A. From this place Mr. Stuart started, accompanied by Thring and Woodforde, to examine the condition of the Finke, and found its bed broad, and filled with white drift sand, but without water. A hole ten feet deep was sunk in the sand, but just as the increasing moisture gave them hope of finding water, the sides gave way, and Thring had a narrow escape of being buried alive. After sinking several other holes, but without success, they turned to another creek, coming more from the westward, and in a short time discovered six native wells near to what was evidently a large camping-place of the natives. The ground for one hundred yards round was covered with worleys, and at one spot they seemed to have had a grand corroberrie, the earth being trodden quite hard, as if a large number had been dancing upon it in a circle. They had left one of their spears behind, a formidable weapon about ten feet long, with a flat round point, the other end being made for throwing with the womera. On the 13th Mr. Stuart and his two companions returned to the camp on the Coglin, after discovering a place about four miles from the six native wells, where sufficient water could be obtained by digging. On the 14th three of the men were sent in advance to dig a hole at this place, and the following day the whole party moved forward to join them. Here the natives annoyed them much by setting fire to the grass in every direction.

Marchant Springs (on the Finke) were reached on February 22nd, and here Mr. Stuart noticed a remarkable specimen of native carving. He says: "The natives had made a drawing on the bark of two trees—two figures in the shape of hearts, intended, I suppose, to represent shields. There was a bar down the centre, on either side of which were marks like broad arrows. On the outside were also a number of arrows, and other small marks. I had a copy of them taken. This was the first attempt at representation by the natives of Australia which I had ever seen."

Following the course of the Finke, they arrived on the 25th at some springs which were rendered memorable by Mr. Stuart's favourite mare Polly. She became very ill, and on the morning of the 26th slipped her foal. Polly had been with her master on all his previous journeys, and was much too valuable and faithful a creature to be left behind; besides, she was second to none in enduring hardship and fatigue. They therefore waited another night to give her time to recover, and Mr. Stuart named the springs Polly Springs in her honour. On the 27th they again moved northwards, still following the course of the Finke, and, after a short journey of ten miles, camped at what were afterwards called Bennett Springs. It is worthy of remark that while the horses were in this water drinking, one of them kicked out a fish about eight inches long and three broad—an excellent sign of the permanency of the water. Here several of the horses were taken violently ill, and the next morning one of them could not be found. Mr. Stuart writes:

"Thursday, 28th February, The Finke, Bennett Springs. Found all the horses but one named Bennett. Sent two of the party out in search of him; at 9 a.m. they returned, having been all round, but could see nothing of him. I then sent out four, to go round the tracks and see if he had strayed into the sand hills. At noon they returned unsuccessful. Sent five men to search, but at 2 p.m. they likewise returned without having discovered him. I then went out myself, and, in half-an-hour, found the poor animal lying dead in a hole, very much swollen. Blood seemed to have come from his mouth and nostrils. He must have died during the night. I am afraid that there is some description of poisonous plant in the sand hills, and that the horses have eaten some of it. As he lay he appeared to have been coming from the sand hills, and making for the water. He seemed to have fallen down three times before he died. I never saw horses taken in the same way before—in a moment they fell down and became quite paralysed. The cream-coloured horse, that was taken so ill last night, must also have eaten the poison. We were upwards of two hours before we could get him right. As soon as he got on his legs, his limbs shook so that he immediately fell down. This he did for more than a dozen times. As we were very much in want of hobble-straps, I sent Mr. Kekwick, with three others, to take Bennett's skin and shoes off. We found no indication of poison on opening him. This is a very great loss to me, for he was one of my best packhorses—one that had been with me before, and that I could depend upon for a hard push."

On the 2nd March, while still following the course of the Finke, they passed two or three holes containing fish about eight inches long, and enclosed by small brush fences, apparently for the purpose of catching fish. They also saw a lot of shields, spears, waddies, etc., which the natives had deposited under a bush. As to the aborigines themselves, although it was evident there were plenty of them about, they never allowed themselves to be seen. There was an abundance of timber which Mr. Stuart says would be well suited for electric-telegraph poles.

Mr. Stuart's journal continues as follows:

Tuesday, 5th March, The Finke. Started at 8.5 a.m., bearing 345 degrees, for the Hugh, with Thring and Lawrence. On arriving there found the water nearly all gone, only a little in a well dug by the natives; cleared it out, but it took us until 12 p.m. to water the four horses. At three miles further, we passed round a high conspicuous table hill, having a slanting and shelving front to the south; this I have named Mount Santo, after Philip Santo, Esquire, M.P. The country passed over to-day has been sand hills, with spinifex, grassy plains, with mulga and other shrubs, and occasionally low table-topped hills, composed of sand, lime, and ironstone, also the hard whitish flinty rock; kangaroo plentiful, but very wild. Wind south-east. The day has been very hot; horses very tired.

Wednesday, 6th March, The Hugh. Started at 8.45 a.m. on a bearing of 209 degrees. At nine miles, finding the water gone that I had seen on my last return, I dug down to the clay, and obtained a little, but not enough for us. Followed the creek up into the gorge, and found it very dry. Our former tracks are still visible in the bed of the creek. No rain seems to have fallen here since last March. I had almost given up all hopes of finding any water, when, at seven miles, we met with a few rushes, which revived our sinking hopes; and, at eight miles, our eyes and ears were delighted with the sight and sound of numerous diamond birds, a sure sign of the proximity of water. At the mouth of a side creek coming from the James range, on the eastern side of the Hugh, found an excellent water hole, apparently both deep and permanent. We saw a native and his lubra at the upper end at a brush fence in the water; they appeared to be fishing, and did not see us until I called to them. The female was the first who left the water; she ran to the bank, took up her child, and made for a tree, up which she climbed, pushing her young one up before her. She was a tall, well-made woman. The man (an old fellow), tall, stout, and robust, although startled at our appearance, took it leisurely in getting out of the water, ascended the bank, and had a look at us; he then addressed us in his own language, and seemed to work himself up into a great passion, stopping every now and then and spitting fiercely at us like an old tiger. He also ascended the tree, and then gave us a second edition of it. We leisurely watered our horses, and he was very much surprised to see Thring dismount and lead the pack-horse down to the water, so much so that he never said another word, but remained staring at us until we departed, when he commenced again. This water being sufficient for my purpose, I will go no further up the creek, but return to the last night's camp. Wind, south-east.

Thursday, 7th March, The Hugh. As my horses are very tired, and the distance between my main camp on the Finke and the water we discovered yesterday being upwards of fifty miles, I will remain here to-day, dig down to the clay, and try if I can obtain enough water for all the party; for, owing to the extreme heat, and the dryness of the feed, many of our weak horses are unable to go a night without water. By 8 p.m. we dug a trench ten feet long, two feet and a half deep, and two feet and a half broad; it is about twelve feet below the level of the creek. We have had a very hard day's work. Wind, south-east. Day very hot.

Friday, 8th March, The Hugh. This morning very cold; wind, still south-east. The trench is quite full; our four horses made very little impression on it. I shall send up and enlarge the trench, so that we may be enabled to water the whole lot. At 6.40 a.m. started back for the camp. At 1.45 p.m. halted to give the horses a little rest. At 2.30 p.m. changed to 184 degrees, and at four miles reached the table hills, but there was no creek, only a number of clay-pans, all quite dry, with stunted gum-trees growing round them. Changed my bearing to Mount Santo, passing a number of clay-pans of the same description; from thence proceeded to the camp; arrived there at sundown, and found all right. Plenty of water; the horses make little impression on it. Wind, south-east.

Saturday, 9th March, The Finke. I shall give Thring a rest to-day, and will send him with two others, and a part of the horses, to-morrow to the Hugh, to make a place large enough to water all. From about 2 a.m. until after sunrise the morning has been very cold. Wind, south-east.

Sunday, 10th March, The Finke. At 7 a.m. despatched Thring, Thompson, and Sullivan, with eleven pack and three riding-horses, to the Hugh to dig a tank. Wind, still south-east; clouds east.

Monday, 11th March, The Finke. Clouds all gone; wind still south-east. I will remain here to-day with the rest of the party, to give the others time to have all ready for us when we arrive. One of the horses missing; found him in the afternoon. Wind variable.

Tuesday, 12th March, The Finke. Started at 8.30 a.m. for the Hugh, course 345 degrees, following our former tracks. The day has been exceedingly hot; wind from east and south-east, with heavy clouds in the same direction. About 3 p.m. missed the party that was behind; they were last seen about one mile and a half back. Thinking that the packs had gone wrong, and that they were remaining behind to repair them, I waited an hour, but finding they did not come up, I sent Ewart back to the place where they were last seen to find out what was wrong; in an hour he returned, and informed me that their tracks were going away to the eastward. As the James range was in sight, and two of the party had been there before, I concluded that they must have lost my tracks and were pushing on for the water. This loss of two hours would make it late before we arrived there, so we hurried on; but within four miles it became so dark, from the sky being overcast with heavy clouds, and the mulga bushes being so thick, that we were in great danger of losing some of our pack-horses, for we could not see them more than ten yards off. I therefore camped until daylight, having to tie the horses during the night. Wind variable.

Wednesday, 13th March, Between the Finke and the Hugh. Started at daybreak; and in a little more than an hour arrived at the Hugh; found that Thring had gone up the creek to the other water, not finding enough here for the horses he had with him. We could only get sufficient for ten of ours. As the fire was still alight, I was led to believe that the other party had arrived here last night, having had two hours more sunlight than we, and that they, seeing Thring's note to me, which he had fastened on a tree, and also the small quantity of water, had watered their horses last night, and gone on this morning, leaving the water that had accumulated during the night for us and our horses; we cleared out the hole in order to obtain sufficient for our other five. At about 10 a.m. had breakfast; before we finished, the other party came in sight; they had lost the tracks, and could not find them again. They made the creek about one mile to the eastward. Unsaddled and gave their horses a rest, and as much water as we could get for the weak ones; those of mine which have had none will have to go without. By 1 p.m. obtained a drink for seven of them. Pushed on to the other water, fifteen miles up the creek; arrived there a little before sundown. The day, although cloudy, has been very hot. Found Thring and his party all right. They had seen no more of our spitting friend. Wind variable, with heavy clouds from east and south-east, but still no rain.

Thursday, 14th March, The Hugh, James Range. As the done-up horses will not be able to travel to-day, I have sent Thring and Wall up the creek to look for other water. Sky still overcast. No rain. Thring and Wall returned in the afternoon, having found water a little below the surface, about nine miles up; a very light shower has fallen. Wind all round the compass.

Friday, 15th March, The Hugh, James Range. A few drops of rain have fallen during the night, but this morning it seems to be breaking up again, which is a great disappointment. Started at 8 a.m., course 10 degrees west of north; passed through the gorge in James range, found all the water gone that I had seen on my journey down; followed up the creek to the native wells that Thring found yesterday. This water is situated about one mile and a half from where the creek enters the gorge in James range, and under a concrete bank on the north side. The natives seem to have quitted this water on hearing us coming, for they have left behind them a large, long, and unfinished spear, two smaller ones, and some waddies, one of which was quite wet, as if the owner had been in the act of clearing out one of the wells when he heard or saw us coming: he also left a shield cut out of solid wood, which I think was, from its lightness, cork-wood. I also observed on one of the gum-trees, marks similar to those which I saw on the Finke, broad arrows and a wavy line round the tree. Still cloudy, but much broken. No rain. Wind, south-east.

Saturday, 16th March, The Hugh, James Range. Rain all gone. Proceeded up the creek, course 30 degrees, to examine the east bend before it enters the Waterhouse range; in about six miles arrived and followed it upwards, pushing on through the gorge to the large water I had previously seen on the north side of the range; found it gone, but water in some native wells in its bed. Proceeded on to the second bend of the creek from Waterhouse range, to a water which I consider to be a spring (it is under conglomerate rock), and am glad to see that there is still a large hole of beautiful water, with bulrushes growing round about it. Camped. This water I have named Owen Springs, after William Owen, Esquire, M.P. Wind variable, from south-east to north-east. Cloudy.

Sunday, 17th March, Owen Springs, The Hugh. During the night we had a few light showers, which will be of great advantage to us, causing the green feed to spring up. The morning still cloudy; wind from the east, with a few drops of rain. Wind still variable—all round the compass.

Monday, 18th March, Owen Springs, The Hugh. Very heavy clouds this morning; and it seemed as if it was setting in for a wet day, but it cleared off, and only a little rain fell. Wind still all round the compass.

Tuesday, 19th March, Owen Springs, The Hugh. Saddled and started for Brinkley Bluff, bearing 349 degrees. After entering the McDonnell range the water is permanent. It has been here for twelve months; no rain has fallen during that time, for my former tracks, both up and down, are as distinct as if they had been made a month ago. At 3.30 p.m. camped at the waterhole about a mile north-west of Brinkley Bluff; it is situated under a rocky cliff. There are some seams of beautiful grey granite crossing the creek, and abundance of marble of all colours, also a little iron and limestone. We found some specimens of the palm tree, but there is neither seed nor blossom at this season of the year. Lawrence got one of the leaves, ate the lower end of it, and found it sweet—resembling sugar-cane; he ate a few inches of it, and in about two hours became very sick, and vomited a good deal during the evening. Wind variable; but mostly south-east, with heavy thunder clouds.

Wednesday, 20th March, Brinkley Bluff, McDonnell Range. About 1 p.m. we were delighted with the sight and feeling of heavy rain. At about 4 the creek came down, and by sunrise it was running at the rate of five miles an hour—a new and delightful sight to behold. At about 9 the clouds were breaking and the rain lighter. We were all truly thankful for this great boon. It is too wet to move to-day; the horses are bogging up to their knees. After sundown we had a heavy thunder storm, accompanied by vivid lightning, and heavy rain from south-east and east. Wind from same direction.

Thursday, 21st March, Brinkley Bluff, McDonnell Range. Rain has continued at intervals during the night; a great deal has fallen. A horse having gone into the creek to drink during the night, one of his hobbles became undone, and got fastened to his hind shoe. He was found this morning up to his body in water, and unable to move. Having relieved him, it was with difficulty he could get out. He is in a tremble all over, and can scarcely walk. The ground is so soft, even on the hills, that we cannot walk without sinking above the ankle. I should gain nothing by starting to-day. It would injure the horses more than a week's travelling.

Friday, 22nd March, Brinkley Bluff, McDonnell Range. About 1 a.m. the rain came down in torrents, and continued until nearly sunrise, from south-east. Wind from same quarter. It is impossible to move to-day. The creek is higher than it has been before, and running with great rapidity. All the horses were found right this morning but the one which got into the creek yesterday. After searching all the hills and the creeks round about, he was found in a small gully by himself.

Saturday, 23rd March, Brinkley Bluff, McDonnell Range. Heavy shower of rain about 4 a.m. this morning. After sunrise it all cleared away and became fine. Started at 8.20 to cross the northern portion of the range by following the creek up. We have had a very hard and difficult journey of it. It is now 4 p.m., and we have arrived at Hamilton Springs. The ground was so soft, even at the top of the ranges, that we had the greatest difficulty in getting the horses through. We did so, however, with the loss of a great number of shoes, and many of the horses were very lame. Wind still south-east.

Sunday 24th March, Hamilton Springs. I am compelled to have some of the horses shod to-day, and also to have a number of saddle-bags mended, which were torn by the scrub yesterday. This afternoon there is a great deal of thunder and lightning in the north and north-east.

Monday, 25th March, Hamilton Springs. Part of the horses missing this morning in consequence of the green feed; did not get a start until 10.20 a.m.; bearing 43 degrees. The country became so boggy after seven miles that we were unable to proceed further than eleven miles. There being no surface water, although the ground was so soft that the horses kept bogging up to their bodies, we were forced to retreat five miles to obtain some for them. Wind south-east, the stormy weather apparently breaking up. Camped at 5 p.m. Latitude, 23 degrees 28 minutes 51 seconds.

Tuesday, 26th March, Scrub North-east of Hamilton Springs. Started at 9 a.m. on a south-south-east course to round the boggy country. At about six miles we were enabled to cross the lower part, and go in the direction of a low range. Camped on the north-east side of it. The last four miles were over fair travelling-country of a red soil, with mulga and other bushes, in some places rather thick, abounding in green grass. We also passed many bushes of the honey mulga, but the season is passed, and it is all dried up. Wind, east. Latitude by Pollux, 23 degrees 24 minutes 51 seconds; by Jupiter, 23 degrees 24 minutes 52 seconds.

Wednesday, 27th March, Low Granite Range in Scrub. More than half of the horses are missing this morning; at noon we have managed to get all but ten; they are scattered all over the place; at 5 p.m. they cannot be found, and the water is nearly all gone, and the country much dried towards Strangway range. I have sent the horses four miles back to a large clay-pan that we saw yesterday, to remain there to-night and in the morning to return. Two of the party to separate from there, and to go in search of the missing horses, which I suppose have gone back to the Hamilton Springs; it is very vexing, some of our best are amongst them. Wind, east.

Thursday, 28th March, Low Granite Range in Scrub. At 11 a.m. the horses were brought back from the clay-pan. Two of the missing ones were found about a mile after they started, making towards where they had camped last night. I think that the other eight must be also in that direction; we find that all the tracks have gone that way; I shall therefore move down to-day to the south end of the swampy country, which I know they cannot cross, and endeavour, if possible, to find them to-night. By 1 p.m. arrived at the end of the swamp; camped, and despatched Thring in one direction and Sullivan in another to try and cut their tracks; at a little before sunset Sullivan returned with three of the missing ones. Five are still wanting. Wind, south-east.

Friday, 29th March, South End of Swamp in Scrub. At sunrise sent Thring and Sullivan again to look for the missing horses; they arrived at 5 p.m. with three of them. If we do not find the other two to-morrow, I shall push on without them, and endeavour to pick them up on our return.

Saturday, 30th March, South End of Swamp in Scrub. Again sent Thring and Sullivan in search of the two remaining horses; at about 11 a.m. they returned with them. I shall now move up to our camp of 25th instant. Camped at some rain water a little south of our former place, where there is plenty of feed for the horses. Wind, south-east; clouds from north-west.

Sunday, 31st March, Rain Water in Scrub. All day the sky has been overcast with clouds from the north-west. Wind from south-east.

Monday, 1st April, Same Place. Started at 7.30 a.m.; course, 330 degrees. At 1 p.m. we came upon a very pretty flat of beautiful grass, with water in the middle of it; and, as the afternoon has every appearance of rain, I have camped—to go on in the rain will only spoil our provisions. We had scarcely got the packs off when it came on heavily, and lasted about an hour: it then ceased until sundown, when it came on again, and continued till 10.30 p.m.

Tuesday, 2nd April, Green Flat in Scrub. Started at 8.20 a.m. on same course, and camped at 1.30 p.m. under a prominent rocky hill, which I ascended and have named Mount Harris, after Peter G. Harris, Esquire, of Adelaide. I obtained bearings of the different points all round. The last seven miles was sandy soil, with spinifex and scrub, which was mostly young cork-tree, and the broad-leafed mallee.

Wednesday, 3rd April, Mount Harris. We have put up a small cone of stones on the top of this mount. Started at 8 a.m. for Anna's Reservoir. Arrived at the creek about two miles south-south-east of it, and, finding it running, camped amongst excellent feed. By keeping to west of my former track I have found the country much opener; but nearly all day the journey has been through spinifex. Wind from west.

Thursday, 4th April, The Wicksteed, Reynolds Range. Started at 7.40 a.m. to cross the range, bearing to Mount Freeling 312 degrees. At 1.30 p.m. crossed the range, and arrived at the creek, camping at the same place as I did on my previous journey, and finding water and feed abundant. I have named this creek the Woodforde, after Dr. Woodforde, of Adelaide. After crossing the range, we found the bean-tree in blossom; it was magnificent. I have obtained a specimen of it; also some beans, a number of which were of a cream colour; we have roasted a few of them, and find that they make very good coffee. Wind, south-east.

Friday, 5th April, The Woodforde, Reynolds Range. Started at 7.30 a.m. Camped at 4.30 p.m. on the Hanson, which is now a running stream. About five miles back we passed a freshly-built native worley. I observed a peculiarity in it which I never noticed in any before—namely, that it was constructed with greater care than usual. It was thatched with grass down to the ground. Inside the worley there was a quantity of grass laid regularly for a bed, on which some one had been lying. Round about the front was collected a large quantity of firewood, as much as would have done for us for a night. Latitude, 22 degrees 5 minutes 30 seconds, bearing to Central Mount Stuart, 25 degrees. Wind, south-south-west.

Saturday, 6th April, The Hanson. Started at 8 a.m., on a course of 46 degrees 30 minutes, to the springs in the Hanson; this course led me through about four miles of very thick mulga. After crossing the central line we arrived on the creek and camped, below the springs, at 1.30 p.m. Bearing to Central Mount Stuart, 251 degrees 20 minutes. Wind variable.

Sunday, 7th April, The Hanson, East-north-east of Centre. Day hot. Wind variable, with a few clouds.

Monday, 8th April, The Hanson, East-north-east of Centre. Five of the horses missing this morning. Started at 9.45, course 45 degrees; camped on the Stirling at 3.50 p.m. Through all the day's journey the country abounded in grass and water. Wind from south.

Tuesday, 9th April, The Stirling, Forster Range. Started at 7.30 a.m., to cross Forster range on the same course. At 10.50 a.m. camped on north side of it, on a large gum creek with water. I have named this the Taylor, after John Taylor, Esquire, of the firm of Messrs. Elder, Stirling, & Co., of Adelaide. This is a most beautiful place, a plain four miles broad between two granite ranges, completely covered with grass, and a gum creek winding through the centre. I made a short journey to-day in consequence of having some of the horses lame, and some weak through the effects of the green grass, and to-morrow's journey will be a long one. Had I gone on to-day, they would in all probability be without water, and would require to be tied up during the night. I shall now be able to get through in one day, and keep them in good condition for the unexplored country, which I expect to commence next Monday.

Wednesday, 10th April, The Taylor. Started at 7.25 a.m. on a course of 11 degrees 30 minutes for Mount Morphett; at 12.30 ascended the summit. On the north side we had some difficulty in getting the horses down; however, we managed without accident. Ran a creek down and found some water; gave the horses a drink; still followed it until it was lost in a grassy plain. Proceeded on to the next hills, passed through a gap, and made for a creek on the north side, in which we found water, and camped at 4 p.m.

Thursday, 11th April, North Side Mount Morphett, Crawford Range. Started at 7.45 a.m. on a course of 10 degrees. The first four miles was over a beautiful grassy plain, with mulga wood, not very thick; it then became more sandy, and covered with gum, cork-trees, and other scrubs, which continued within a mile of where we camped, in a small, but beautiful grassed plain; no water. Latitude 20 degrees 38 minutes 33 seconds. Wind, south-east.

Friday, 12th April, Grassy Plain. Started at 6.15 a.m., same course. At 1 p.m. arrived at the Bonney; it is now running—green feed abundant. As some of the horses are still very lame, I will rest them to-morrow and Sunday, and start into the unexplored country on Monday morning. Wind from south-east; a few clouds from north-east.

Saturday, 13th April, The Bonney. Sent Thring down the creek to see what its course is, and if the country gets more open; the men mending saddle-bags, cleaning and repairing saddles, shoeing horses, etc. While I and Woodforde were endeavouring to get a shot at some ducks on the long water holes, a fish, which he describes as being about two feet long, with dark spots on either side, came to the surface; he fired at it, but was unsuccessful in killing it. A little before sundown Thring returned; he gave a very bad account of the creek; it was a dry deep channel. Wind, variable; cloudy.

Sunday, 14th April, The Bonney. Wind from every quarter, with clouds; a few drops of rain fell about the middle of the day; after sundown much lightning in the south-west.

Monday, 15th April, The Bonney. Cloudy; wind still variable. Mount Fisher, bearing 120 degrees. Started at 7.15 a.m., bearing 290 degrees; at 11.40 changed to 264 degrees, to some rising ground; at 12.45 p.m., after crossing stony hills, we crossed a gum creek on the west side, with long reaches of water in it running north-west, which I supposed to be the Bonney; but as there appeared to be more and larger gum-trees farther on, I continued, to see if there were not another channel. Proceeded three miles over low limestone rises, with small flats between, on which was growing spinifex, and the gum-trees which I had seen—exactly the same description of country from which I was forced to return through want of water on my former journey from Mount Denison to north-west. I therefore returned to the creek, which I find to be the Bonney, now much smaller, but containing plenty of water—followed it down to north-north-west for about one mile, and then camped. The water is in long reaches, which I think are permanent.

Tuesday, 16th April, The Bonney. Still cloudy. Started at 8 a.m. on a bearing of 380 degrees. At 11.15 changed to 40 degrees, with the intention of cutting the McLaren. Camped at 3.40 p.m. Three miles from our start the creek spreads itself over a large grassy plain, thickly studded with gum-trees, covered with long grass, and a great number of white ants' nests of all sizes and shapes, putting one in mind of walking through a large cemetery. In many places it was very boggy. We followed it for ten miles, but it still continued the same; I could not see more than one hundred yards before me, the gum-trees, and sometimes a low scrub, being so thick. Not seeing anything of the McLaren coming into the plain, I changed my course to cut it and run it down, as I think that it will form a large creek where they join. In three miles we got out of the plain upon a red sandy soil, with spinifex, and scrubs of all kinds, in some places very thick, and difficult to get the horses through. When we were in the gum plain the atmosphere was so close and heavy, and the ground so soft, that the sweat was running in streams from the horses; and when we halted for a few minutes they were puffing and blowing as though they had just come in from running a race. I continued the second course for fourteen miles, but saw nothing of the McLaren; it must have joined the plain before I left it. Thus ends the Bonney and the McLaren. We passed over several quartz and ironstone ranges of low hills crossing our course, and camped under a high one, without water. Wind south-east. Cloudy.

Wednesday, 17th April.[15] Quartz Hill, West Mount Blyth. Started at 7.25 a.m. on a bearing of 70 degrees. We again passed quartz hills running as yesterday; the spinifex still continuing, with a little grass, until we came within a mile of the hills in the Murchison range; finding some water, I camped, and gave the horses the rest of the day to recruit. Last night after sundown, and during the night, we had a few slight showers of rain, and a great deal of thunder and lightning, mostly from south-west. About 11 to-day the clouds all cleared away. About a mile before camping, we observed the ground covered with numerous native tracks; also that a number of the gum-trees were stripped of their bark all round.

Thursday, 18th April, West Mount Blyth. Started at 7.40 a.m., same bearing, across the Murchison range, in which we found great difficulty. On the north-east side of Mount Blyth we found a large gum creek of permanent water, and camped. I have named this Ann Creek. I then rode to the highest point of the range, taking Thring with me, to see if there is any rising ground to north-west by which I may cross the gum plain. I could see no rise, nothing but a line of dark-green wood on the horizon. We had great difficulty in getting to the top, the rocks being so precipitous. In coming down the eastern side we were gratified by the sight of a beautiful waterfall, upwards of one hundred feet high, over columns of basaltic rock, its form, two sides of a triangle, the water coming over the angle. Wind, south-east.

Friday, 19th April, Ann Creek. Started at 7.45 a.m., on a course of 324 degrees, towards Mount Samuel. After sundown arrived at Goodiar Creek; one of the horses done up; had to leave him a little distance back; he is unwell. On leaving the Murchison range we crossed a number of quartz reefs and hills running east and south-west. Wind, south-south-east.

Saturday, 20th April, Goodiar Creek. Three horses missing this morning, in consequence of the scarcity of feed. The horse left behind last night has been brought in; he looks very bad indeed. About 11 a.m. the other horses were found, brought in, and saddled, and we proceeded on a north-north-west course for Bishop Creek, but found the sick horse too ill to proceed further than Tennant Creek, where we camped, there being plenty of water and feed. Two natives were seen by Masters this morning when in search of the horses—he could not get them to come near him. Wind, south-west.

Sunday, 21st April, Tennant Creek. Wind from south-west; a few clouds from east.

Monday, 22nd April, Tennant Creek. Started at 7.30 a.m., course 21 degrees, for Bishop Creek, and at twelve miles made it. I find that two of the horses are so weak that they are unable to go any further without giving in, I have therefore camped, giving them the remainder of the day to recruit. Native fires are smoking all around us, but at some distance off. Wind, east.

Tuesday, 23rd April, Bishop Creek. It is late before we can get a start to-day,

---

[15] The Journal of this Expedition, as published by the Royal Geographical Society, commences here.

in consequence of one of the horses concealing himself in the creek. He is an unkind brute, we have much trouble with him in that respect; he is constantly hiding himself somewhere or other. Started at 9.30 a.m., on a course of 17 degrees, to cross Short range. Found plenty of water in Phillips Creek; the grass on its banks, and on the plains where it empties itself, is splendid, two feet and a half long, fit for the scythe to go into, and an abundant crop of hay could be obtained. We then crossed the range a little north of where I passed before, and found some slight difficulty. After descending, we struck a small creek which supplies Kekwick Ponds, and is a tributary to Hayward Creek; found plenty of water and camped at 3 p.m. Feed abundant. Wind, south-east.

Wednesday, 24th April, Hayward Creek. Started at 7.40 a.m.; course 17 degrees. At 9.30 changed to 14 degrees 30 minutes west of north, and at 12.30 arrived at Attack Creek; camped at the same place that I did on my former journey. Tracks of natives about, but we have seen none of them. I kept about a mile to the west of my former track, and found the country much more open. The banks of both creeks for two or three miles are splendidly covered with grass, in some places over the horses' heads. Four of the horses are ill, and looking very bad indeed. Wind, south-west.

Thursday, 25th April, Attack Creek. Started at 7.50 a.m., on a course of 294 degrees, to the top of the range, which I have named Whittington Range, after William S. Whittington, Esquire, of Adelaide. At six miles reached the top. At 9.50 changed to north-west, and at 11.30 struck a large gum creek running east, with large water holes in it. At about two hundred yards crossed it again, running to the west, and shortly afterwards crossed it again, running to the east. I have called it Morphett Creek, after the Honourable John Morphett, Chief Secretary. We then ascended another portion of the range, and continued along a spur on our course. This range presents quite a new feature, in having gums growing on the top and all round it; it is composed of masses of ironstone, granite, sand, and limestone, and in some places white marble. Thinking that the creek we had passed might break through a low part of the range, which I could see to the north-west, at ten miles I changed to west, and crossed to the other range, but found the dip of the country to the south. We could find no water; traced the creek to the south-east for two miles, found some water and camped. The range is very rough and stony, covered with spinifex; but the creeks are beautifully grassed. Native smoke to east. This is one of the sources of Morphett Creek, and flows to the east; it is as large, if not larger, than Attack Creek, and, in all probability, contains water holes quite as fine to the eastward. Latitude, 18 degrees 50 minutes 40 seconds.

Friday, 26th April, Morphett Creek. At 8 a.m. started on a course of 300 degrees to cross the north-west part of the range. Camped upon a plain of the same description as John Plain, that I met with on my former journey to the north-east of Bishop Creek, a large open plain covered with grass, and with only a few bushes on it. The journey to-day has been very rough and stony. Not a drop of water have we passed to-day, nor is there the appearance of any on before us. I shall be compelled to fall back to-morrow to the water of last night.

Four of the horses, I am afraid, will not be able to get there. I must try more to the north, and endeavour to get quit of the plains, and get amongst the creeks. There is no hope of success on this course. Latitude, 18 degrees 38 minutes. Wind, east.

Saturday, 27th April, Grassy Plains. Started at 7.10 a.m., course 110 degrees, to the other side of the plain. At three miles came upon a small creek running towards the north; I followed it down to the north. At three miles came upon a fine large creek, coming from the south-east, with plenty of water. Returned to the party, took them down to the large creek on north course, and at three miles camped. Two of the horses are nearly done up. Wind, south-east. Latitude, 18 degrees 35 minutes 20 seconds.

Sunday, 28th April, Tomkinson Creek. Sent Thring down to examine and see how the creek runs. I have named it after S. Tomkinson, Esquire, Manager of the Bank of Australasia, at Adelaide. We have found many new plants and flowers, also some trees, one of which grows to a considerable size, the largest being about a foot in diameter. The fruit is about the size and colour, and has the appearance of plums; the bark is of a grey colour; the foliage oval, and dark-green. Another is more of a bush, and has a very peculiar appearance; the seed vessel is about the size of an orange, but more pointed. When ripe it opens into four divisions, which look exactly like honeycomb inside, and in which the seeds are contained; they are about the size of a nut, the outside being very hard. The natives roast and eat them. The leaves resemble the mulberry, and are of a downy light-green. We have obtained a few of the seeds of it. The bean-tree does not seem to grow up here. Mr. Kekwick, in looking for plants this morning, discovered one which very much resembles wheat in straw (which is very tough), ear, and seed. It grows two feet high. The seed is small, but very much like wheat both in shape and colour. At about 3 p.m. Thring returned, having run the creek out into a large grassy plain. The course of this creek is west-north-west for about nine miles; it then turns to west, and empties itself into the plain. There is plenty of water about, but where it empties itself it becomes quite dry. The native companion, the emu, and the sacred ibis are on this creek. The country is splendidly grassed. We have got to the north side of the Whittington range. I shall have to leave my two done-up horses here, and will get them when I return. The hills and rocks are of the same description as the first part of the range. Wind, south. Sun hot, but the nights and mornings are very cold.

Monday, 29th April, Tomkinson Creek. Had a late start this morning in consequence of my having to take a lunar observation. Started at 10.30 a.m. At 2.10 p.m. reached the top of a high hill; from this we could see a gum creek. Started at 2.30 to examine it; found water, and camped at 4. I have named the hill Mount Primrose, after John Primrose, Esquire, of North Adelaide. This water will last us six or eight weeks. The country passed to-day has been mostly stony rises of the same description as the other parts of the range. The valleys have a light sandy soil, nearly all with spinifex and scrub. The view from the top of Mount Primrose is not extensive, except to the west and south-west, which appears to be thick wood or scrub. Near the top we met with the Eucalyptus Dumosa. Wind, south-east. Latitude, 18 degrees 25 minutes.

Tuesday, 30th April, Carruthers Creek. The creek in which we are now camped I have named Carruthers Creek, after John Carruthers, Esquire, of North Adelaide. Started at 8.50 a.m. At 1.50 p.m. found a creek running from the range, with a splendid hole of permanent water situated under a cliff, where the creek leaves the range; it is very deep, with a rocky bottom. From the top of the range the country seems to be very thick, which I am afraid is scrub; no high hills visible. To the north of this the range appears to cease; I wish it had continued for another sixty miles. The country passed to-day has been stony rises coming from the range, very rough and rocky indeed. My horses' shoes are nearly all gone; I am obliged to let some go without—they have felt the last four rough days very much. Spinifex, scrub, and stunted gums all the day, with occasionally a few tufts of grass; this is very poor country indeed. Smoke of native fires still in south-east. The hills of the same formation as those we first came upon in entering the ranges from Attack Creek. I have named this creek Hunter Creek, after Mr. Hunter, of Messrs. Hunter, Stevenson, and Co., of Adelaide. Camped. The horses seem very tired. Wind, east. Latitude, 18 degrees 17 minutes.

Wednesday, 1st May, Hunter Creek. Started at 8 a.m., course, 305 degrees. At 8.45 crossed the Hunter going south-west; it came round again and continued crossing our course thirteen times in nine miles, after which it was lost in a large grassy and gum plain. At 5.15 camped. The plain in which the creek loses itself bears south-west; the banks are beautifully grassed, but about a mile on either side the soil is sandy, with spinifex and scrub, which continued for nine miles; we then entered upon a scrub and grassy plain. Here I noticed a new and very beautiful tree—in some instances a foot in diameter—with drooping branches. Its bark was grey and rough, and it had a small dark-green leaf, shaped like a butterfly's wing. Not finding a creek, nor the least indication of a watercourse, and the scrub becoming very thick, I changed to north, to see if I could find any water; but at three miles we lost the gums, the new tree taking their place, and becoming very thick scrub with plenty of grass, but no signs of a watercourse. I again changed to east in the hope of cutting one in that direction. At one mile and a half again came upon small gums; and at three miles, seeing neither creek nor any hope of getting water, camped. The horses very tired. Wind light from west-north-west. Latitude, 18 degrees 3 minutes 19 seconds.

Thursday, 2nd May, Large Scrubby and Grassy Plain. Started at 10 a.m. in consequence of some of the horses having strayed a long way to the east during the night; course, 143 degrees 30 minutes, back to Hunter Creek. I have taken a different course to see if there is any creek that supplies this plain with water. For about nine miles we passed over a splendidly grassed plain, with gum-trees, the new tree, and a number of all sorts of bushes. One part for about three miles is subject to inundation, and the Eucalyptus Dumosa grows thickly on it. We then passed over about two miles of spinifex and grass, and again entered the grassy plain, which continued to Hunter Creek. During the whole day we have not seen the shadow of a creek or watercourse. If there had been any sign of a watercourse, or if I could have seen any rising ground near our course, I would have gone on another day. I sent Wall to the top of the highest tree to see if there was anything within view; he could see nothing but the same description of plain. If my horses can travel to-morrow, I will try a course to the north, and run down the creek, to see if there is one that will lead me through this plain. If I could get to some rising ground, I think I should be all right; but there is none visible except the end of the range, which is lost sight of to the north-east. Wind again south-east, with a few clouds. Latitude, 18 degrees 13 minutes 40 seconds.

Friday, 3rd May, Hunter Creek. Started at 8.40 a.m.; course, north. At 11.15 (nine miles), came upon a creek; bed dry and sandy; searched for water, and, at three quarters of a mile to east, found a nice hole; watered the horses and proceeded on the same course—starting at 12. At 3.20 p.m. changed to 20 degrees north of east; the first ten miles were over a plain of gums covered with grass two feet long; we had then six miles of spinifex, and a thick scrub of dwarf lancewood, as tough as whalebone. After that we entered upon another gum plain, also splendidly grassed, which continued for four miles, when the gums suddenly ceased, and it became a large open plain to north, as far as I could see. Seeing no appearance of water, I changed my course to 30 degrees north of east, to some high gums; and, at one mile, not finding any, I camped without it. This seems now to be a change of country; there is no telling when or where I may get the next water on this course, so that I shall be compelled to go towards the range to-morrow to get some, and have a long day's journey to the new country. The wind has been from east all day. Latitude, 17 degrees 56 minutes 40 seconds.

Saturday, 4th May, Sturt Plains. Started at 7.15 a.m., course east, to find water. At 3.20 p.m. came upon a little creek and found a small quantity of water, which we gave to the horses. Started again at 9 p.m., course south-east, following the creek to find more; at a mile and a half found water which will do for us until Monday morning. I proceeded to the top of the range to obtain a view of the country round, but was disappointed in its height; from the plain it appeared higher than it really is. This range I have named Ashburton Range, after Lord Ashburton, President of the Royal Geographical Society. The point upon which I am at present is about three miles east of our camp; the view from south to north-west is over a wooded plain; from north-west to north is a large open plain with scarcely a tree upon it. On leaving our last night's camp, we passed over three miles of the plain, which is subject to inundation. There are numerous nasty holes in it, into which the horses were constantly stumbling. It is covered with splendid grass, and is as fine a country as I have ever crossed. These plains I have named Sturt Plains, after the venerable father of Australian exploration and my respected commander of the expedition in 1845. Ashburton range is composed of sandstone and ironstone, granite, and a little quartz; it is very rough and broken. Native tracks about here. Wind, south-east. This creek I have named Watson Creek, after Mr. Watson, formerly of Clare.

Sunday, 5th May, Watson Creek, Ashburton Range. Sent Thring to the north along the range to see if there is permanent water; at eight miles he returned, having found plenty. One large hole is about a mile from here; in another creek it is apparently permanent, having a rocky bed. A flight of pelicans over head to-day; they seem to have come from the north-west, which course I will try to-morrow. Wind, south-east. Latitude, 17 degrees 58 minutes 40 seconds.

Monday, 6th May, Watson Creek, Ashburton Range. Started at 8.20 a.m., course 300 degrees, to cross Sturt Plain. At eleven miles arrived at the hill which I saw from Ashburton range. It turned out to be the banks of what was once a fresh-water lake; the water-wash is quite distinct. It had small iron and limestone gravel, with sand and a great number of shells worn by the sun and atmosphere to the thinness of paper, plainly indicating that it is many years since the water had left them. Judging from the water-marks, the lake must have been about twelve feet deep in the plain. The eucalyptus is growing here. We then proceeded over another open part of it, for about two miles, when the dwarf eucalypti again commenced, and continued until we camped at twenty-one miles; the horses quite worn out. This has been the hardest and most fatiguing day's work we have had since starting from Chambers Creek; for, from the time we left in the morning until we camped, we have had nothing but a succession of rotten ground, with large deep holes and cracks in it, caused at a former period by water, into which the poor horses have been constantly falling the whole day, running the risk of breaking their legs and our necks, the grass being so long and thick that they could not possibly see them before they were into them. I had a very severe fall into one of these holes; my horse came right over and rolled nearly on top of me. I was fortunate enough to escape with little injury. Some of the shells resemble the cockle shell, but are much longer, many of them being three or four inches long; the others are of the shape of periwinkles, but six times as large. Both sorts are scattered over the plain, which is completely matted with grass. The soil is a dark rich alluvial, and judging from the cracks and holes, some of which are of considerable depth, they are splendid plains, but not a drop of surface water could we see upon them, nor a single bird to indicate that there is any. It was my intention at starting to have gone on thirty miles, but I find it quite impossible for the horses to do more; it would be madness to take them another day over such a country, when from the highest tree we can see no change. If I were to go another day and be without water, I should never be able to get one of the horses back, and in all probability should lose the lives of the whole party. If I could see the least chance of finding water, or a termination of the plain, I would proceed and risk everything. I see there is no hope of my reaching the river by this course. I believe this gum plain to be a continuation of the one I met with beyond the Centre, and that it may continue to the banks of the Victoria. The features of the country are nearly the same. The absence of all birds has a bad appearance. Day very hot. Wind, south-east. Latitude, 17 degrees 49 minutes.

Tuesday, 7th May, Sturt Plains. Before sunrise this morning I sent Wall up a tree to see if any hills or rising grounds would be visible by refraction. To the west, with a powerful telescope he can just see the top of rising ground. As the grass is now quite dry, the horses feel the want of water very much; many of them are looking wretched, and I hardly think will be able to reach it. However reluctant, I must go back for the safety of the party. At 3 p.m. arrived at the creek which Thring found about one mile to the north of my former camp, with the loss of only one horse; we had to leave him a short distance behind, he would not move a step further, although during a great part of the journey he had been carrying little or nothing. This water will last two months at least; feed good. It is inside the first ironstone rise in Ashburton range, in a gum creek which empties itself into the plains. This creek I have named Hawker Creek, after James Hawker, Esquire, of her Majesty's Customs at Port Adelaide. The day has been very hot. Wind, south-east. Latitude, 17 degrees 58 minutes.

Wednesday, 8th May, Hawker Creek, Ashburton Range. I have sent Masters back to bring up the horse we left behind. Sturt Plains have been at one time the bed of a large fresh-water lake; our journey of the 6th instant was over the middle of it, and we were not at the end of it when I was forced to return; the same rotten ground and shells continued, although we had got amongst the eucalypti. I shall give the horses a rest to-day, and to-morrow will take the best of them (those that I had out on my former journey), and endeavour to cross the plain to the rising ground seen yesterday morning; I shall take Thring and Woodforde, with seven horses and one week's provisions. I may be fortunate enough to find some water, but from the appearance of the country I have little hope. I shall, however, leave nothing untried to accomplish the object of the expedition. In the morning the horse we left behind could not be found; sent Masters and Sullivan in search of him; in the afternoon they returned with him looking miserable. He had wandered away beyond the other camp.

Thursday, 9th May, Hawker Creek, Ashburton Range. Started at 7 a.m., with Thring and Woodforde, and seven horses, following our tracks through the rotten ground to the first eucalypti, for about twelve miles, as it made it lighter for the horses, the tracks being beaten to that place. Changed our course to 282 degrees, still journeying over Sturt Plains; at twenty-seven miles arrived at the end of the portion of them that had been subject to inundation, but there are still too many holes to be pleasant. I certainly never did see a more splendid country for grass; in many places for miles it is above the horses' knees. We entered upon red sandy soil, with spinifex and grass, from which we changed our bearing. The country became thickly studded with eucalypti, in one or two places rather open, but generally thick. After the twenty-seven miles we again met with the new small-leafed tree, the broad-leafed mallee, the eucalypti, and many other scrubs. At sundown we camped; distance, thirty-three miles, but not a drop of water have we seen the whole day, or the least indication of its proximity. I hope to-morrow we may be more fortunate, and find some. Wind, south.

Friday, 10th May, Sturt Plains. This morning there are a few birds about. Started at 8.15 a.m., same course; at 10.30 arrived on first top of rising ground seen from the camp of 7th instant, which turns out to be red sand hills covered with thick scrub. Changed our course to north-west, and at 11.15 arrived at the highest point; the view is very discouraging—nothing to be seen all round but sand hills of the same description, their course north-north-east, and south to west. No high hills or range to be seen through the telescope. We can see a long distance, apparently all sand hills with scrub and stunted gums on them. The first ridge is about two hundred feet above Sturt Plains, but further to the west they are much lower, and become seemingly red sandy undulating table land; but further to the west they are much lower. There is no hope of reaching the Victoria on this course. I would have gone on further to-day had I seen the least chance of obtaining water to-night; but during the greater part of yesterday and to-day we have met with no birds that frequent country where water is. Both yesterday and to-day have been excessively hot, and the country very heavy. From this point I can see twenty-five miles without anything like a change. To go on now with such a prospect, and such heavy country before me, would only be sacrificing our horses and our own lives without a hope of success—the horses having already come forty-five miles without a drop of water, and over as heavy a country as was ever travelled on. I have therefore, with reluctance, made up my mind to return to the camp and try it again further north, where I may have a chance of rounding the sand hills; the dip of them from here seems to be south-south-west. Turned back, and at eighteen miles camped on Sturt Plains, where there is green grass for the horses. Wind, south.

Saturday, 11th May, Sturt Plains. At dawn of day started for the camp; arrived at 2 p.m. It was fortunate I did not go on further, for some of the horses were scarcely able to reach it; a few more hours and I should have lost half of them. The day has been so hot that it has nearly knocked them all up. Found the rest of the party all right at the camp. We had a job to keep the horses from injuring themselves by drinking too much water. I gave them a little three separate times, tied them up for twenty minutes, and then gave them a good drink, and drove them off to feed. They took a few mouthfuls of grass, and were back again almost immediately, and continued to do so nearly all the afternoon. They drank an immense quantity. Wind, south.

Sunday, 12th May, Hawker Creek, Ashburton Range. My old horses that were out with me before look very well this morning, but the others, whose first trip of privation this has been, are looking very bad indeed. They could not have gone another night without water; it has pulled them down terribly. Yesterday, while Masters was looking for the horses, he saw what appeared to him to be a piece of wood stuck upon a tree, about two feet and a half long, sharp at both ends, broad at the bottom, and shaped like a canoe. Having pulled it down, he found it to be hollow. On the top of it were placed a number of pieces of bark, and the whole bound firmly round with grass cord. He undid it, and found the skull and bones of a child within. Mr. Kekwick brought it to me this morning for my inspection. It certainly is the finest piece of workmanship I have ever seen executed by natives. It is about twelve inches deep and ten wide, tapering off at the ends. Small lines are cut along both sides of it. It has been cut out of a solid piece of wood, with some sharp instrument. It is exactly the model of a canoe. I told him to do it up again, and replace it as it was found. If it is here when I return, I will endeavour to take it to Adelaide with me. Wind, variable. A few clouds about.

Monday, 13th May, Hawker Creek, Ashburton Range. Started at 8 a.m., course 360 degrees. At five miles crossed the large gum-tree creek, with water, that Thring found; proceeded along the side of Sturt Plains. At ten miles ascended the north point of Ashburton range; descended, and the country became red sand with spinifex, gum-tree, the new tree, and other shrubs very thick; at fifteen miles, gained the top of another stony rise; followed three creeks down in search of water; found a little, but not sufficient for us; followed it still further down, leading us to the south for about six miles, but could find no more. I thought it best to return for water to the large creek, which I have named Ferguson Creek, after Peter Ferguson, Esquire, of Gawler Town. From the top of the range the view is limited. To the north and north-east are stony rises, at about nine miles distant; from north to west are Sturt Plains, in some places wooded; to the north they are open for a very long distance; the country in the hills is bad, but in the plains is beautiful. I am afraid, from the view I have of the country to the north, that I shall again meet with the same description of sand hills that I came upon on my last western course. Wind east-south-east, blowing strong. Latitude, 17 degrees 53 minutes 20 seconds.

Tuesday, 14th May, Ferguson Creek. Started at 8.30 a.m., on a north course, to the place I turned back from yesterday; arrived at noon; changed course to 345 degrees. Started again at 12.20. At 1 p.m. crossed a gum creek that has the appearance of water. At 1.40 changed course to 260 degrees, and came upon two large water holes, apparently very deep, situated in the rocks—they are seemingly permanent. Camped. I named this creek Lawson Creek, after Dr. Lawson, J.P., of Port Lincoln. A number of natives have been camped about them. We found another canoe, of the same description as the one in which the bones of the child were found—it is broken and burned, and seems to have been used as a vessel for holding water. Wind south-east, blowing strong. Mornings and evenings very cold. Latitude, 17 degrees 43 minutes 30 seconds.

Wednesday, 15th May, Lawson Creek. Started at 8.10 a.m.; went a mile west to clear the stones; changed course to 340 degrees. At 2.45 p.m. changed again to 45 degrees. Camped at 4.15. The first twelve miles is poor country, being on the top of stony rises, with eucalypti, grass, and scrubs. After descending from the rises, we crossed a wooded plain, subject to inundation; no water. The trees are very thick indeed—they are the eucalyptus, the Eucalyptus Dumosa, the small-leaved tree, another small-leaved tree much resembling the hawthorn, spreading out into many branches from the root; it rises to upwards of twenty feet in height. We have also seen three other new shrubs, but there were no seeds on them. After crossing the plain we got upon red sandy rises, very thick with scrub and trees of the same description. We continued on this course until 2.45 p.m.; then, as there is an open plain in sight, with rising ground upon it to north-east, and as this scrubby ridge seems to continue, without the least appearance of water, I have changed to north-east. Crossed the plain, which is alluvial soil, covered with grass, but very dry. At 4.15 camped on north-east side, without water. I would have gone on to the rise, but I feel so ill that I am unable to sit any longer in the saddle. I have been suffering for the last three days from a severe pain in the chest. Wind, east. Latitude, 17 degrees 16 minutes 20 seconds.

Thursday, 16th May, Sturt Plains. Sent Thring to see if there is a creek or a sign of water under the rise. At 8.20 a.m. he returned, having found no water. It is a low sandy rise, covered with a dense scrub. Started at 8.20 a.m.; course, east. At three miles I was forced to return; the scrub is so dense that it is impossible to get through. Came back two miles; changed to 20 degrees west of south to get out of it. At two miles gained the plain, then changed to the east of south at 10.45. At 2 p.m. there is no hope of a creek or water. Changed to south-west. At two miles and a half struck our tracks and proceeded to Lawson Creek. We found the open parts of the plain black alluvial soil so rotten and cracked, that the horses were sinking over their knees; this continued for six miles. It is covered with long grass and polygonum; also a few eucalypti scattered over it. The scrub we were compelled to return from was the thickest I have ever had to contend with. The horses would not face it. They turned about in every direction, and we were in danger of losing them. In two or three yards they were quite out of sight. In the short distance we penetrated it has torn our hands, faces, clothes, and, what is of more consequence, our saddle-bags, all to pieces. It consists of scrub of every kind, which is as thick as a hedge. Had we gone further into it we should have lost everything off the horses. No signs of water. From south to west, north and north-east nothing visible but Sturt Plains, with a few sand rises having scrub on them, which terminate the spurs of the stony rises. They are a complete barrier between me and the Victoria. I should think that water could be easily obtained at a moderate depth in many places on the plains. If I had plenty of provisions I would try to make it by that way. The only course that I can now try is to the north-east or east, to round the dense scrub and plains. At sundown arrived at Lawson Creek. The horses, owing to the dryness of the grass, drank a great quantity of water; they are falling off very much. Wind, south-east.

Friday, 17th May, Sturt Plains. I must remain here to-day to mend saddle-bags, etc. I have sent Thring to north-east to see if the stony rises continue in that direction. He has returned and gives a very poor account of the country. He crossed them in about six miles, and again came upon the plain that we were on yesterday, extending from north-east to south. Nothing but plains. To the north is the dense scrub, thus forming a complete stop to further progress. From here I fear it is a hopeless case either to reach Victoria or the Gulf. The plains and forest are as great a barrier as if there had been an inland sea or a wall built round. I shall rest the horses till Monday, and will then try a course to the north-west, and another to north-east. I have not the least hope of succeeding without wells, and I have not sufficient provisions to enable me to remain and dig them. It is a great disappointment to be so near, and yet through want of water to be unable to attain the desired end. Wind, south-east.

Saturday, 18th May, Lawson Creek. Resting horses, etc. Wind, south-east.

Sunday, 19th May, Lawson Creek. Wind, south-east.

Monday, 20th May, Lawson Creek. Started at 7.25 a.m., course 15 degrees, with Thring, Woodforde, and seven horses. The first four miles was over the stony rises; the next three, sandy table-land, with spinifex, eucalyptus, and scrub. Crossed part of Sturt Plains, open and covered with grass. Five miles of it were very heavy travelling-ground, very rotten, and full of holes and cracks. At about thirty miles camped on the plains. We have seen no birds, nor any living thing, except kites and numerous grasshoppers, which are in myriads on the plains. From this place to the east, and as far as south-south-west, there is no rising ground within range of vision—nothing but an immense open grassy plain. The absence of birds proclaims it to be destitute of water. We have not seen a drop, not a creek, nor a watercourse during the whole day's journey. To-morrow I shall again try to get through the scrub. On leaving the camp this morning, I instructed Kekwick to move the party about three miles down the creek to another water hole, the feed not being good. Wind, east.

Tuesday, 21st May, Sturt Plains, East. Started at 7.10 a.m. Passed through a very thick scrub seven miles in extent. We again entered on another portion of the open plains at ten miles from our last night's camp. Nothing to be seen on the horizon all round but plains. Changed to 300 degrees, to where I saw some pigeons fly. At two miles came across their feeding-ground; skirted the scrub until we cut our tracks. No appearance of water. This is again a continuation of the open portion of Sturt Plains; they appear to be of immense extent, with occasional strips of dense forest and scrub. We had seven miles of it this morning as thick as ever I went through; it has scratched and torn us all to pieces. At my furthest on the open plain. I saw that it was hopeless to proceed, for from the west to north, and round to south-south-west, there is nothing to be seen but immense open plains covered with grass, subject to inundation, having an occasional low bush upon them. I think with the aid of the telescope I must have seen at least sixty miles; there is not the least appearance of rising ground, watercourse, or smoke of natives in any direction. The sun is extremely hot on the plain. Having no hope of finding water this morning, I left Woodforde with the pack and spare horses where we camped last night, as the heat and rough journey of yesterday have tired them a great deal; so much so, that I fear some of them will not be able to get back to water. Returned to where I had left him, and followed our tracks back to the open plain. After sundown camped among some scrub. Wind, south-east.

Wednesday, 22nd May, Sturt Plains. After sunset we saw a number of turkeys flying towards the stony rises where our main camp is; they appear to come from the north-west. Upwards of fifty passed over in twos and threes; and this morning we observed them going back again. Two of the horses which had been short hobbled walked off during the night, following our tracks. Saddled and followed, overtaking them in three miles and a half, standing under the shade of a tree. Unhobbled and drove them on before us. At 12 o'clock arrived at Lawson Creek. Had great difficulty in preventing the horses from drinking too much, and, as there are other holes down the creek, I gave them a little at a time at each. Found that Kekwick had moved with the party. Followed them, and at three miles and a half west-south-west arrived at their camp, and allowed the horses to drink as much as they chose. Poor brutes! they have had very hard work, eighty miles over the heaviest country, under a burning sun, without a drop of water. Three of them were those I had on my former journeys; I could depend upon them; the rest were the best I could pick from the other lot. They have all stood the journey very well, but could not have done another day without water. Natives seem to have been about this water lately, but we have not seen one since leaving our spitting friend on the Hugh. Wind, east.

Thursday, 23rd May, Lawson Creek. Started 7.45 a.m., course 315 degrees, with Thring, Woodforde, and seven fresh horses. At fourteen miles came across a splendid reach of water, about one hundred and fifty yards wide, but how long I do not know, as we could not see the end of it. It is a splendid sheet of water, and is certainly the gem of Sturt Plains. I have decided at once on returning, and bringing the party up to it, as it must be carefully examined, for it may be the source of the Camfield, or some river that may lead me through. On approaching it I saw a large flock of pelicans, which leads me to think that there may be a lake in its vicinity. There are mussels and periwinkles in it, and, judging from the shells on the banks, the natives must consume a large quantity. The gum-trees round it are not very large. The first ten miles of that part of the plain travelled over to-day is full of large deep holes and cracks, black alluvial soil covered with grass, with young gum-trees thicker as we approached the water. This I have named Newcastle Water, after his Grace the Duke of Newcastle, Secretary for the Colonies. Duck, native companion, white crane, and sacred ibis abound here. Returned to bring the party up to-morrow. Wind, south-east.

Friday, 24th May, Lawson Creek. Started at 8 a.m. for Newcastle Water; arrived at noon. Camped. Sent Kekwick to north-east and Thring to west to see the length of it; I have had the depth tried. It is about six feet deep ten yards from the bank, and in the middle seventeen feet. I should say it was permanent. Thring found it still the same at three miles west. Kekwick returned after following it for four miles. At two miles there is a break in it. At four miles it is more of a creek coming from north-east. Gum-trees much larger. Woodforde succeeded in catching four fish about ten inches long, something resembling the whiting. I had one cooked for tea; the skin was as tough as a piece of leather, but the inside was really good, as fine a fish as I have ever eaten. To-morrow I shall follow the water to the west; its bed is limestone. Wind, south-east, with a few clouds. Latitude, 17 degrees 36 minutes 40 seconds.

Saturday, 25th May, Newcastle Water, Sturt Plains. Started at 7.50 a.m. and followed the water nine miles round. It still continued, but became a chain of ponds. As I could see some rising ground north-north-east about four miles distant, I camped the party and took Thring with me to see what the country was before us. At four miles we found that the first part of the rise was stony, but on the top it was sandy table-land, covered with thick scrub. The view is obstructed to the east-north-east to north by it; but to the north-west and west there is an appearance of rising ground, thickly wooded, about twenty miles off. Wind, west. Latitude, 17 degrees 30 minutes 30 seconds.

Sunday, 26th May, Newcastle Water, Sturt Plains. This morning we were visited by seven natives, tall, powerfully-made fellows. At first they seemed inclined for mischief, making all manner of gestures and shaking their boomerangs, waddies, etc. We made friendly signs to them, inviting them to come nearer; they gradually approached, and Kekwick and Lawrence got quite close to them; in a short time they appeared to be quite friendly. I felt alarmed for the safety of J. Woodforde (who had gone down the water in search of ducks, and in the direction from which they had come), and endeavoured to make them friends by giving them pieces of handkerchiefs, etc. During the time we were talking with them I heard the distant report of his gun; at the same time Thring and Masters returned from collecting the horses that were missing. I told them to remain until the natives were gone, as I wished to keep them as long as possible to give Woodforde a chance of coming up before they left us; shortly afterwards they went off apparently quite friendly. Sent Thring and Wall to round up the horses which were close at hand, and while they were doing so the natives again returned, running quite close up to the camp and setting fire to the grass. It was now evident they meant mischief. I think they must have seen or heard Woodforde, and have lit the grass in order to engage our attention from him. I felt very much inclined to fire upon them, but desisted, as I feared they would revenge themselves on him in their retreat. They did very little injury by their fire, which we succeeded in putting out. By signs I ordered them to be off, and after much bother they left us, setting fire to the grass as they went along. I now ordered Thring and Wall to go with all speed to protect Woodforde. In about twenty minutes he came into the camp. After leaving us they had attacked him, throwing several boomerangs and waddies at him; he had only one barrel of his gun loaded with shot; they all spread out and surrounded him, gradually approaching from all sides. One fellow got within five yards of him, and was in the act of aiming his boomerang at him. Seeing it was useless to withhold any longer, while the black was in the act of throwing he gave him the contents of his gun in his face, and made for the camp. In a short time Thring and Wall returned at full speed; they had passed where he was, and hearing the report of his gun, made for the place, overtook the blacks, gave chase and made them drop the powder-flask and ducks (which Woodforde had laid down before firing when they attacked him); knowing them to be his, they gave up the chase to look for him, but seeing nothing of him, and two of the natives supporting one apparently wounded, they returned to the camp, where they saw him all safe, relating his adventure, his shot-belt still missing. I sent Thring and him to look for it, and to bring up the missing horses which they had seen. Wind variable. Cloudy.

Monday, 27th May, Newcastle Water, Sturt Plains. Started at 8.10 a.m., course 335 degrees. At 10.20 changed to north; at 1.20 p.m. changed to 90 degrees; and at one mile found water; gave the horses some, and proceeded north-north-east; at 3.40 changed to 90 degrees to some gums: at one mile and a half camped. The gums turn out to be thick wood. I went north-north-west this morning, with the expectation of meeting with water, or rather a chain of ponds; at four miles, I could see nothing of them; and, as we were getting into a very thick scrub of lancewood, I changed to north; and at ten miles on that course, still seeing nothing of them, I changed to east; at one mile came upon them, found water, and followed them; their course now, 20 degrees; at one mile found another pond; in a short time, lost the bed of them in a thick wooded plain. Found a native path running nearly in my course; followed it, thinking it would lead me to some other water, but in a few miles it became invisible. I continued on the same course for nine miles, and found myself on Sturt Plains, with belts of thick wood and scrub; to the north, nothing visible but open plains; to north-east, apparently thick wood or scrub; to north-west and west, apparently scrubby sand hills. The ponds seem to drain this portion of the plains. Changed to east, to what seemed to be large gum-trees, thinking there might be a creek; arriving there, I found them to be stunted gums on the edge of the plain. There is no hope of succeeding in this quarter. Camped without water. Wind, east. Latitude, 17 degrees 12 minutes 30 seconds.

Tuesday, 28th May, Sturt Plains, North. Fourteen of the horses missing this morning before sunrise. From the highest tree nothing is to be seen from east to north and north-west but immense open grassy plains, without a tree on them; no hope of water. I must go back to the ponds and try again to the westward. Did not find the horses until 9.30 a.m., and started at 10. I observed very large flocks of pigeons coming in clouds from the plains in every direction towards the ponds. Some time afterwards we saw them coming back and flying away into the plains as far as the eye can reach, apparently to feed. Arrived at the water at 1.30 p.m. Wind, east-north-east.

Wednesday, 29th May, Chain of Ponds. Started at 7.20 a.m. with Thring, Woodforde, and Wall, and nine horses, to follow a native track, which is leading to the westward. At 9.20 made the track; its course, west-north-west. At twenty-eight miles camped without water. The track led us into very thick wood and scrub, and at five miles became invisible. I still continued on the same bearing through the scrub. We have again met with the mulga—a little different from what we have seen before, growing very straight, from thirty to forty feet high, the bark stringy, the leaf much larger and thicker. Amongst it is the hedge-tree. We had seven miles of it very dense, when we again met with an open plain. At three miles entered another dense wood and scrub, like that passed through in the morning. To-day's journey has been over plains of grass, through forest and scrub, without water. In the last five miles we passed through a little spinifex, and the soil is becoming sandy. Wind, south.

Thursday, 30th May, Sturt Plains. As I can see no hope of water, I will leave Woodforde and Wall with the horses, take Thring with me, and proceed ten miles, to see if there will be a change in that distance. Went into a terrible thick wood and scrub for eleven miles and a half, without the least sign of a change—the scrub, in fact, becoming more dense; it is scarcely penetrable. I sent Thring up one of the tallest trees. Nothing to be seen but a fearfully dense wood and scrub all round. Again I am forced to retreat through want of water. The last five miles of the eleven the soil is becoming very sandy, with spinifex and a little grass. It is impossible to say in which way the country dips, for, in forty-five miles travelled over, we have not seen the least sign of a watershed, it is so level. Returned to where I left the others, followed our tracks back, and at eleven miles camped. Horses nearly done up with heavy travelling and the heat of the sun, which is excessive. It is very vexing and dispiriting to be forced back with only a little more than one hundred miles between Mr. Gregory's last camp on the Camfield and me. If I could have found water near the end of this journey, I think I could have forced the rest. It is very galling to be turned back after trying so many times. Wind, east.

Friday, 31st May, Sturt Plains. Not having sufficient tethers for all the horses, we had to short hobble two, and tie their heads to their hobbles; and, in the morning, they were gone. I suppose they must have broken their hobbles or fastenings; they will most likely make on to our outward tracks. I have sent Thring and Woodforde to follow them up, while Wall and I, with the other horses, proceed on our way to the camp. In two hours they made the tracks before us, and I then pushed on as hard as I could get the horses to go; being very anxious about the safety of the party—for, on the first day that I left them, at about seven miles, we passed fourteen or fifteen natives going in the direction of their camp; I also observed, this morning, that they had been running our tracks both backwards and forwards. At three o'clock we arrived, and found all safe; they have not been visited by them, although I observed the prints of their feet in our tracks, a short distance from the camp. It was as much as some of our horses could do to reach the camp. The day has been excessively hot; wind from north-north-east, with clouds. Latitude, 17 degrees 7 minutes.

Saturday, 1st June, Chain of Ponds. I must rest the horses to-day and to-morrow, for they look very miserable; our longitude is 133 degrees 40 minutes 45 seconds. Before leaving the Ponds I shall try once more to the westward—starting from a point three miles west of my first camp on them. To try from this, for the Gulf of Carpentaria, I believe to be hopeless, for the plain seems to be without end and without water. If I could see the least sign of a hill, or hope of finding water, I would try it; but there is none—if there is a passage it must be to the south of this. Wind variable, with clouds.

Sunday, 2nd June, Chain of Ponds. The day has again been very hot. Wind variable.

Monday, 3rd June, Chain of Ponds. Started back to the commencement of the Chain of Ponds, and camped. During the day the sky has been overcast with heavy clouds. Wind, south-east.

Tuesday, 4th June, Chain of Ponds. Last night one of the horses was drowned in going down to drink at the water hole. He went into a boggy place, got his hind foot fastened in his hobbles, from which he could not extricate himself, and was drowned before we could save him. This is another great loss, for he was a good pack-horse, and was one that I intended taking on my next trip to the westward. At about 8 p.m. it began to rain, and continued the whole night, coming from the east and east-south-east. It still continues without any sign of a break. The ground has become so soft that when walking we sink up to the ankle, and the horses can scarcely move in it. At sundown there is no appearance of a change. It has rained without intermission the whole of last night and to-day. I do not know what effect this will have on my further progress, for now it is impossible to travel. The horses in feeding are already sinking above their knees. Wind and rain from east and east-south-east.

Wednesday, 5th June, Chain of Ponds. There is a little sign of a break in the clouds this morning. The rain has continued the whole night. Ground very soft; it has become about the thickness of cream. The horses can scarcely get about to feed. Sundown: It has been showery all day; sky overcast; clouds and rain from same direction, south-east. In the afternoon some natives made their appearance at about six hundred yards' distance. As the rain had damped the cartridges I caused the rifles to be fired off in that direction; and, as the bullets struck the trees close to them they thought it best to retreat as fast as possible, yelling as they went.

Thursday, 6th June, Chain of Ponds. During the night it has been stormy, with showers of rain, and is still the same this morning. Sundown: Still stormy, with a few drops of rain. Wind, east.

Friday, 7th June, Chain of Ponds. During the night the rain ceased, and this morning is quite bright. Ground so soft that it is impossible to travel. Latitude, 17 degrees 35 minutes 25 seconds. Sent Thring some miles to the west, to see in what state the country is, if fit for us to proceed, and if he can see any water that I could move the party to, for I do not like this place. If more rain falls it will lock us in all together—neither do I like leaving the party with so many natives about. At one o'clock he returned. The ground was so heavy that he had to turn at five miles. He could see no water, but a number of native tracks going to and coming from the west. I shall be obliged to leave the party here, and on Monday try another trip to the west. If I find water I shall return and take them to it. The day has been clear, but at sundown it is again cloudy. Clouds from north-west. Wind from east.

Saturday, 8th June, Chain of Ponds. This morning it has again cleared off, and there is every appearance of fine weather. If it hold this way I shall be able to travel on Monday. Sundown: A few clouds. Wind, south-east.

Sunday, 9th June, Chain of Ponds. The day has again been fine. Wind, still south-east.

Monday, 10th June, Chain of Ponds. Started at 7.55 a.m., course 275 degrees, with Thring, Woodforde, and Wall, nine horses, and fourteen days' provisions. The first five miles were over a grassy plain, with stunted gum and other trees. It was very soft, the horses sinking up to their knees. We met with a little rain water at three miles, where the soil became sandy; continued to be more so as we advanced, with lancewood and other scrubs growing upon it. At fourteen miles gained the top of a sand rise, which seems to be the termination of the sand hills that I turned back from on my west course south of this. From here the country seems to be a dense forest and scrub; no rising ground visible. Camped at 5 p.m., distance thirty-two miles. The whole journey from the sand hills has been through a dense forest of scrubs of all kinds—hedge-tree, gum, mulga, lancewood, etc. We have had great difficulty in forcing the horses through it so far; they are very tired. It is the thickest scrub I have yet been in. Ground very soft; heavy travelling, with the exception of the last five miles, where little rain seems to have fallen. I am afraid this will be another hopeless journey. I fully expected to have got water to-night from the recent rains, but there is not a drop. The country is such that the surface cannot retain it, were it to fall in much larger quantities. I shall try a little further on to-morrow. I had a hole dug, to see if any rain had fallen, and found that it had penetrated two feet below the surface, below which it is quite dry. Wind, east.

Tuesday, 11th June, Dense Forest and Scrub. Leaving Woodforde, Wall, and the pack-horses, I took Thring with me, and proceeded on the same course to see if I could get through the horrid forest and scrub, or meet with a change of country, or find some water. At two miles we came upon some grass again, which continued, and at another mile the forest became much more open and splendidly grassed, which again revived my sinking hopes; but alas, it only lasted about two miles, when we again entered the forest thicker than ever. At eleven miles it became so dense that it was nearly impenetrable. The horses would not face it; when forced, they made a rush through, tearing everything we had on, and wounding us severely by running against the dead timber (which was as sharp as a lancet) and through the branches. I saw that it was hopeless to force through any further. Not a drop of water have we seen, although the ground is quite moist—the horses sinking above the fetlock. The soil is red and sandy; the mulga from thirty to forty feet high and very straight; the bark has a stringy appearance. There is a great quantity of it lying dead on the ground, which causes travelling to become very difficult. I therefore returned to where I left Woodforde and Wall, and came back ten miles on yesterday's journey, and camped. This morning, about 5.30, we observed a comet bearing 110 degrees; length of tail, 10 degrees, and 10 degrees above the horizon. Wind, south-east.

Wednesday, 12th June, Western Dense Forest and Scrub. Proceeded to camp and found all well. This is the third long journey by which I have tried to make the Victoria in this latitude, but have been driven back every time by the same description of country and the want of water. There is not the least appearance of rising ground, or a change in the country—nothing but the same dismal, dreary forest throughout; it may in all probability continue to Mr. Gregory's last camp on the Camfield. My farthest point has been within a hundred miles of it. I would have proceeded further, but my horses are unable to do it; they look as if they had done a month's excessive work, from their feet being so dry, the forest so thick, and the want of water. Thus end my hopes of reaching the Victoria in this latitude, which is a very great disappointment. I should have dug wells if my party had been larger, and I had had the means of conveying water to those engaged in sinking the wells. I think I could accomplish it in that way; but by doing so, I should have to divide the party into three, (one sinking, one carrying water, and one at the camp), which would be too small a number where the natives appear to be so hostile. I have not the least doubt that water could be obtained at a moderate depth, near the end of my journeys, amongst the long thick timber, which seems to be the lowest part of the country. I had no idea of meeting with such an impediment as the plains and heavy scrub have proved to be. For a telegraphic communication I should think that three or four wells would overcome this difficulty and the want of water, and the forest could be penetrated by cutting a line through and burning it. In all probability there is water to be found nearer than this in the Camfield, Mr. Gregory's last camp, somewhere about its sources, which might be thirty miles nearer. Wind, south-east. Country drying up very fast.

Thursday, 13th June, Chain of Ponds. To-day I shall move the camp to the easternmost part of Newcastle Water, and now that rain has come from the east, I shall try if I can cross Sturt Plains, and endeavour to reach the Gulf of Carpentaria. My provisions are now getting very short. We are reduced to four pounds of flour and one pound of dried meat per man per week, which is beginning to show the effects of starvation upon some of them; but I can leave nothing untried where there is the least shadow of a chance of gaining the desired object. Started at 9.40 a.m. At three miles and a half passed our first camp of Newcastle Water. At eight miles and a half camped at the last water to the eastward. The ground is firmer than I expected, travelling good. The large part of the water is reduced two inches since 24th ultimo. The late rains seem to have no effect on it. Wind, south-east.

Friday, 14th June, East End of Newcastle Water. Started with Thring, Woodforde, and Wall, with one month's provisions and ten horses, at 7.45 a.m.; course, 60 degrees. At two miles crossed our former tracks, on the top of the sandy table land, and after leaving it we again got on the open plains, black alluvial soil, covered with grass, with deep holes and cracks into which the horses were continually falling on their noses, and running the risk of breaking our necks. These plains have swallowed up every drop of rain that has fallen. The extent of the plain is seven miles. We then entered a thick wooded country, of the same description as the western forest, being equally thick, if not thicker, and as difficult to penetrate. This continued for thirteen miles, when we met with another small plain about half a mile wide, but opening out wider to north-west and south. Not a drop of water have we seen since leaving Newcastle Water, a distance of about thirty miles, except a little rain water about three miles east of it. The plains are quite dry, scarcely showing that rain has fallen. Camped. The horses have had a hard day's work and are very tired. I wish I could have found water for them to-night. Latitude, 17 degrees 26 minutes 20 seconds. Wind, south-east.

Saturday 15th June, North-east Small Plains, Sturt Plains. Started at 7.30 a.m.; course, 60 degrees, through another ten miles of very thick forest, the thickest we have yet seen. At eleven miles came again upon the large open grassy plain, at the point where I turned on the 21st ultimo. I expected to have found some rain water here, this being the only place in all the plain I have seen that is likely to retain it. Sent Thring and Woodforde in different directions, while I proceeded in another, to see if we could find any, but not a drop could we see. It has been all swallowed up by the ground, which is again dry and dusty. It must take an immense quantity to saturate it, and leave any on the surface; and if that were to be the case, the country would become so soft it would be quite impassable. I am again forced to turn; it is quite hopeless to attempt it any farther. It would be sacrificing our horses, and, perhaps, our own lives, without the least prospect of attaining our end. If I could see rising ground, however small, or a change in the country to justify my risking everything, I would do so in a moment. I only wish there was. I have tried my horses to their utmost. Even my old horses that are inured to hardship are unable to be longer than three days without water, owing to the heat of the sun, the dryness of the feed, and the softness of the country. We saw a few cockatoos and pigeons. There might be water within a short distance, but none can we see or find; for on my course 20 degrees west of north I passed within two miles of Newcastle Water, where the main camp is now, but could not see it. It would require a long time to examine this country for water. There are so many clumps of trees, and strips of scrub on the plain, where water might be, that it would take upwards of twelve months to examine them all. At sundown camped fifteen miles from the main camp. Horses look very bad. It has been very heavy travelling, over rotten ground, and tearing through thick wood and scrub, which has skinned our legs from the knees to the ankles and caused no little pain. Wind, variable.

Sunday, 16th June, Sturt Plains East. Proceeded to the camp, where I found all well. No natives had been near them. This is very disheartening work. I shall proceed to the south, and try once more to round that horrid thick western forest; it is now my only hope; if that fail I shall have to return. I am doubtful of the water in Ashburton range, if no rain has fallen there; those hills are the last of the rising ground within range of vision, which ends in about latitude 17 degrees 14 minutes. From south-south-east round the compass to south-south-west nothing but dense forest and Sturt Plains. Wind, south-east.

Monday, 17th June, Newcastle Water East. Returned to the Lawson and camped. Little rain seemed to have fallen there. I kept a little to west of my former tracks to see the nature of the large open plain. It is completely matted with grass, having large deep holes and cracks, and is as dry as if no rain had fallen for months. Wind, south-east.

Tuesday, 18th June, Lawson Creek. Proceeded to Hunter Creek. Tracks of natives upon ours to Hawker Creek. Light winds, variable.

Wednesday, 19th June, Hawker Creek. Although the water holes in this creek are full from recent rains, the water is very hard, evidently showing it must come from a spring in the hills. Proceeded to the Hunter along the foot of the hills, and at nine miles crossed the large gum creek, where I watered the horses on my north course; this I have named Powell Creek, after J.W. Powell, Esquire, of Clare. At twenty miles crossed another gum creek, which I have named Gleeson Creek, after E.B. Gleeson, Esquire, J.P., of Clare. Camped on the Hunter. Between this and Hawker Creek we crossed eleven gum creeks with water in them. The country passed over is not so good, being close to the hills: it is scrubby, and generally covered with spinifex. Wind, south-east.

Thursday, 20th June, Hunter Creek. Three horses missing; could not be found until too late to reach the other water to-night. Wind, calm.

Friday, 21st June, Hunter Creek. Proceeded to the water under Mount

Primrose, over stony hills, the highest of which I have named Mount Shillinglaw, after —— Shillinglaw, Esquire, F.R.C.S., of Melbourne, who kindly presented me with Flinders' Charts of North Australia. The gum creek on which we are now camped I have named Carruthers Creek, after John Carruthers, Esquire, of Adelaide. Calm.

Saturday, 22nd June, Carruthers Creek. Proceeded to Tomkinson Creek, where I left the two horses; I will there rest the horses a day, and have those shod which I intend to take with me. The last two days have been over very stony country, which has made some of the horses quite lame. I am now running short of shoes. We can see nothing of the two horses about our old camp. Light wind from north-east, with a few clouds. Very hot in the middle of the day; evenings and mornings cold.

Sunday, 23rd June, Tomkinson Creek. Sent Thring and Woodforde down the creek, and Masters up into the open plain, to see if they could find the horses on their tracks. In the afternoon they returned unsuccessful, except Masters, who had seen their tracks when the ground was boggy. Recent tracks of natives were also seen. If they have not been frightened away, they will not be far off. I have instructed Sullivan to follow their tracks, and try to find them during my absence. Wind, north-east, with a few clouds. The sun is very hot in the middle of the day.

Monday, 24th June, Tomkinson Creek. Started with Thring, Masters, and Lawrence, and ten horses, with fourteen days' provisions, at 7.40 a.m.; course, 270 degrees east. We crossed the plain and the creek several times. At 12.20, fifteen miles, ascended a stony rise, and saw that the creek emptied itself into an open grassy plain, about two miles north of us. Proceeded on the same course over a gum plain covered with grass for five miles. The country then became sandy soil, slightly undulating, with ironstone, gravel, spinifex, gums, and occasionally a little scrub, which continued throughout the day. Camped without water. Very little feed for the horses, it being nearly all spinifex. Distance, twenty-eight miles. Wind, west; a few clouds.

Tuesday, 25th June, Spinifex and Gum Plain West. Started at 7.40 a.m. on the same course, 270 degrees. Camped at twenty-seven miles. The country travelled through to-day is bad—red sandy light soil, covered with spinifex, slightly undulating, and having iron gravel upon it. Scarcely a blade of grass to be seen. Some gum-trees, and a low scrub of different sorts. I seem to have got to the south of the dense forest, but into a poorer country. Not a drop of water or a watercourse have we seen since we left Tomkinson Creek. We have crossed two or three low rises of ironstone gravel. Not having the dense forest to tear through has induced me to go on all day in the hope of meeting with a change, but at the end of the day there seems as little likelihood as when we first came upon it, and it may continue to the river. I am again forced to return disappointed. There is no hope of making the river now; it must be done from Newcastle Water with wells. I wish that I had twelve months' provisions and convenience for carrying water, I should then be enabled to do it. Wind, east.

Wednesday, 26th June, Spinifex and Gum Plain. Started at 7 a.m. back towards Tomkinson Creek. At dusk found some water on the small plain into which the creek empties itself. Camped. Distance travelled to-day, forty miles. One of the horses completely done up. I am fortunate in finding this water, for another night without it and I should have lost some of them. I am also glad we had a cool day—only two hours' heat. The horses have travelled one hundred miles without water, and the country being sandy, made it very heavy walking for them. Wind, east.

Thursday, 27th June, Tomkinson Creek. Started for the camp, and arrived at noon. Sullivan had gone after the horses, and lost himself for three days and two nights. Not making his appearance the first night, Kekwick sent Woodforde in search of him from south-east to north. Not returning the second night, Kekwick and Woodforde went out in another direction to try if they could cut his tracks, but were again unsuccessful. At about 3 a.m. he came into the camp perfectly bewildered, and did not seem to recognise anyone. From what we can learn from him he must have gone to the south instead of the east, where the tracks of the two horses were seen. On the first night he came close to the camp—saw the other horses feeding, but could not find them. He can give no account of where he went the next day and night; on the third day he cut my outward tracks to the west, and the horse brought him to the camp. I observed his horse's tracks upon ours this morning, about ten miles down the creek, and could not imagine how they came there. Woodforde found the two horses he went in search of within three miles of the camp—they had not left the creek. The cream-coloured one had improved very much; but Reformer still looks miserable—I think he must be ill. Wind, north, with a few clouds coming from the same direction.

Friday, 28th June, Tomkinson Creek. Shoeing horses and preparing for another start. I shall try once more to make the Gulf of Carpentaria from this. There may be a chance of my being able to round Sturt Plains to the east or north-east. Wind, varying from south-east to north.

Saturday, 29th June, Tomkinson Creek. Shoeing horses, etc. Wind, south-east. Clouds all gone.

Sunday, 30th June, Tomkinson Creek. Wind, north-east.

Monday, 1st July, Tomkinson Creek. Started at 8.10 a.m., course 54 degrees, with Thring, Woodforde, and Masters. At 11.20 (eleven miles), top of a high hill, which I named Mount Hawker, after the Honourable George C. Hawker, Speaker of the House of Assembly, S.A. At 12.45, four miles, struck a large creek; its course a little east of north, which I have named McKinlay Creek, after John McKinlay, Esquire. The first part of the journey was over stony undulations, gradually rising until we reached the top of Mount Hawker, the view from which was not very extensive on our course, being intercepted by stony spurs of the range nearly the same height, about eight hundred feet, and very rocky and precipitous. They are composed of sandstone, quartz, iron, limestone, and hard white flinty rocks. The sandstone predominates. We descended with great difficulty, crossed McKinlay Creek, and at five miles ascended another high hill, which I have named Mount Hall, after the Honourable George Hall, M.L.C. From this our view is most extensive, over a complete sea of white grassy plains. At about fifteen or twenty miles south-east are the terminations of other spurs of this range; beyond them nothing is visible on the horizon but white grassy plains. To the east and north-east the same. To the north apparently a strip of dense scrub and forest, which seems to end about north-east, beyond which, in the far distance, we can see the large grassy plain I turned back from on the 21st of May and 15th of June. No rising ground visible except the hills of Ashburton range to north-west and south-east. Descended towards the plains over stony rises, with gum-tree, lancewood, and other scrub and spinifex. At five miles reached the plain. It is of the same description as the other parts I have been over. No appearance of water. It is hopeless to proceed further; it will only be rendering my return more difficult, by reducing the strength of my horses, without the slightest hope of success. All hope of gaining the Gulf without wells is now gone. I have therefore turned back to a small plain (four miles), searched round it, and in one of the small creeks found a little rain water, at which I have camped. Wind, south.

Tuesday, 2nd July, Loveday Creek. This creek I have named Loveday Creek, after R.J. Loveday, Esquire, Lithographer to the South Australian Government. Returned towards the camp. On reaching McKinlay Creek I was informed by Woodforde that Masters had remained behind, about six miles back, and had not yet come up. This is against my strict orders which are that no one shall leave the party without informing me, that I may halt and wait for them. I have sent Thring back to one of the hills to fire off a gun, and see if he is to be seen, as I have left my outward tracks to avoid crossing Mount Hall—and the tracks are very difficult to be seen over such stony country. I am afraid that he is lost. In an hour and a half, Thring returned; he can see nothing of him. He cut our former tracks, but can see nothing of his on them. My conjectures, I fear, are too true. If he has missed the tracks, it is a thousand chances to one if he is ever found again. To track a single horse is impossible. I proceeded towards Mount Hawker, and camped on my outward tracks, at a remarkable gorge that we had come through. Sent Thring back to the top of Mount Hall to raise a smoke, to remain there some time, and see if he comes up; if not, he is to proceed to our last night's camp, there to remain all night, in case he should go there—while I and Woodforde raised another smoke on top of Mount Hawker. A little after 2 p.m. Thring returned with him. He found him on a hill near Mount Hall, looking for the tracks. He was quite bewildered, and in a great state of excitement. I am most thankful that he is found. The account that he gives is, that his horse slipped the reins out of his hand, and that he was unable to catch him for some time, and when he did so, he was unable to find our tracks, or to track his own horse back, and he became quite confused. He seems to be most thankful for his narrow escape. As it is too late to reach the camp, I shall remain here to-night. Wind, west.

Wednesday, 3rd July, Under Mount Hawker. Proceeded to the camp on the

Tomkinson. Found all right, with the exception of one of the horses (Reformer), which cannot be found. He is one of the two that I left here formerly, and was looking so ill when I found him. He was last seen on Monday night, when he looked miserable. I have sent three men in search of him. Wind, variable.

Thursday, 4th July, The Tomkinson. Started at 8.20 a.m., course 300 degrees, with Woodforde, Thring, and Masters, ten horses, and a month's provisions, to try once more to make the Victoria. Between my first and last attempts, I may succeed. I am very unwilling to return without trying all that is in my power. At three miles we left the plains, and proceeded over stony rises for two miles. The country then became sandy, with gum, spinifex, and lancewood scrub, not difficult to get through. There is no grass. At twenty-five miles came to a little, and, as I am not sure of coming upon any more soon, I camped. We have seen no water since leaving the creek. Latitude, 18 degrees 25 minutes 40 seconds. Wind, south-east.

Friday, 5th July, Spinifex and Gum Plains. Started at 7.50 a.m., course 360 degrees, to find water. At 9.10 (five miles), struck a creek with water; followed it down, course 285 degrees, and at eight miles camped on the last water. The banks in places have good feed upon them, but there is a great deal of spinifex and scrub. The creek is getting narrower, and, as the horses had but little to eat last night, I shall give them the remainder of the day here, for there is no telling when they will get another good feed. Day exceedingly hot, horses covered with sweat. This I have named Burke Creek, after my brother explorer, Richard O'Hara Burke, Esquire, of Melbourne. On camping I saw a remarkable bird fly up; I sent Woodforde to try and shoot him, which he did. It was of a dark-brown colour, and spotted like the landrail; the tail feathers were nine in number, and twelve inches long. I have had it skinned, and will endeavour to take it to Adelaide. Thring, Woodforde, and Masters cooked the body, and ate it. They had scarcely finished, when, in a moment, they were seized with violent vomiting, but in a few minutes they were all right again. Wind, calm. Latitude, 18 degrees 19 minutes 30 seconds.

Saturday, 6th July, Burke Creek. Started at 7.45 a.m., same course, to follow the creek (285 degrees). At three miles it was lost in a grassy gum plain; changed to 300 degrees. On this course the plain continued for three miles; it then became sandy soil, with spinifex, gums, and scrub. Crossed a low sand hill at fourteen miles; descended into another low grassy plain subject to inundation, which, I suppose, receives Hunter Creek. It continued for two miles, at the end of which we again ascended a sandy rise, on the top of which the country became a sandy table land, and continued so the rest of the day's journey. Camped without water, and with very little grass. The table land was spinifex, gums, and scrub, in some places very difficult to get through. Distance, thirty miles. Wind, south-east. Latitude, 18 degrees 7 minutes 5 seconds. At 7 p.m. I observed the comet, 5 degrees above the horizon, bearing 15 degrees west of north, the nucleus more hazy, and the tail much longer. Calm.

Sunday, 7th July, West Sandy Table Land. Started at 8 a.m. on the same course. At 3 p.m. we got into dense scrub, and, as I could see some distance on before, being on one of the slight undulations, I felt there was not the slightest hope of obtaining water; there was no change, no rising ground visible. It would be hopeless to continue such sandy country, as it can never hold water on the surface. We dug five feet, in one of the small plains, but came to the clay without finding water, or even moisture. There is not a mouthful of grass for the horses to eat; the whole of the journey, with the exception of the small grassy plains, is spinifex, gums, and scrub. I shall have to retreat to the last plain we passed through to get feed for the horses, which are looking very bad. The travelling has been heavy tearing through thick scrub, which in some places has been burned: this makes it very rough for them. I must now give up all hope of reaching the Victoria, and am unwillingly forced to return, my horses being nearly worn out. Wind, variable. Distance, twenty-five miles.

Monday, 8th July, Small Grass Plain in Scrub. Started at break of day and continued until 4.30 p.m. Meeting with a little grass, camped; some of the horses unable to go further. Wind, south.

Tuesday, 9th July, Sandy Table Land. Started at sunrise and arrived at Burke Creek. At 11 a.m. turned the horses out to feed for two hours, and proceeded up the creek to where I first struck it. Camped. At a little more than a mile down the creek from here, there is a course of concrete ironstone running across it, which forms a large pond of water nearly a mile in length, apparently deep and permanent. Wind, west.

Wednesday, 10th July, Burke Creek. Shortly after sunrise proceeded toward the main camp, and arrived there at 3 p.m. Found all well. The natives have been about. They attacked Wall while in search of the missing horse; he and his horse narrowly escaped being hit by their boomerangs. The missing horse cannot be found. I suppose that he has crept into some bushes and died; for, the night before he was missed, he left the other horses and came to the camp fire; he appeared to be very stupid, and for some time they could not get him away; when they did so, he went off reeling. Wind, south-west.

Thursday, 11th July, Tomkinson Creek. Shoeing horses, and repairing saddles and bags to carry our provisions back. We have now run out of everything for that purpose, and are obliged to make all sorts of shifts. The two tarpaulins that I brought from Mr. Chambers's station for mending the bags, are all used up some time ago, and nearly all the spare bags; the sewing-twine has been used long since, and we are obliged to make some from old bags. We are all nearly naked, the scrub has been so severe on our clothes; one can scarcely tell the original colour of a single garment, everything is so patched. Our boots are also gone. It is with great reluctance that I am forced to return without a further trial. I should like to go back, and try from Newcastle Water, but my provisions will not allow me. I started with thirty weeks' supply at seven pounds of flour per week, and have now been out twenty-six, and it will take me ten weeks before I can reach the first station. The men are also failing, and showing the effects of short rations. I only wish I had sufficient to carry me over until the rain will fall in next March. I think I should be able to make both the Victoria and the Gulf. I had no idea when starting that the hills would terminate so soon in such extensive level country, without water, or I should have tried to make the river, and see what the country was, when I first saw the rising grounds from Mount Primrose, which are the sand and iron undulations passed over on my southernmost western journey. Before I went to Newcastle Water they completely deceived me; for from the top of the mount they had the appearance of a high range, which I was glad to see, thinking that if the range I was then following up should cease, or if I could not find a way into the river further north, I would be sure to get in by that distant range, which caused me to leave the Newcastle Water country sooner than I should otherwise have done; and now I have not provisions to take me back again. From what I have seen of the country to the west and south of Newcastle Water, I am of opinion that it would be no use trying again to make the river, for I believe no water can be obtained by sinking. To the west and north-west of Newcastle Water the country is apparently lower, and I think that water could be obtained at a moderate depth. It is the shortest distance between the waters; but the greatest difficulty would be in getting through the dense forest and scrub, but that, I should think, could be overcome. It certainly is a great disappointment to me not to be able to get through, but I believe I have left nothing untried that has been in my power. I have tried to make the Gulf and river, both before rain fell, and immediately after it had fallen; but the results were the same, UNSUCCESSFUL. Even after the rain I could not get a step further than before it. I shall commence my homeward journey to-morrow morning. Wind, south. The horses have had a severe trial from the long journeys they have made, and the great hardships and privations they have undergone. On my last journey they were one hundred and six hours without water.

On Friday, July 12th, Mr. Stuart quitted Tomkinson Creek to return to Adelaide, and on the following Friday reached Ann Creek on the north side of the Murchison range. On the 30th, the party proceeded across the Centre, and camped south-west of it, on the Hanson. The nights now became very cold, and there was usually white frost on the grass, and ice in the buckets every morning. On August 6th they camped under Brinkley Bluff, and remained there until the 8th. The Hamilton was reached on the 23rd, and here the natives again showed some hostility, contenting themselves, however, with yelling and howling, and endeavouring to set fire to the grass, in which they were happily unsuccessful. On Saturday, August 31st, they arrived at Mr. Levi's station, where all of them "were overjoyed at once more seeing the face of a white man." They were received with great kindness and attention. After remaining there three days they proceeded by way of Louden Spa, William Springs, Paisley Ponds, and Hamilton Springs to Chambers Creek, where they arrived on September 7th and remained until the 10th.

The last entry in Mr. Stuart's journal is as follows:

Sunday, 15th September, Moolooloo. I shall leave to-morrow for Port Augusta, and proceed by steamer for Adelaide, leaving the party to be brought into town by Mr. Kekwick.

I cannot close my Journal without expressing my warmest thanks to my second in command, and my other companions; they have been brave, and have vied with each other in performing their duties in such a manner as to make me at all times feel confident that my orders were carried out to the best of their ability, and to my entire satisfaction; and I also beg to tender my best thanks to the promoters (Messrs. Chambers and Finke) and the Government, for the handsome manner I was fitted out.

JOHN McDOUALL STUART,
Leader of the Expedition.

JOURNAL OF MR. STUART'S SUCCESSFUL EXPEDITION ACROSS THE CONTINENT OF AUSTRALIA FROM DECEMBER, 1861, TO DECEMBER, 1862

Mr. Stuart made his public entry into Adelaide on Monday, 23rd September, and reported himself to the authorities. Almost at the same time the Victorian Government obtained their first traces of the survivors of the ill-fated expedition under Burke and Wills.[16] The South Australian Government had such confidence in Mr. Stuart that, on his expressing his readiness to make another attempt to cross the continent, they at once closed with his offer, and in less than a month (on October 21st) the new expedition started from Adelaide to proceed to Chambers Creek, and get everything in order there for a final start. Mr. Stuart accompanied them for a few miles to see that everything went on well, when, one of the horses becoming restive, he advanced with the intention of cutting the rope which was choking the animal; the horse reared and struck him on the temple with its fore foot, knocking him down and rendering him insensible. The brute then sprang forward and placed one of his hind feet on Mr. Stuart's right hand, and, rearing again, dislocated two joints of his first finger, tearing the flesh and nail from it, and injuring the bone to such an extent that amputation of the finger was at first thought unavoidable. By careful treatment, however, it was unnecessary to resort to such a course, and in five weeks the leader was able to start to overtake his party, some of whom were to remain at Moolooloo until he joined them.

In no way discouraged either by his own unlucky accident and previous want of success, or by the melancholy end of his brother explorers, Burke and Wills, Mr. Stuart arrived at Moolooloo on Friday, December 20th, and at Finniss Springs on the 29th. The names of the party were as follows:

John McDouall Stuart, Leader of the Expedition.
William Kekwick, Second Officer.
F.W. Thring, Third Officer.
W.P. Auld, Assistant.
Stephen King.
John Billiatt.
James Frew.
Heath Nash.
John McGorrerey, Shoeing Smith.
J.W. Waterhouse, Naturalist to the Expedition.

Besides these, there were at starting, Woodforde and Jeffries; but at Finniss Springs, the latter struck one of his companions, and, on being called to account by his leader, refused to go any further. As to the former, when quitting Mr. Levi's station on January 21st, it was arranged, in order to lighten the weak horses, that the great-coats of the party should be left, but Woodforde objected to this, and said he would not go unless he had his great-coat with him. Mr.

[16] The news of their death reached Melbourne on November 2nd.

Stuart had very properly decided not to take any man who refused to obey orders, and he therefore started without him. The next day Woodforde rejoined the party near Milne Springs, but did not accompany them many days longer; for on February 3rd, shortly after starting, he asked McGorrerey to hold his gun while he returned to get something he had left behind at the previous night's camp. About an hour afterwards, McGorrerey discovered a piece of folded-up paper on the nipple of the gun, and on examination this proved to be an insolent note, addressed to his leader, stating that he had gone back, taking with him a horse, saddle, bridle, tether-rope, and sundry other things not belonging to him. Mr. Stuart had been much dissatisfied with his conduct for some days, and had made up his mind to send him back, believing that he was doing everything in his power to discourage the party, and bring his leader's authority into contempt.

At Marchant Springs, where they arrived on February 15th, they began to experience annoyance from the natives. On the 17th, as Auld was approaching the water-hole, a native who was there called to some others who were posted in trees, and shortly afterwards a great cloud of smoke was seen to windward, coming towards the camp. It was evidently their intention to attack the exploring party under cover of the smoke, "but Thring, while looking for the horses, came suddenly on three of them concealed behind a bush, armed with spears and boomerangs; he did not perceive them until within twelve yards of them. They immediately jumped up, and one of them threw a boomerang at him, which fortunately missed both him and his horse. He was obliged to use his revolver in self defence," but with what result Mr. Stuart does not state.

The excessive heat of the weather now proved a great hindrance to the expedition. They had already lost so many horses that a large part of their provisions, etc. had to be abandoned on various occasions. On February 23rd, Mr. Stuart writes:

"Before reaching this place (the Hugh) five other horses gave in, and were unable to proceed further. I cannot understand the cause of the horses knocking up so much; every one of them has fallen off the last week. Whether it is the excessive heat or the brackish water of the Finke, I am unable to say. Last night I tried some citric acid in the water of the Finke, and it caused it to effervesce, showing that the water contained soda." It was afterwards ascertained that the horses were suffering from worms, which may partially account for their failing strength.

After leaving the Hugh, on February 25th, they were again annoyed by the natives. When about half-way through the gorge, they "set fire to the grass and dry wood across the creek, which caused a dense smoke to blow in our faces. I had the party prepared for an attack. After passing through the smoke and fire, three natives made their appearance about twenty-five yards off, on the hill side, armed with spears and shields, and bidding us defiance by placing the spears in the womeras, and yelling out at the highest pitch of their voices. I ordered Auld to dismount and fire a shot a little distance on one side of them, to let them know what distance our weapons carried. The ball struck the rock pointed out to him to aim at, and stopped their yelling, but seemed to have no other effect. I again ordered him to fire at the rock on which the middle one of the three was standing; the shot was a good one, for the ball struck the desired spot, and immediately had the effect of sending them all off at full speed."

Again, on March 5th, while crossing the plains under Mount Hay, they came suddenly on three natives armed with long spears and shields, who ran off into the scrub. A short distance further, while watering the horses at some rain water, these three natives returned, accompanied by four others, and made signs of hostility, by yelling and shaking their spears, and performing other threatening antics while widely separating themselves in a half-circle. Mr. Stuart says: "I had the party prepared to receive an attack; but when they saw us stationary they approached no nearer. I ordered some of the party to fire close to them, to show them we could injure them at a long distance, if they continued to annoy us, but they did not seem at all frightened at the report of the rifles nor the whizzing of the balls near to them, since they still remained in a threatening attitude. With the aid of a telescope we could perceive a number of others concealed in the belt of scrub. They all seemed fine muscular men. There was one tall fellow in particular with a large shield and a very long spear (upwards of twelve feet), which he seemed very anxious to use if he could have got within distance. We crossed the creek, and had proceeded a short distance across the plain, when they again came running towards us, apparently determined to attack; they were received with a discharge of rifles, which caused them to retire and keep at a respectful distance. Having already wasted too much time with them, I proceeded over the plain, keeping a sharp look-out; should they threaten us again, I shall allow them to come close, and make an example of them. It is evident their designs are hostile. Before entering the scrub we could see no signs of them following. About sundown, arrived at Mount Harris without further annoyance."

A week later (on March 12th) the Centre was passed; and on the 17th, while going from Woodforde Creek through the bad country towards the Bonney, Thring met with an awkward accident, which his leader thus describes: "Being anxious to keep my old tracks through the scrub, as it does not wear the saddle-bags so much as breaking through a new line, I missed them about two miles after starting, in consequence of the earliness and cloudiness of the morning. I sent Thring in search of them, and he, on finding them a short distance off, fired his revolver to let us know that the tracks were found. The young horse he was riding stood the first report very well. I, not hearing the report, was moving on, which caused him to fire again, whereupon his horse backed and threw him with violence to the ground on his chest. He feels his chest is much hurt by the fall. The horse then returned on the tracks at full gallop," and was not recovered until shortly before sundown.

The party camped at Attack Creek on Friday, March 28th, and at Tomkinson Creek on the 31st. On April 3rd, while crossing the Gleeson, Kekwick's horse fell back with him in ascending the bank, and broke the stock of his gun, but he himself escaped unhurt. On Saturday, April 5th, they camped at the east end of Newcastle Water, and the following day, "at about 9 o'clock a.m. Kekwick, in endeavouring to shoot some ducks, went towards some native smoke, and was met by two natives, who ran away. In an hour afterwards, five natives came within a hundred yards of the camp, and seemed anxious to come up to it, but were not permitted. Two hours afterwards we were again visited by fifteen more, to some of whom a present was made of some looking-glasses and handkerchiefs; at the same time they were given to understand that they must not approach nearer to the camp, and signs were made to them to return to their own camp, which they shortly did. In the afternoon we were again visited by nineteen of them, who approached within a hundred yards of the camp, when they all sat down and had a good stare at us, remaining a long time without showing any inclination to go. At length some of them started the horses which were feeding near the water, and made them gallop towards the camp. This so frightened the natives that they all ran away, and we were not troubled with them for the rest of the evening."

The next day the camp was moved to the north end of Newcastle Water, where they remained for a week resting horses and repairing bags, saddles, etc. The Journal then continues as follows:

Monday, 14th April, North End of Newcastle Water. Leaving Mr. Kekwick in charge of the party, started with Thring and Frew at 7.15 a.m., on a northerly course, in search of water; and at six miles, on the edge of the open plains, found some rainwater, sufficient for a few days. Proceeded across the plain on the same course; but at three miles saw something like a watercourse, and changed my course to 20 degrees, to see what it was. At two miles I struck a dry course running south-west. Followed it up towards the small rise without finding any water. Three miles further on the same course I ascended a low stony rise, from which I could see nothing but a thick forest of tall mulga and gums. I changed to a northerly course, and, at 4.20 p.m., camped in a forest without water. Wind, south-east.

Tuesday, 15th April, Sturt Plains, Forest. Proceeded on a course of 250 degrees, and at five miles again struck the open plains, and changed to 180 degrees. At one mile I found a fine water hole three feet deep and about forty feet in diameter, the edge of which was surrounded with conglomerate ironstone rock; watered the horses, and proceeded on a southerly course, through grassy plains with stunted gum-trees, to the first water I found yesterday, and camped. The plains and forest are of the same description as I have already given, only that the plains have not quite so many holes in them, and the forest in many places is covered with ironstone gravel. I shall try a course to the north of west to-morrow, to see if I can find water. Wind variable.

Wednesday, 16th April, Frew's Water Hole, Sturt Plains. Started at 7.45 a.m., on a course of 302 degrees, keeping along the edge of the open plain. I have made many twistings and turnings, but my general course is north-west for ten miles. Seeing a small rise on the open plain, a little to the north of west, I changed to 275 degrees; and at two miles came on some fine ponds of water about one mile and a half long, twenty feet broad, and three feet and a half deep.

I examined them on both sides, to see if they would do for a permanent camp for the party as it is a point nearer; and I think I may depend on the water lasting two months without any more rain. I shall camp here to-night, and try another day to-morrow to the westward, and endeavour to make the Victoria, for I can see but little chance of making the Adelaide. By my journal of the 14th, everything is quite dry and parched up; no rain seems to have fallen there for a long time. The last two days have been excessively hot. The further to the west the hotter I find it. The natives seem to be numerous, for their smoke in the scrub is to be seen in every direction. I name these ponds after John Howell, Esquire, of Adelaide.

Thursday, 17th April, Howell Ponds. Started at 7 a.m., on a bearing 10 degrees north of west. At twelve miles crossed the open plains, and entered a thick forest of gums and other trees and shrubs. Seeing that there is no chance of finding water to-day, returned to the ponds. The open plains seem to tend more to the north-west; I shall examine them when I bring the party up to the ponds. Distance, fifteen miles. Wind, south-east.

Friday, 18th April, Howell Ponds. Started for the camp on the Newcastle Water. On my arrival, I found the party all right, but very anxious about me, as I had been absent longer than I intended. No natives had been near them during my absence at this time; smoke was seen all around. Weather hot during the day, but cold at night and in the morning. Wind, south-east.

Saturday, 19th April, North End of Newcastle Water. I shall remain here till Monday, in order to take some lunar observations, as I am not quite certain that my longitude is correct. Wind, south-east.

Sunday, 20th April, North End of Newcastle Water. Wind from the east; blowing strongly during the day, but it dropped a little before sundown, allowing the mosquitoes to annoy us very much.

Monday, 21st April, North End of Newcastle Water. Some of the horses having strayed some distance made it 10 o'clock a.m. before I could get a start. Proceeded through six miles of forest and scrub to the water that I found on the 14th instant; from thence I changed to 301 degrees 30 minutes for nine miles, and then to 275 degrees, and at two miles camped at the ponds I had discovered on the 16th. Native smoke all around us. The day has been very hot, and the flies a perfect nuisance. Wind, south-east.

Tuesday, 22nd April, Howell Ponds. Preparing for a start to-morrow to the north-west in search of water. Wind, south-east.

Wednesday, 23rd April, Howell Ponds. Leaving Mr. Kekwick in charge of the party, I started with Thring and Frew at 8.5 a.m., on a course of 284 degrees. At 9.55 (seven miles) changed to 320 degrees. At 11.20 (four miles and a half) crossed the open plain, changing to 40 degrees to avoid the scrub. At one mile and a half changed to west. At one mile changed to north-west. At 2.20 (five miles) changed to 45 degrees. At 3 o'clock (two miles) changed to north. At 3.25 one mile and a half changed to north-west. At 3.45 camped without water. I have skirted the border of the forest land in the hope of finding water, but am disappointed. I have not seen a drop since I started. The plains are covered with beautiful grass, two or three feet high. There are a great many different kinds of birds about, and native smoke all round. I have searched every place where I thought there was the least chance of finding water, but without success. The day has been exceedingly hot. With such hot weather as this I dare not attempt to make the Victoria. The horses could not stand a hundred and forty miles without water. Those I have had with me to-day seem to have suffered enough, and would not stand another two days without. I must therefore return to the camp to-morrow. Wind, calm.

Thursday, 24th April, Sturt Plains. Returned to the camp and found all right. The day has been excessively hot. We have seen nothing new during the journey—the same open plains, with forest between.

Friday, 25th April, Howell Ponds. Leaving Mr. Kekwick in charge of the party, started at 8.20 a.m., with Thring and Frew and fresh horses, on a northerly course, in hopes of better success in that direction: course 360 degrees for twenty-two miles; grassy plains, covered in many places with stunted gums, and a new tree with a small green leaf. After that, we entered again a thick forest, and scrub almost impassable. At twenty-eight miles, seeing no prospect of getting through it, I returned two miles to a small open space, where I could tether the horses. I have not seen a drop of water this day's journey. The forest is so very thick, and so many twistings and turnings are required to pass through it, that, although I travelled thirty miles, I don't believe I made more than fifteen miles in a straight line. The day again exceedingly hot, with a few clouds. A few birds were seen during this day's journey, but no pigeons, which are the only sign we have now of being near water. Wind variable.

Saturday, 26th April, Dense Forest. Returned to the camp. The horses felt the heat and the want of water very much. In the forest the heat was almost suffocating. I hope it will rain soon and cool the ground and replenish the ponds, which are drying up fast. There have been a few clouds during the day, but after sundown they all cleared away. Wind, south-east.

Sunday, 27th April, Howell Ponds. A few clouds have again made their appearance, but still no rain. There has not fallen a drop of rain since I left the Woodforde, which was on the 9th of March. Wind, south-east. Latitude, 17 degrees 5 minutes 16 seconds.

Monday, 28th April, Howell Ponds. Leaving Mr. Kekwick in charge of the party, started with Thring and King, on a course of 338 degrees, to try and find an opening in the dense forest and scrub, as well as water. At ten miles we crossed the open plain, with stunted gum-trees and long grass. At this point we met with a small ironstone rise, about twenty feet in height. On ascending I was again disappointed in finding before me a dense forest and scrub. Proceeding in our course, it became thicker than any which I had ever encountered before, and was almost impassable. Still continued, and for a short distance in some places it became more open. A little before sundown I camped on the edge of a stunted gum-tree plain. There are a few slate-coloured cockatoos and other birds, which lead me to hope that, in the morning, I may drop across some water. Wind variable, with a few clouds during the day.

Tuesday, 29th April, Sturt Plains. Started on an easterly course, following the flight of the birds; but at five miles crossed the open gum plain, and again encountered the thick forest. Examined every place I could see or think of where water was likely to be found, but was again disappointed—not a drop was to be seen. Changed my course, so as to keep on the plain; at four miles again crossed, and again met the dense forest, but still no water. Changed to south-east, and at ten miles found ourselves on a large stunted-gum plain. Changed to a little east of south, and arrived at the camp without seeing a drop of water. Wind variable, with heavy clouds from the east.

Wednesday, 30th April, Howell Ponds. I feel so unwell to-day that I am unable to go out, besides I shall require my compass case and other things mended; they got torn to pieces in the last journey by the forest and the scrub. Yesterday's clouds are all gone, and have left us no rain. Another hot day. Wind, east.

Thursday, 1st May, Howell Ponds. Leaving Mr. Kekwick in charge of the party, started with King and Thring to the water hole that I discovered on the 15th ultimo; arrived in the afternoon and camped. This water hole I have named Frew's Water Hole, in token of my approbation of his care of, and attention to, the horses. This waterhole is about twenty feet below the plain, surrounded by a conglomerate ironstone rock. Since my last visit it is only reduced two inches, and is still a large body of clear water from the drainage of the adjacent country; it will last much longer than I anticipated. I shall use my best endeavours to-morrow to find an opening in the thick scrub from north to north-west. The course of the forest seems to run a little west of north, and I am afraid the open plains are surrounded by it; however, I shall try to get through it if I possibly can. Wind, south-east. Day excessively hot.

Friday, 2nd May, Frew's Water Hole. Started at half-past seven o'clock a.m. Course, 335 degrees. At ten miles, a dense forest and scrub. Changed to 10 degrees east of north. At half a mile struck a water-shed, and followed it north for two miles. Found a little rainwater in it, and at two miles further arrived at its source. At three miles further on the same course changed to 30 degrees east of north. At three miles and a half again changed to 320 degrees, and at about a mile and a half struck some fine ponds of water. At two miles further, arrived at what seemed to be the last water, a small shallow pond. Examined around the plain to try and find others, but without success. A little before sundown, returned to the last water and camped. The first part of the day's journey was over a stunted-gum plain, covered with grass. At ten miles we again met with thick forest and scrub. I then changed my course to get out of it, and struck the small water-shed running to the east of south. Following it generally for two miles on a northerly course, we met with a little rain water. Continued the same course through a thick forest and scrub for three miles and a half to get through it if possible. At this point it becomes denser than ever. Sent Thring to climb to the top of a tree, from which he saw apparently a change in one of the low scrubby rises, which appeared to be not so thickly covered with scrub as the others. I directed my course to it, 30 degrees east of north, to examine it. I observe that there is some sandstone in the low scrubby rises, which leads me to hope that I may not be far from a change of country. On this last course we travelled three miles, through a dense thicket of hedge-tree, when I observed some large gum-trees bearing 320 degrees, and decided to examine them before leaving the rise. As I approached nearer to it I again sent Thring to climb a tree to see if there was any change. He could see nothing but the same description of forest and scrub. The change that he saw from the other tree was the shade of the sun on the lower mulga bushes, which caused him to suppose that it was more open country. Not seeing any opening in that direction, I changed to the gum-trees. At a mile and a half was delighted at the sight of a chain of fine water holes; their course north-west to south-east, the flow apparently to south-east. I followed one pond, which was about half a mile long and appeared to be deep. A number of smaller ones succeeded. They then ceased, and I crossed a small plain, which shows signs of being at times covered with water. Observing some green and white barked gum-trees on the west side of it, I went to them, and found a small watercourse with small pools of water, which flowed into the plain coming from the north-west. Following it a little further, we met with some more water. A short distance above this it ceased in the dense forest which seems to surround these ponds. I shall endeavour to force my way through it to-morrow to the west of north. Wind, south-east, with a few clouds in the same direction. These ponds I name King's Ponds, in token of my approbation of his care of, and attention to, the horses, and his readiness and care in executing all my orders. Wind, south-east, with a few clouds in the same direction.

Saturday, 3rd May, King's Chain of Ponds. Started at twenty minutes past seven a.m., on a course of 350 degrees. At twenty-four miles changed to 45 degrees; at three miles and a half changed to north; at two miles and a half camped. At two miles from our last night's camp found an easy passage through the forest; the rest of the twenty-four miles was over a well-grassed country, well wooded with gum and some new trees that I had found last year, and occasionally a little scrub, in some places thick for a short distance. On my first course, before changing, I was crossing low ironstone undulations, which caused me to think I was running along the side of one of the scrubby rises. I therefore changed to 45 degrees east of north to make the plain—if there is any—the scrub being so thick that I cannot see more than fifty yards before me. At three miles and a half I found that I was travelling over the same description of small rises and getting into much thicker scrub. I again changed to north, to see if that would lead me into a plain. At two miles and a half it was still the same, and apparently a thick forest and scrub before us. I camped a little before sundown at a small open place to tether the horses. I have not seen a drop of water during the whole journey, nor any place likely to retain it, with the exception of a small flat about six miles from the last camp. The day very hot. Wind, south-east, with a few clouds.

Sunday, 4th May, Dense Forest. Returned to King's Ponds. This country seems but little frequented by the natives, as we have seen no recent tracks of them. There are a number of cockatoos and other birds about. We have seen no other game, except one wallaby and one kangaroo. There are plenty of old emu tracks about the ponds. Wind, variable. Cloudy.

Monday, 5th May, King's Ponds. Returned to Frew's Water Hole and camped. Before sundown the sky became overcast with clouds. Wind variable.

Tuesday, 6th May, Frew's Water Hole. Towards morning we had a few drops of rain. Returned to the camp and found all well. Yesterday they were visited by a few natives who seemed to be very friendly; they called water ninloo: they were armed with spears, about ten feet long, having a flat sharp flint point about six inches long, with a bamboo attached to the other end. They pointed to the west as the place where they got the bamboo and water also, but they seemed to know nothing of the country north of this; they were tall, well-made, elderly men. After talking for some time they went away very quietly. To-day they have set fire to the grass round about us, and the wind being strong from the south-east it travelled with great rapidity. In coming into the camp, about three miles back, I and the two that were with me narrowly escaped being surrounded by it; it was as much as our horses could do to get past it, as it came rolling and roaring along in one immense sheet of flame and smoke, destroying everything before it.

Wednesday, 7th May, Howell Ponds. Resting. The natives have not again visited us, but their smoke is seen all around. I shall start to-morrow on a course west of north, to try and make the Victoria by that route. I shall take some of the waterbags with me to see how they answer. Wind, south-east. Clouds all gone.

Thursday, 8th May, Howell Ponds. Leaving Mr. Kekwick in charge of the party, started with Thring and McGorrerey, also with King and Nash, who are to bring back the horses which carry the waterbags, whilst I with Thring and McGorrerey proceed on a west course. Started at half-past eight a.m., keeping the former tracks made on my previous journey to the westward, to where we met with the thick forest. About a mile beyond, struck a native track, followed it, running nearly north-west, until nearly three o'clock p.m., when we came upon a small water hole or opening in the middle of a small plain, which seems to have been dug by the natives, and is now full of rain water. This is apparently the water that the natives pointed to, for their tracks are coming into it from every direction. This opening I have named Nash Spring, in token of my approbation. I am very much disappointed with the water-bags; in coming this distance of twenty-one miles they have leaked out nearly half. Wind, south east.

Friday, 9th May, Nash Spring. Sent King and Nash with the horses that carried the water-bags back to the depot, while I and the other two, at twenty minutes to eight o'clock a.m. proceeded on a bearing of 290 degrees, following one of the native tracks running in that direction. At about a mile they became invisible; for that distance I observed that a line of trees was marked down each side of the track by cutting a small piece of bark from off the gum-trees with a tomahawk. This I had never seen natives do before; the marks are very old. At eighteen miles and a half struck another track (the trees cut in the same way) crossing our course; followed it, bearing 10 degrees east of north, and at about two miles came on a native well with moisture in it. Followed the valley on the same course, but seeing no more appearance of water, I again changed to my original course, and, at a quarter to four o'clock, finding that I was again entering the dense forest and scrub, I camped at a good place for feed for the horses, but no water. The whole of the day's journey has been through a wooded country, in some places very thick, but in most open; it is composed of gums, hedge-trees, and some new trees—the gums predominating; there were also a few patches of lancewood scrub. For the first eighteen miles the soil was light and sandy, with spinifex and a little grass mixed. At the end of eighteen miles I again got into the grass country, with occasionally a little spinifex. Wind, south-east. Cold during the night and morning.

Saturday, 10th May, The Forest. Started at five minutes to seven o'clock a.m. (same course, 290 degrees). Almost immediately encountered a dense forest of tall mulga, with an immense quantity of dead wood lying on the ground. It was with the greatest difficulty that the horses could be made to move through it. At a mile it became a little more open, which continued for six miles. At seven miles I thought, from the appearance of the country, that it was dipping towards the north-north-west; I therefore changed my course to north-west, and in less than a mile again entered a dense forest of tall mulga, thicker than I had yet been into. Continued pushing, tearing, and winding into it for three miles. The further I went the denser it became. I saw that it was hopeless to continue any further. We were travelling full speed, and making little more than a mile an hour throughout the ten miles gone over to-day. The country is a red light soil and covered with abundance of grass, but completely dried up. No rain seems to have fallen here for a length of time. We have not seen a bird, nor heard the chirrup of any to disturb the gloomy silence of the dark and dismal forest—thus plainly indicating the absence of water in and about this country. I therefore retraced my steps towards Nash Springs; passed our last night's camp, and continued on till sundown, one of the horses being completely knocked up. Camped without water. Wind, south-east.

Sunday, 11th May, The Forest. This morning the horse that was so bad last night was found dead, which puts us in a very awkward position—without a pack-horse. We had to leave behind the pack-saddle, bags, and all other things we could not carry with us on our riding-horses. Proceeded to Nash Spring, which we reached after two o'clock p.m., with another of the horses completely knocked up. It was with difficulty that he reached it. I suppose the days being so extremely hot, and the feed so dry that there is little nourishment in it, is the cause of this, as they were horses that had been out with me on my last year's journey, and had suffered from want of water a longer time than on this occasion. I am nearly in a fix with a long journey before me, the horses unable to do more than two nights without water, and the water-bags losing half their contents in one day's journey. To make the Victoria through the country I have just passed into would be impossible. I must now endeavour to find a country to the northward and make the Roper. I am very vexed about the water-bags turning out so badly, as I was placing great dependence on them for carrying me through. I must try and push through the best way possible. Wind, south-east.

Monday, 12th May, Nash Spring, West Forest. Proceeded very slowly with the knocked-up horse to the Depot; he appears to be very ill, and is looking very bad this morning. Arrived there and found all right; they had been visited by the natives twice during my absence. They appeared to be very friendly, and were hugging Frew and King, for whom they seemed to have taken a great fancy; they were old, young, and children. Some pieces of white tape were given to them, which pleased them much. They still pointed to the west, as the place where the large water is, and made signs with a scoop to show that they have to dig for it in going through; which I am now almost sure is the case from what I saw of the country in my last journey in that direction. In upwards of fifty miles we did not see the least signs of a watercourse—nor could I discover any dip in the country; it has the same appearance all round; one cannot see more than half a mile before one, and in many places only a few yards. I have been deceived once or twice by what appeared to be a dip in the country, but it turned out to be only lower trees and scrub than what we were travelling through. With a small party I might make the Victoria from here, but there is every chance of losing the horses in doing so; and I should be in a sad predicament to be there without horses, and without the possibility of receiving supplies from the party at the Depot; I should have to perish there. Therefore, I consider it would be folly and madness to attempt it, and might be the cause of sacrificing the lives of both parties. Had the feed been green, or had it any substance in it, I would have tried, but every blade of grass is parched and dried up as in the middle of summer, and the horses have not the strength nor endurance to undergo much privation, of which I have had a proof in the journey I have just taken. After resting a day or two to recover the horses, and get ourselves a little refreshed, I shall move the party up to King's Ponds, and try to push through wherever I can find an opening. Day very hot. Wind, south-east. A few clouds came up from that quarter after sundown.

Tuesday, 13th May, Depot, Howell Ponds. Resting ourselves and horses. Day again hot, with a few clouds round the horizon. The natives had again set fire to the country all around, which increases the heat. I wish it would come on to rain, and put out their fires, and fill the ponds, which are shrinking a great deal more than I expected. Wind, south-east. Clouds.

Wednesday, 14th May, Depot, Howell Ponds. As I don't feel well enough to-day, I shall remain here, and start to-morrow morning. This morning, while Thring was collecting the horses, he came on a place where the natives had been encamped a day or two before, and there saw the remains of the bones of one of them that had apparently been burnt; this is another new feature in their customs. Wind, south-east.

Thursday, 15th May, Depot, Howell Ponds. Started with the party across the plain to Frew's Water Hole, course 15 degrees east of north; found the plain burnt for ten miles. The fire has been so great that it has burned every blade of grass, and scorched all the trees to their very tops. I was very fortunate the other day in having escaped it; nothing could have lived in such a fire, and had we been caught in it we must have perished. Wind, south-east. Clouds all gone, Latitude, 16 degrees 54 minutes 7 seconds.

Friday, 16th May, Frew's Water Hole. Started at fourteen minutes past eight o'clock a.m., course 345 degrees, for King's Chain of Ponds. Arrived at about half-past three o'clock p.m. In coming through, one of the horses separated from the rest and bolted off into the dense forest, tearing everything down before him. We got him in again, but with a broken saddle, and the top off one of the bags, which we afterwards recovered. Arrived at the ponds without any further accident. Wind, north-east. Very hot, and a few clouds. Latitude, 16 degrees 38 minutes 53 seconds.

Saturday, 17th May, King's Chain of Ponds. Sent King and Thring to follow round the flat to see where the ponds go to. About noon they returned, and reported that the water loses itself in a flat, which is surrounded by thick forest and scrub. This certainly is a very pretty place, and a great pity it is not more extensive. It reminds me much of the park land found by Captain Sturt in 1845, where he had his second depot, named Fort Grey. Wind, south-east, with a few clouds.

Sunday, 18th May, King's Chain of Ponds. In the afternoon the sky became cloudy, and at sundown was quite overcast; the day exceedingly hot, and the wind nearly calm. The clouds came from the north-west, and the little wind there is from the south-east.

Monday, 19th May, King's Chain of Ponds. As the sky is overcast with clouds, so that I cannot see the sun, and as it is nearly impossible to keep a straight course in such thick country without it, I shall remain here to-day, and if it should break up I shall endeavour to take a lunar observation. At 9 o'clock a.m. it cleared up, which enabled me to take one. The remainder of the day very hot. Wind variable, with clouds from every direction; towards sundown it again settled in the south-east, and all the clouds disappeared without any rain falling.

Tuesday, 20th May, King's Chain of Ponds. Leaving Mr. Kekwick in charge of the party, I started with Thring, King, and Auld, at half-past nine a.m., on a northern course; at one o'clock p.m. changed to 65 degrees, to what appeared to be a bare hill. At a little more than a mile struck a small watercourse running towards the north; followed it, and at about two miles and a half came on some ponds of water, but not so large as those at our depot; at present they are not more than three feet and a half deep. Examined around the wooded plain to see if there was any larger body of water, but could see none. This plain is covered with small gums, having a dark bluish-green leaf with a grey-coloured bark; there are also a few white ones around the ponds of water, which abound with grass. Before reaching the plain we crossed what seemed to be elevated sandy table land, extending about nine miles, covered with spinifex and dark-coloured gum-trees; we also passed two or three narrow belts of tall mulga and hedge-trees which grow on the stony rises, about twenty feet high. These ponds I name Auld's Ponds, in token of my approbation of his conduct. Wind, south-east. Latitude, 16 degrees 28 minutes 16 seconds.

Wednesday, 21st May, Auld's Chain of Ponds. Started at twenty minutes past eight o'clock a.m., course north. The morning was so thick, with a heavy fog, that I did not get a start till late. At three miles I found another chain of ponds, but not so large; these I name McGorrerey Ponds. Proceeded on the same course and passed through some thick belts of hedge-tree and scrub; the country then opened and became splendidly grassed, with gums and other trees. We also saw, for the first time, a new gum-tree, having a large broad dark-green leaf, and the bark of a nankeen colour, which gave a very pretty effect to the country. At seventeen miles, not finding any water, and having passed five deep holes surrounded with ironstone conglomerate rock similar to Frew's Water Hole, but without any water in them, and to all appearance the dip of the country being to the north-east, I have changed my course to that direction, again travelling over a splendidly grassed country for ten miles, occasionally meeting with low stony rises of ironstone and gravel, at the foot of which were some more deep holes without water. In the last three miles we had to get through a few patches of scrub; the grass is all very dry. No rain seems to have fallen here for a long time. At sundown camped without water. Day very hot. Wind variable, with a few clouds. Latitude, 16 degrees 8 minutes 39 seconds.

Thursday, 22nd May, Fine Grass Country. Returned to McGorrerey Ponds. Day very hot, and the horses much distressed for want of water; they have the appearance of being half-starved for a month, and have taken an immense quantity of water, having gone to it about four or five times in an hour. As I am not satisfied that these ponds cease here, I shall try again to-morrow a little more to the east. Wind, south-east.

Friday, 23rd May, McGorrerey Ponds. Gave the horses a little time to feed after daylight in consequence of their having been tethered during the night; the country is so thickly wooded that I dare not trust them in hobbles the whole night, as, if they were lost sight of there would be great difficulty in finding them here. There is still the appearance of a small creek, which I shall follow until it runs out or trends too much to the east. Started at half-past eight o'clock a.m., course 20 degrees east of north, following the small creek about two miles; it seems to be getting larger, with occasionally a little water in it. We have also seen, on both sides of us, ponds with water surrounded by gum-trees; these ponds, when full, must retain water for a long time. We have also seen a new tree growing on the banks of the creek, with a large straight barrel, dark smooth bark, with bunches of bright yellow flowers and palmated leaves. At a mile and a half further the creek is improving wonderfully. We have now passed some fine holes of water, which will last at least three months; at five miles the water is becoming more plentiful and the creek broader and deeper, but twisting and turning about very much, sometimes running east and then turning to the west and all other points of the compass. Having seen what I consider to be permanent water, I shall now run a straight course, 20 degrees east of north, and strike it occasionally to see if the water continues. I have named these Daly Waters, in honour of his Excellency the Governor-in-Chief. Within a hundred yards the banks are thickly wooded with tall mulga and lancewood scrub; but to the east is open gum forest, splendidly grassed. Proceeded, occasionally touching the creek, and always found fine reaches of water, which continued a considerable way. At thirteen miles they become smaller and wider apart; at fifteen miles the creek seems to be trending more to the eastward, its bed is now conglomerate ironstone, and, as this appears to be about the last water, I shall give the horses a drink and follow it as far as it goes. In a short distance it has become quite dry, with a deep broad course upwards of twenty yards wide. At seventeen miles it separated into two channels, and at a quarter of a mile the two channels emptied themselves into a large boggy swamp, with no surface water. I examined the swamp, but could see no outlet. The country round about is thickly timbered with gum and other trees. Returned to the last water and camped. I shall return to the Depot and bring the party up here. Wind, south-east; a few clouds at sunset.

Saturday, 24th May, Chain of Ponds, Large Creek. Followed my tracks back to Auld's Chain of Ponds, and had difficulty in doing so, the ground being so hard that the hoofs of the horses scarcely left any impression on it. This would be a fearful country for any one to be lost in, as there is nothing to guide them, and one cannot see more than three hundred yards around, the gum-trees are so thick, and the small belts of lancewood make it very deceptive. Should any one be so unfortunate as to be lost, it would be quite impossible to find them again; it would be imprudent to search for them, for by so doing the searchers would run the risk of being lost also. Arrived at Auld's Ponds and camped. Wind, south-east. A few clouds.

Sun day, 25th May, Auld's Chain of Ponds. Proceeded to the Depot, where I arrived in the afternoon and found all well. No natives have been near them, although some of their smoke has been seen at a short distance from the Depot. Yesterday we hoisted the Union Jack in honour of her Most Gracious Majesty's birthday, that being the only thing we had to commemorate this happy event, with our best wishes for her long and happy reign. Wind, south-east.

Monday, 26th May, Chain of Ponds. Removed the party on to Auld's Chain of Ponds.

Tuesday, 27th May, Auld's Chain of Ponds. Proceeded with the party to the fourth chain of ponds and creek. This water has every appearance of being permanent, and I hope I may fall in with such another in the next degree of latitude. It may be from this that the Wickham receives a supply of water when this overflows. Wind, south-west. Latitude, 16 degrees 14 minutes 31 seconds.

Wednesday, 28th May, Daly Waters, Fourth Chain of Ponds and Creeks. Sent Thring and King to round the swamp into which this creek flows, to see if there is any outlet to the eastward of this within two miles. There are other ponds and a creek, which also empties itself into a swamp a little to the eastward of the one into which this one empties itself. In the afternoon they returned, having found a small watercourse forming the north-west side of the swamp; followed it, running nearly 10 degrees east of north. In about one mile and a half they came upon a large swamp covered with water, but shallow. They then proceeded seven miles on a north-east course; then meeting with some white-barked gum-trees, appearing to run to the north-west, followed them for three miles, crossing a gum and grass plain. Observing some native smoke to north-east, they returned. Wind, south-east.

Thursday, 29th May, Daly Waters. Leaving Mr. Kekwick in charge of the camp, at half-past seven o'clock proceeded with Thring, Auld, and Frew down the creek to examine the swamp found yesterday. It is about 30 degrees east of north, about three miles from the Depot at Daly Waters. The water does not appear to be deep, but covers a large area; there were a few pelicans and other water-birds on it. From this we proceeded, on a course 20 degrees east of north, to search the flat where Thring and King saw the smoke yesterday. At eighteen miles from Daly Waters, having crossed the gum plain without meeting with any water, and being on apparently higher ground than the plain, I changed my course to 90 degrees east of north. At two miles and a half again crossed the plain, and got upon low rising ground of ironstone and gravel, but still no water. Changed to former bearing of 20 degrees east of north, and at seven miles came upon a dry swamp, covered with long blue grass and deep holes, but still no water could we find. Proceeded another mile, and finding I was getting on rising ground, and the horses having done a long and heavy day's journey, camped without water. After leaving the swamp with the water (which was very boggy all round it), the country became similar to that of Sturt Plains surrounding Newcastle Water, being so full of deep holes that we were in danger of getting our necks broken, and also the horses' legs. The soil is good, and completely covered with grass and stunted gum-trees. In rainy weather it seems to be covered with water. There is no watercourse, or any appearance of which way the water flows. A number of various kinds of birds were about. Wind variable, but mostly from south-west. Latitude, 15 degrees 56 minutes 11 seconds.

Friday, 30th May, North-north-east of Blue-Grass Swamp. Wishing to see a little more of the country further on and to find where the birds get their water, I proceeded with Thring, leaving the other two behind with the horses, three miles and a half on the same course, following their flight. In half a mile came again upon the stunted gum plain, splendidly grassed to above the horses' knees. Can find no water, although the birds are still round about us. The same description of country continues from the swamp with the water to beyond this, consisting of small undulations of gravel and ironstone. Retraced my steps to where I had left the other two, and proceeded towards the Depot at nine miles. The country was in a blaze of fire to the east of us. I am very thankful there was scarcely a breath of wind, which enabled us to pass within a quarter of a mile of it: had there been a strong wind we should have been in great danger, the grass being so long and thick. Returned to the Depot after six p.m., being all very tired with the shaking we have had the last two days by the horses falling into the holes nearly every step, and they also are nearly exhausted; twelve hours in the saddle over such a country is no easy task. It was my intention to have come back more to the east, but having seen the smoke I saw we should be in the middle of the fire, and so changed my intention. Wind, south-west. Very hot.

Saturday, 31st May, Daly Waters. As there are no appearances of rain, the weather very hot, and I have a good deal of work in plans, etc. to bring up, I shall remain here until Monday. I feel this heavy work much more than I did the journey of last year; so much of it is beginning to tell upon me. I feel my capability of endurance beginning to give way. There are a number of small fish in this water, from three to five inches long, something resembling a perch; the party are catching them with hooks; they are a great relish to us, who have lived so long upon dry meat. Any change is very agreeable. Wind variable.

Sunday, 1st June, Daly Waters. The day has been as hot as if it were in the middle of summer. Surely we must get a change soon. Wind variable, with a few light clouds. Mr. Waterhouse has shot two new parrots.

Monday, 2nd June, Daly Waters. Leaving the party in charge of Mr. Kekwick, I started at twenty minutes past seven (course north), with Thring, Auld, and Frew. Camped at 4.20. The whole day's journey has been through a splendid grass country, and open forest of gum-trees and other shrubs, some of them new to us. Here again we have also met with the bean-tree, the blossoms of a bright crimson, and at this season they seem to shed their leaves. The country passed over consisted mostly of undulations of ironstone and gravel, with a brown-coloured rock occasionally, between which were broad valleys of a light-coloured soil, all cracked and having many deep holes, which, being hidden with the long grass, caused the horses to tumble into them, and made it very fatiguing both to them and us. I have been constantly in the hope all day of coming upon some water, but have been disappointed. After rain this country can be passed over with the greatest facility, for we have passed holes that will hold water for a long time. The dip of this country is now to the eastward. To-day I think I have been running along where the dip commences from the table land. It was my intention to have tried a journey to the north-west; but, from what I have seen of the country to-day, and on my other journeys to the north, as well as Mr. Gregory's description of it on the other side, I am led to believe that it would be hopeless to expect to find water there. To try it will only be losing time, and reducing the strength of my horses. I must now try on a north-east course towards the Gulf of Carpentaria. I do not wish to go east if I can help it; but I must go where the water leads me. During the day's journey we passed through three narrow belts of hedge-tree scrub, which was very thick. There does not seem to be so much of that as we get to the north, neither is there so much of the tall mulga. We have not seen a drop of water since we left the camp. Camped without it. Wind, south. Day very hot. Latitude, 15 degrees 50 minutes 20 seconds.

Tuesday, 3rd June, Gum Forest. Fine country. I sent Thring on three miles to see if there was any change, there being a number of birds about that frequent the place where water is. I think there may still be a chance of falling in with some. He has returned and can see none. Country the same as that travelled over yesterday. Returned to the Depot. Arrived a little before sundown, and found all well. Wind light; south. Day again very hot.

Wednesday, 4th June, Daly Waters. Preparing for a start to-morrow to the north-east. I shall take the water-bags; they may retain as much as will suffice for a drink night and morning for four horses. I shall proceed to the blue-grass swamp that I found in my last north-north-east course, trace that down as far as it goes, and, should there be no water, shall strike for the sources of the Wickham River. Wind, south-south-east.

Thursday, 5th June, Daly Waters. Started at a quarter to eight with Thring and Auld, taking all the water-bags full, also King and Billiatt to take back the horses that carry the water. I have chosen King for this purpose, as being the next best bushman to Thring, and one in whom I can place the greatest dependence to execute any charge I may give him with care and faithfulness. At four o'clock arrived at the blue-grass swamp. Changed my course to 70 degrees east of north, following down the middle of it, which contains a great number of large deep holes in which water has been, but are now quite dry. Followed it until it spread itself over the plain, causing a great number of deep cracks and holes completely covered with grass, gums, and other trees, too thick to get an easy passage through. At sundown camped on the plain without water. A few hours before sundown the sky had a very peculiar appearance to the eastward, as if a black fog were rising, or smoke from an immense fire at a long distance off, but it was too extensive for that. At sundown it assumed a more distinct aspect in the shape of black clouds coming from that direction. Wind, south-east.

Friday, 6th June, Plain East of Blue Swamp. Sent King and Billiatt back with the horses, while I proceeded with the other two on a course 70 degrees east of north. At a mile and a half came suddenly upon a scrubby ironstone rise about twenty feet high. After passing over a rotten plain, full of holes and covered with grass and stunted gum-trees, proceeded to the top, from which we had a good view of the surrounding country—to all appearance one of the blackest and most dismal views a man ever beheld; even the splendid grass country I had been coming through has the same appearance. The cause of it is the trees being so thick, and some of them of a very dark colour, that nothing but their tops can be seen, which gives it the appearance of being a dense scrub. To the west there is an appearance of a scrubby rise—the one on which I have been on my other journeys to the north. No hills visible; all appears to be a level country. Proceeded down the gradual slope, crossing two other lower ironstone undulations, meeting occasionally with small rotten plains with holes, and covered with grass. At five miles the ground became firmer; at seven miles met with what seemed to be a water-shed. After a long search found that the flow of the water was to the west of north; traced it a short distance to the south-east and found a small shallow pool of water and gave our horses a drink; and wishing to take advantage of anything that may take me to the north-west, I turned and traced it down; passed three ponds with some water in them, and at three miles came upon a fine large one two and a half feet deep; followed it still on, but was disappointed on finding it terminate in a dry swamp, all cracked and full of holes; circled round it to see if the creek took up again, but could see no appearance of any. As this last pond will do for the party, I will return and bring them up, for there is a slight appearance of rain, and I wish to get them on as far as possible before the winter rain comes on. Returned to our last night's camp, where we arrived at sundown. Wind, south-east, with few clouds.

Saturday, 7th June, Plain East of Blue Swamp. Returned to the Depot; found all well. Clouds all gone, but the wind blowing strong from the south-east.

Sunday, 8th June, Daly Waters. Strong winds still from south-east, and sometimes from the south. Day very hot.

Monday, 9th June, Daly Waters. Last night, a little after sundown, Mr. Waterhouse was seized with a violent pain in the stomach, which was followed by a severe sickness, and continued throughout the night; this morning he is a little better. I think it was caused by eating some boiled gum which had been obtained from the nut-tree Mr. Kekwick discovered last year. When boiled it very much resembles tapioca, and has much the same taste. I also ate some of it yesterday, which occasioned a severe pain in the stomach, but soon went off. Some of the others also felt a little affected by it, but none so bad as Mr. Waterhouse; on others it had no effect whatever, and they still continue to eat it. Mr. Waterhouse looks so ill that I think it desirable not to move the party to-day, and trust by to-morrow he will be quite well. Light wind from the south-east, with a few clouds.

Tuesday, 10th June, Daly Waters. As Mr. Waterhouse is better, I shall move the party to-day. Started at half-past eight a.m., following my former tracks. At half-past four p.m. camped at the blue-grass swamp; twenty-six miles without water. The horses will require to be watched during the night. Wind, south-east. Day very hot. Latitude, 15 degrees 56 minutes 31 seconds.

Wednesday, 11th June, Blue-Grass Swamp. Started at seven o'clock; course, 70 degrees east of north. At three miles crossed the ironstone rise, and at eleven miles changed to north, to cut the chain of ponds, which I have named Purdie Ponds, in honour of Dr. Purdie, of Edinburgh, M.D. At one mile and three quarters, on the last course, camped on the largest pond. The country that we have gone over, although there are a number of holes and cracks in it, is really of the best description, covered with grass up to the horses' bodies. We have passed several new trees and shrubs. The bean-tree is becoming more numerous here. At this season and in this latitude it sheds its leaves; the flower is in full bloom without them. The course of the ironstone rise seems to be north and south. Wind, south-east. Weather a little cooler, but clouds all gone. Latitude, 15 degrees 52 minutes 58 seconds.

Thursday, 12th June, Purdie Ponds. Preparing for another start to-morrow with the water-bags. It takes two men nearly half a day to fill them. The orifices for filling them are a great deal too small; they ought to be at least two inches in diameter. The American cloth with which they are lined is useless in making them watertight, and is a great annoyance in emptying them, for the water gets between it and the leather. It takes a long time to draw through again, and does not answer the purpose it was intended for. A piece of calico would have done far better. It is very vexing to bring things so far, and, when required, to find them nearly useless. Wind, south-east. Cloudy. Nights cold, but the day hot.

Friday, 13th June, Purdie Ponds. Leaving Mr. Kekwick in charge of the party, started at fifteen minutes past eight with Thring and Auld, also with King and Billiatt, who were to bring back the horses carrying the water-bags. Proceeded on a north course, and at seven miles crossed what seemed to be a water-shed, seemingly running to the west of north. Halted the party, and sent Thring a short distance to see if the flow was in that direction. In a quarter of an hour he returned and informed me that it was, but only very slightly so. Changed to north-north-west to follow it. It gradually assumed the appearance of a small creek. At two miles came upon three small pools of water. I now resolve to follow it down and see where it goes to. I should think there must be more water further on. Its course is west of north. Continued to follow it down, winding and twisting about very much to almost every point of the compass. At seven miles from the pools found a little more water, but not a drop between. Allowed the horses to drink what there was, and proceeded down it. I sent Thring to follow it on one side, while I and the rest of the party kept on the other. By this we were enabled to cut off the bends and see all the creeks, so that no water could escape us. Twice it became very small, and I was afraid we were going to lose it altogether, but it commenced again and became a fine creek. Not a drop of water. At a quarter to five camped without it. Stony rises are now commencing, which are covered with gum and other trees, also a low scrub. They are very rough and running nearly west and south. The one on the west is a continuation of the one I crossed in coming to Purdie Ponds. The general flow of the creek is north. Some of the new trees are growing very large on its banks. The cabbage-tree is growing here also. This is the first time I have met with it, sometimes growing to the height of fifteen feet. All along the banks of the creek, and apparently for some distance back, is covered with an abundance of grass, but all dried up. In some places both horse and rider were completely hidden by it. Wind, south-east—few clouds. Latitude, 15 degrees 30 minutes 27 seconds.

Saturday, 14th June, River Strangways. Named after the Honourable H.B. Templar Strangways, Commissioner of Crown Lands, South Australia, and who, since his taking office, has done all in his power to promote exploration of the interior. Sent King and Billiatt back with the horses to the camp at Purdie Ponds, whilst I proceed with the further examination of the creek. I find it now running to the east of north, and the stony rises are closing upon it at two miles and a half. They begin to assume the shape of hills, which causes the travelling to be rather rough. At three miles and a half the hills run close to the creek, and are precipitous; the bed is very rough and stony—so much so that I could not take the horses down it. Ascended a hill near the creek to see what it and the country ahead was like; the hills being so rough that I could not get the horses close enough to see if there was any water, dismounted and scrambled to the top of the precipices; was delighted to see below me a large hole of water. Sent the horses across a gully to another hill still higher, while I descended into the creek; found the bed very rough, having large masses of sandstone and ironstone, which rendered it impassable for the horses. Found the water to be deep and beautifully clear; proceeded down a little further, and saw another large one. The hills close to the creek are very precipitous, and we shall have difficulty in getting the horses down to water; the hills, where they come close to the creek, are covered with spinifex. I shall therefore require to camp the party at the mouth of the gorge, where there is plenty of feed. The hill I had sent the horses to was so rocky they were unable to cross it, and there being higher hills still on ahead, I have left the horses with Auld, and, taking Thring with me, have walked to the top of it to see what course the creek was taking, but they are all so much of the same height and appearance that I can scarcely tell in which direction it runs. There is an appearance of a large creek coming in from the westward, and higher hills towards the north. I shall return and send the party on to this permanent water, and try to find an easy road over the ranges for them. I would have gone on to-day, but my horses are without shoes, and some of them are already lame, and the shoes I brought with me are nearly all exhausted; we have not been using any since shortly after leaving South Australia. Returned to our last night's camp, where we had left the canvas tank with some water that the horses did not drink in the morning; gave them what remained, and proceeded up the creeks to the last water we saw yesterday, where we arrived at sundown and camped. Wind, south.

Sunday, 15th June, River Strangways. Returned to the Depot at Purdie Ponds; found all well. Wind, south-east. Cool.

Monday, 16th June, Purdie Ponds. It was late before the horses could be found. Proceeded to the first pool of water in the River Strangways, distance about ten miles, and camped. Wind, south-east.

Tuesday, 17th June, River Strangways. Proceeded down the creek to the gorge and camped; day very hot. We had some difficulty in finding a way down for the horses to drink, it being so very rough and stony, but at last succeeded. On the west side there is a layer of rocks on the top of the hard sandstone, black and rugged, resembling lava; spinifex close to the creek. Wind, south-east.

Wednesday, 18th June, Gorge, River Strangways. I shall require to have some of the horses shod for further exploration, and shall therefore remain here to-day to get that done. I sent Thring and King a little way down the creek to see what the country is, and if there is any more water. They went about nine miles, but could see no more. In some places the country is sandy, and in others stony and grassy. Mr. Kekwick has discovered four new trees that we have not seen before, and several new shrubs. Some of the party succeeded in catching a few fine large fish, some of them weighing two pounds and a half. Some were of the perch family, and others resembled rock cod, with three remarkable black spots on each side of their bodies. There are also some small ones resembling the gold fish, and other small ones with black stripes on their sides, resembling pilot fish. Wind, south-east. Latitude, 15 degrees 30 minutes 3 seconds.

Thursday, 19th June, Gorge, River Strangways. Leaving Mr. Kekwick in charge of the party, started with Thring, Auld, and King, to look for water. No rain seems to have fallen here for a long time back; the grass is quite dry and withered. At 8.15 proceeded down the river, and, to avoid the hills, I went about a mile to the west, and found a very passable road; for about two miles we had sandy soil and spinifex mixed with grass, also a few stony rises of lime and sandstone. The country after that again became excellently grassed, the soil light and a little sandy. No water in the bed, which appears to have a very rapid fall; its general course is about north-north-east. At twelve miles, seeing a stony hill of considerable elevation, I left the bed, and went towards it. At the base of it was a deep creek; I was pleased to see a fine supply of water in it. I immediately sent Thring back to guide the party up here to-morrow, whilst I with the two others proceeded with the examination of the river further down. After following it for about ten miles through a beautifully grassed country, passing occasionally sandstone rises, with apparently scrub on their tops, camped at the base of one of them.

Friday, 20th June, First Camp North of Gorge. Returned to the other water, and at noon met the party and brought them on to this water. We have passed a few stringy-bark trees. In the bed of the river there is growing some very large and tall timber, having a dark-coloured bark, the leaf jointed the same as the shea-oak, but has not the acid taste: the horses eat it. There are also some very fine melaleuca-trees, which here seem to displace the gums in the river. We have also passed some more new trees and shrubs. Frew, in looking about the banks, found a large creeper with a yellow blossom, and having a large bean pod growing on it. I shall endeavour to get some of the seed as we go on to-morrow. I shall now move on with the whole party, and I trust to find water in the river as long as I follow it; its banks are getting much deeper and broader, and likely to retain water; it is dreadfully slow work to keep searching for water. Before this I could not do otherwise, in consequence of the season being so very dry. Since the commencement of the journey the only rain that we have had to have any effect upon the creeks was at Mr. Levi's station, Mount Margaret. Since then we have had only two or three showers, which have had no effect upon the creeks. Light winds, south-east. Latitude, 15 degrees 15 minutes 23 seconds.

Saturday, 21st June, First Camp North of Gorge. It was late before we could get a start, in consequence of our not being able to find two of the horses which separated from the rest during the night. Started, following the river down; it frequently separating into two or three channels, and again joining. Numerous small watercourses are coming in on both sides, from east and west; it winds about a great deal—its general course to-day for nine miles has been nearly north-north-east. We passed a number of large lagoons, nearly dry, close to the stony hills: when full they must retain water for a long time. There is very little water in the main channel. At nine miles I found a large and excellent pool of water in one of the side creeks; it will last some time. It being now afternoon, and there being a nice open plain for the horses, I have camped. The river is now running through stony hills, which are very rough, composed of hard sandstone mixed with veins of quartz, some of which are very hard, much resembling marble with crystalline grains in it. We are now passing a number of stringy bark along with gum and other trees, Mr. Kekwick still finding new shrubs. After we had camped, taking Thring with me, I ascended a hill a little way from the camp, but was disappointed in not having an extensive view. To the north, which is now apparently the course of the river, there seems to be an opening in the range of stony hills. The dip of the country seems to be that way.

At 33 degrees east of north from the camp, about eight miles distant, there is a high wooded tent-hill on the range; this I have named Mount Muller, after my friend the well-known botanist of Victoria. All round about are rough stony hills with grassy valleys between, having spinifex growing on their sides and tops. The valley through which the main channel flows is good soil, and covered with grass from two to four feet high. Towards the north-west the hills appear to be very rugged. Wind south-east, with a few clouds. Latitude, 15 degrees 10 minutes 40 seconds.

Sunday, 22nd June, Rock Camp, River Strangways. A few heavy clouds about. We are now in the country discovered by Mr. Gregory. There is a great deal of very good timber in the valley, which is getting larger and improving as we advance. It is still very thick—so much so, that the hills cannot be seen until quite close to them. Wind variable. Latitude, 15 degrees 10 minutes 30 seconds.

Monday, 23rd June, Rock Camp, River Strangways. This morning the sky is overcast with light clouds coming from the south-east. Started at eight o'clock, still following the river, which winds about very much; its general course 10 degrees east of north. At nine miles the channel became much smaller, and shortly afterwards separated into numerous small ones, and was apparently lost to me. I continued a north course, and at twelve miles struck a creek coming from the south-east; at two miles from this creek found another large one coming from the south-west, with shea-oak in it, which makes me suppose it is the River Strangways, and that it formed again and joined this one. At the junction were numerous recent fires of the natives; there must have been a great many of them, for their fires covered the ground, also shells of the mussel which they had been eating. Searched for water, and found a little, but not sufficient for my horses, and too difficult to approach; the course of the river is still to the north. One mile and a half from the junction found enough water that will do for me at night. As there seems to be so little water, and this day being exceedingly hot and oppressive, I have camped. The country travelled over to-day has been of the same description, completely covered with long grass; the soil rich, and a great many of the cabbage-tree growing about it. Wind variable. Latitude, 14 degrees 58 minutes 55 seconds.

Tuesday, 24th June, Mussel Camp, River Strangways. With the sun there came up a very thick and heavy fog which continued for about two hours; it then cleared off and the day became exceedingly hot. The river, after rounding the hills (where we were camped), ran nearly east for three miles, meeting there a stony hill which again throws it into a northerly course. I ascended the hill, but could see nothing distinctly, the fog being so thick. Descended and pursued the bed, which separated frequently into many channels, and at ten miles it spread into a large area, and its courses became small with no water in them. The grass above our heads was so high and thick that the rear-party lost me and could not find the rocks; by cooeing I brought them to me again. Before I had heard them I had sent Thring back to pick up their tracks and bring them to the clear ground I was on with the rest of the party, but they arrived before he made up to them. The scrub is also very thick close to the river. Mr. Kekwick found cane growing in the bed, and also brought in a specimen of a new water-lily—a most beautiful thing it is; it is now in Mr. Waterhouse's collection. At twelve miles, finding some water, the horses being tired in crossing so many small creeks, and working through the scrub and long grass, I camped at the open ground. The country gone over to-day is again splendidly grassed in many places, especially near the river; it has very lately been burned by the natives. There are a great number of them running along the banks; the country now seems to be thickly inhabited. Towards the east and the north-east the country is in a blaze; there is so much grass the fire must be dreadful. I hope it will not come near us. The day has been most oppressively hot, with scarcely a breath of wind. Latitude, 14 degrees 51 minutes 51 seconds.

Wednesday, 25th June, River Strangways. Two of the horses having separated from the others, and crossing the river, quite hidden in the long grass, it was late before they were found. Started at nine o'clock; course about 70 degrees east of north, following the channel. I expect, in two or three miles, to meet with the Roper. At three miles struck a large sheet of deep clear water, on which were a number of natives, with their lubras and children; they set up a fearful yelling and squalling, and ran off as fast as they could. Rounded the large sheet of water and proceeded along it. At a mile, three men were seen following; halted the party, and went up to them. One was a very old man, one middle-aged, the third a young, stout, well-made fellow; they seemed to be friendly. Tried to make them understand by signs that I wished to get across the river; they made signs, by pointing down the river, by placing both hands together, having the fingers closed, which led me to think I could get across further down. They made signs for us to be off, and that they were going back again. I complied with their request, and after bidding each other a friendly good-bye, we followed down the banks of the river, which I now find is the Roper. At seven miles tried to cross it, but found it to be impossible; it is now divided into a number of channels, very deep and full of running water. Proceeded further, and tried it at several places, but with the same result. At twelve miles, camped close to a steep rocky hill on the north side of the river. Searched all round for a crossing, but was unable to find one. To the eastward the country is all on fire. The banks of the river are thickly lined with cabbage-trees, also the cane, bamboo, and other shrubs. Two small turtle-shells were picked up by the party at the native camp. The country is still of the same fine description. We are now north of Mr. Gregory's tracks. Latitude, 14 degrees 5 minutes. Wind variable.

Thursday, 26th June, Roper River. As I cannot find a crossing, I shall have to return to my last camp and try to cross there. Arrived and camped. Day again oppressively hot. Almost immediately on leaving our camp this morning I observed native tracks on ours close to it. They must have followed us up last night, although we saw nothing of them. They are not to be trusted: they will pretend the greatest friendship one moment and spear you the next. They have been following us to-day, but keeping on the other side of the river and setting fire to the grass as they go along. I wish it would rain and cause the grass to become green, so as to stop them burning, as well as to give me some fresh food for the horses, for they now begin to show the want of it very much; it is so dried up that there is little nourishment in it. Some of them are beginning to look very poor and are much troubled with worms. My journeys have been very short last week, in consequence of my being so weak from the effects of scurvy and a severe attack of dysentery, for I have scarcely been able to endure the motion of horseback for four hours at a time; but having lately obtained some native cucumbers, I find they are doing me a deal of good, and hope by next week to be all right again. Wind, south. Latitude, 14 degrees 51 minutes 51 seconds.

Friday, 27th June, West Roper River. Started on a course of 320 degrees, crossing the river, and at three miles and a half again struck the Roper, running. Followed it up, coming nearly from the west, but winding about very much, and having many branches, which makes it very difficult for me to get the turns correctly. It is a splendid river. We have passed many brooks and deep reaches of water some miles in length, and the country could not be better: it is really magnificent. At 2.30 I was informed that we were short of a horse. Sent Messrs. Kekwick and Thring back to see where he was left. We have had to cross so many boggy, nasty places, with deep water and thick scrub, that he must have been missed at one of these. The general course of the river to-day has been 280 degrees. Distance, fifteen miles. Messrs. Kekwick and Thring are returned. They found the horse bogged in a side creek. It was so thick with cabbage-tree that they passed in searching for him two or three times. They had great difficulty in getting him out, but at last succeeded, and arrived at the camp before dark. A short time before that, another horse got into a very deep and rapid channel of the river, the top of the banks projecting so much that he could not get out, and the gum-trees having fallen across both above and below him, he was completely fixed. We endeavoured to get him out, but it got so dark that we could not see him, and the rope breaking that we were pulling him out by, he got his head under water, and was drowned in a moment. We then found that the cause of the rope breaking was that he had got one of his hind feet entangled in a sunken tree. It being now so dark we can do no more to-night, and have left him in the water until daylight. Wind, south-east. Latitude, 14 degrees 47 minutes 26 seconds.

Saturday, 28th June, Roper River. As I shall be short of meat, I remain here to-day to cut up the horse and dry him. The water of this river is most excellent; the soil is also of the first description; and the grass, although dry, most abundant, from two to five feet high. This is certainly the finest country I have seen in Australia. We passed three rocky hills yesterday, not high, but having grass up to their tops, round which the river winds at their base, forming large and long reaches of water. On the grassy plains it forms into different channels, and is thickly timbered with shea-oak, gum, cabbage-trees, and other trees and shrubs. Wind variable.

Sunday, 29th June, Roper River. We are all enjoying a delightful change of fresh meat from dry. It is a great treat, and the horse eats remarkably well, although not quite so good as a bullock. At sundown the meat is not all quite dry, but I think we shall be able to preserve the greater part of it. The natives are still burning the grass round about us, but they have not made their appearance either yesterday or to-day. Wind variable.

Monday, 30th June, Roper River. Started at 8.10, course west, following up the river, which winds about very much from north-west to south, and at last to south-east. When coming close to where the grass was on fire, finding a good ford, I crossed the party to the north-east side. At fifteen miles came upon a large reedy swamp through which the river seemed to flow, and again at twenty miles came upon the river running into the swamp, and coming from the north-north-west. Although travelling twenty miles we have not made more than ten miles in a straight line; the general course is west. The country is of the same excellent description. We have passed the stony rises on the north side of the river, which are covered with grass to their tops. After crossing the river I ascended another of the same kind. To the south are a few hills scattered over the grassy plains, with lines of dark-green trees between them, showing that they are creeks flowing into the river whose junctions we have been crossing to-day; the same to the south-west, and at west 20" south the distance appears level, with a single peak just visible. To the north-west seemingly stony hills; to the north the same; to the east I could see nothing, for the smoke conceals from me the country; it is all on fire. The river is still running very rapidly, and as this is a different branch from those previously discovered, I have named it the River Chambers, after my late lamented friend, James Chambers, Esquire, whose zeal in the cause of Australian exploration is already well known. A short time before sundown a number of natives were seen approaching the camp. We were immediately prepared for them. I sent Mr. Kekwick forward to see what their intentions were—friendly or hostile. I immediately followed. On reaching them they appeared quite friendly. There were three men, four lubras, and a number of children. One, an old man, presented a very singular appearance—his legs being about four feet long, and his entire height seven feet, and so remarkably thin that he appeared to be a perfect shadow. Mr. Kekwick having a fish-hook stuck in his hat, which immediately caught the tall old fellow's eye, he made signs of its use, and that he would like to possess it. I told Mr. Kekwick to give it to him, which seemed to please him much. After examining it he handed it over to a young man, seemingly his son, who was a fat, stout fellow, and who was laughing nearly all the time. The other was a middle-aged man of the ordinary height. The women were small, and very ugly. Wind, south-east. Latitude, 14 degrees 47 minutes 24 seconds.

Tuesday, 1st July, Reedy Swamp, River Chambers. Before sunrise the natives again made their appearance, sixteen in number, with small spears. Sent Mr. Kekwick to see what they wanted. On his coming up to them they put two fingers in their mouths, signifying that they wanted more fish-hooks, but we had no more to spare. They remained looking at us until the horses were packed and started. After Thring and Frew had brought in the horses, they rode up to where they were. They (the natives) did not fancy being too near the horses, but having dismounted, it gave them confidence, and they returned again. Thring opened the lips of one of the horses, and showed them his teeth, the appearance of which did not suit their taste. Some of them thought the further off they were from such weapons the better, and ran off the moment they saw them. Others remained, but kept at a respectful distance. Thring pulled a handful of grass, and it amused them much to see the horses eating it. After starting they followed us for some miles, when Mr. Waterhouse, observing a new pigeon, shot it. They, not liking the report of the gun, went off, and we saw no more of them. Started at 8.20, following the river on a course 30 degrees east of north. After a mile it gradually came round to the south-east, and was a running stream in that direction. As that course would take me too much out of my road, I changed my bearing to north-west, to an opening between the hills. After passing a number of fine ponds, many of them with water in them, came upon a large creek, having long reaches of water in it, but not running. It winds about a great deal. Its general course to-day has been west-north-west. The reedy swamp must be a mass of springs, which causes the Roper to run with such velocity. A little after one o'clock camped. The journey to-day has been rough, having so many small creeks to cross, and the day being excessively hot, the horses seem fagged. They have been covered with sweat since shortly after starting until now, and as some of the drowned horse is not quite dry, I have halted earlier than I intended. The country gone over to-day is of the same kind, beautiful soil, covered with grass. We occasionally met with stony hills coming down to the creek, also well grassed and timbered to their tops. Wind west, with heavy clouds from the south-east. Latitude, 14 degrees 41 minutes 39 seconds.

Wednesday, 2nd July, West-north-west of Reedy Swamp, River Chambers. Started 7.40, following the river up until ten o'clock. We kept nearly a north-west course: it then went off to the south-west; as that would take me too much out of my course, I kept the north-west course, crossing the saddle of broken hills, amongst which we have now got; and at twelve again met the river, now coming from the north through the hills, following it still, having plenty of water. At a very large water hole surprised some natives, who ran off at full speed when the rear of the party was passing their camp. One stout fellow came running up, armed with spears, and loaded with fish and bags filled with something to eat. Mr. Kekwick rode towards him. The native held up a green bough as a flag of truce, and patting his heart with his right hand, said something which could not be understood, and pointed in the direction we were going. We then bade him good-bye, and proceeded on our journey. At one o'clock the river suddenly turned to the east, coming from very rough hills of sandstone and other rocks. At one mile and a half on that course it was coming from the south of east, which will not do for me. Changed to the north, and got into some terrible rough stony hills with grassy valleys between, but not a drop of water. It being now after two o'clock, too late to encounter crossing the table land, I again changed my course to south-east for the Chambers, and at 5.3 camped at a large water hole at the foot of a stony rise lined with cabbage (palm) trees. The country although rough is well grassed to the top of the hills, with an abundance of permanent water in the river. I am sorry it is coming from the south-east, and have been in hopes it would carry me through this degree of latitude. To follow it further is only losing time; I shall therefore take to the hills to-morrow. Frew, on coming along, picked up a small turtle alive. Light wind from the south-east; heavy clouds from the south-west. Latitude, 14 degrees 32 minutes 30 seconds.

Thursday, 3rd July, River Chambers. Started at 8.10 o'clock, north-west course. At one mile and a half again struck the river coming from the west-north-west; left it and followed its north-west course: and at another mile again came upon it with plenty of water. Saw four natives, who ran off the moment they saw us. Followed the river, the hill coming quite close to it, very steep and rocky, composed of a hard sandstone, and occasionally a little ironstone. At nine miles again left the river, finding it was coming too much from the eastward; crossed the saddle of the two spurs again; came upon a creek, which I think is the river; ran it up to the west for about a mile, but no appearance of water; left it, and ascended a very rough rugged hill. In the creek we have just left there is a deal of limestone. Crossed three more small spurs and small creeks, but not a drop of water. It being now afternoon, and wishing to see from what direction the river is coming, I changed to north-east, but found that I was still among the rough hills; I then went east for a short distance, and made the river, now quite dry, and having a sandy bed. Followed it up, but saw there was no hope of water; turned, and traced it down to try and find water. After following it for three miles, came upon a fine permanent hole of water, a short distance from where we left in the former part of the day. If it would only rain and put some water in the deep dry holes that are in the other creeks crossed to-day, I should then be enabled to steer a straight line for the Adelaide. It is very tedious and tiresome having to look for water every day. We have now reached to the top of one of the tributaries of the Chambers. This is apparently the last water. It seems to take its rise in a grassy plain to the east of this. The valley through which the creek flows is well grassed, but the sides and the tops of the hills are spinifex mixed with grass. All the small valleys are well grassed. Wind, south-east. Latitude, 14 degrees 26 minutes 50 seconds.

Friday, 4th July, Last Water Hole in the Chambers. Started at 8.10, course north-west, following up the river to its sources. At four miles ascended a rise, which was very rough, composed of sandstone, ironstone, and limestone, with ironstone gravel on the top. Descended on the other side, and at about five miles came upon a nice running stream, but very rough and stony round about it. After crossing several stony rises, in which we had some difficulty in getting our horses over, arrived at a nice broad valley with a creek running through it, course north-west. At a mile it received a large tributary from the east of north, and the bed seems sandy; melaleuca and gum-trees in it; also the bean-tree. The valley is covered with grass from two to four feet high. There is a ridge of rough sandy stone hills, with occasional ironstone on each side, from the direction it was at first taken. I thought I was fortunate in meeting with one of the sources of the Alligator or Adelaide River. After following it for five miles, sometimes going west and south, it went through a stony gorge, and seemed to run to the south, which is a great disappointment. I ascended one of the hills to view the country, but could see very little, it being so thickly wooded. To the north is the appearance of a range running to the east and west that I must endeavour to cross to-morrow if I do not find another creek running to the north-west. There is one benefit I shall derive from following down this creek a day; it will enable me to round the very rough sandstone range that runs on the north side of the creek. It is so rough that I could not take the horses over it. Camped at the gorge of this creek, which I suppose, from the course it is now taking, to be another tributary of the Chambers. The gorge is impassable for horses. It has a very picturesque appearance; immense masses of rock—some thousands of tons in weight—which had fallen from the top of the cliff into the bed of the creek. Mr. Kekwick found a number of new plants, among them a fine climbing fern. Light winds, east. Plenty of permanent water in the creek. Latitude, 14 degrees 25 minutes 8 seconds.

Saturday, 5th July, Gorge on another West Branch of the River Chambers. Started 8.15; course, 5 degrees west of north. After travelling two miles over stony rises we ascended a low table land with coarse grass and a little spinifex; at six miles came up to a high stony tent-hill, which I ascended and named Mount Shillinglaw. All round are stony hills and grassy valleys—dip of the country seemingly to the south. There is apparently a continuous range in the distance to the north-west, the Chambers range. Changed my course to 325 degrees, and at four miles struck another large branch coming from the north-east, and running apparently south—plenty of water in it. This I named the Waterhouse, in honour of Mr. H.W. Waterhouse, naturalist to the expedition. Some of the horses are become so lame on account of the stones they will not be able to travel another day. I have camped early to have them shod, for on Monday I intend taking a north-west course to strike the source of the Adelaide. The country on the last course is again of the very best description and well grassed. The hills are stony, but abound with grass; they are composed of sandstone, ironstone, and occasionally a little limestone; the trees are the same as those on the Roper. Wind, south-east. Latitude, 14 degrees 18 minutes 30 seconds.

Sunday, 6th July, The Waterhouse River. Day again very hot. There is another branch a short distance off, which seems to come from the north-west; I shall follow it to-morrow if it continues the same course. I think these creeks we are now crossing must be the sources of the Adelaide flowing towards the dry river seen by Mr. Gregory running towards the north-west. Wind light; sky cloudy.

Monday, 7th July, Waterhouse River. Started at 7.55; course, north-west. At four miles the creek was coming from the west, north-east, and east; I therefore left it, crossed two low stony rises, and again struck another creek coming from the north-east, with plenty of water; followed it for a short distance to the west, found it so boggy and the body of water so large that I could not get the party round the stony hills. Returned about half a mile, and crossed the stony rise, and again struck it. At eight miles came upon a number of springs coming from the stony rises. Ascended one of the rises, which are not high, and found myself on a sandy table land, which continued for six miles, having coarse grass and spinifex growing on it. Towards the last two miles it again became well grassed. The timber is stringy-bark, some splendid trees; amongst them gums and a number of pines, also very fine. The cabbage palm still growing in the creeks in great numbers, some of them very tall, with several branches on the top. The first eight miles was again over a splendid country, and the last three of the same description. A stony hill being in my course, I proceeded to the top of it, from which I had a good view of the country before me. This hill I named after Lieutenant Helpman. At 10 degrees south of west are two remarkable isolated table hills, Mount Levi and Mount Watts, beyond which is the Chambers range to the north-west; my view in other directions is obstructed by other hills, but to the west about one mile and a half is seemingly a creek, to which I shall go, and if there is water I shall camp. Proceeded and found it a fine creek with plenty of water; followed it about one mile to the north-west, when it became dry. There it seems to come from the south. There are a great number of cabbage palms on its banks. I hope it will soon come round to the north-west and continue on that course. Light winds, variable. Latitude, 14 degrees 9 minutes 31 seconds.

Tuesday, 8th July, Water Creek in Stony Rises. Started at 7.40 a.m., course north-west; followed the creek a little way, but found it was running too much to the west of my course; left it and proceeded to the north-west, crossing some stony rises, now composed of granite and ironstone, with occasionally some hard sandstone. Crossing three small creeks running to the west, at six miles came upon a large one with broad and long sheets of permanent water coming from the north-north-east, and apparently running to the south-west. This I have named the Fanny, in honour of Miss Fanny Chambers, eldest daughter of John Chambers, Esquire. In a small tree on this creek the skull of a very young alligator was found by Mr. Auld. The trees in this creek are melaleuca and gum, with some others. Proceeded across the creek, still going north-west; ascended two stony rises, and got upon low table land with spinifex and grass, passing two stony hills, one on each side of my course. At eighteen miles struck the head of a small creek flowing nearly on my course; followed it down in search of water, now through a basaltic country. At two miles came upon another large creek, having a running stream to the south of west, and coming from the north of east. Timber, melaleuca, palm, and gum, with some of other descriptions. This I have named the Katherine, in honour of the second daughter of James Chambers, Esquire. The country gone over to-day, although there is a mile or two of light sandy soil, is good for pasturage purposes; in the valley it is of the finest description. Light winds, variable. Latitude, 13 degrees 58 minutes 30 seconds.

Wednesday, 9th July, The Katherine. Started at five minutes to eight o'clock, crossing the Katherine, and proceeded on a north-west course over a basaltic country, splendidly grassed. At five miles I ascended a high hill, which I named Mount Stow, but was disappointed in the view. West-north-west course, over a great number of rises thickly timbered with gum. At 20 degrees north of west is a high bluff point of the range; the country on that bearing does not seem to be so rough. No more visible but the range to the west and the hill between.

Descended, and changed my course to the bluff point. At one mile and a half crossed a creek with water in it, coming from the north-east, and running to the south-west. At three miles further arrived at the bluff. The basaltic country has now suddenly changed to slate, limestone, sandstone, and a hard white stone. Crossed three stony rises, and got upon a white sandy rise, with large stringy-bark trees growing upon it; and there seemingly being a creek at the foot of it, from the number of green gums and palm-trees, I went down to it, and found it to be springy ground, now quite dry, although the grass was quite green. Proceeded on the westerly course, expecting to meet with a creek; found none, but large springs coming from sandy rises. Having found water at thirteen miles, and being so very unwell that I cannot proceed, I have been compelled to camp. There is an immense quantity of water coming from these springs; the soil round them is of the best deep black alluvial. About a mile to the west is a strong stream running to the south-west from them. I have called them Kekwick Springs, in honour of my chief officer. Wind light and variable. Latitude, 13 degrees 54 minutes 12 seconds.

Thursday, 10th July, Kekwick's Large Group of Springs. Started at eight o'clock; crossed the springs without getting any of the horses bogged. Proceeded on a north-west course, but at a mile and a half again came upon springs and running water; the ground too boggy to cross it. Changed to north; at three miles and a half on the course changed to north-west. Ascended some very rough stony hills, and got on the top of sandy table land thick with splendid stringy-bark, pines, and other trees and shrubs, amongst which, for the first time, we have seen the fan palm, some of them growing upwards of fifteen feet high; the bark on the stem is marked similar to a pineapple's; the leaf very much resembles a lady's fan set on a long handle, and, a short time after it is cut, closes in the same manner. At half-past one crossed the table land—breadth thteen miles. The view was beautiful. Standing on the edge of a precipice, we could see underneath, lower down, a deep creek thickly wooded running on our course; then the picturesque precipitous gorge in the table land; then the gorge in the distance; to the north-west were ranges of hills. The grass on the table land is coarse, mixed with a little spinifex; about half of it had been burnt by the natives some time ago. We had to search for a place to descend, and had great difficulty in doing so, but at last accomplished it without accident. The valley near the creek, which is a running stream, is very thickly wooded with tall stringy-bark, gums, and other kinds of palm-trees, which are very beautiful, the stem growing upwards of fifty feet high, the leaves from eight to ten feet in length, with a number of long smaller ones growing from each side, resembling an immense feather; a great number of these shooting out from the top of the high stems, and falling gracefully over, has a very pretty, light, and elegant appearance. Followed the creek for about two miles down this gorge, and camped on an open piece of ground. The top course of the table land is a layer of magnetic ironstone, which attracted my compass upwards of 20 degrees; underneath is a layer of red sandstone, and below that is an immense mass of white sandstone, which is very soft, and crumbling away with the action of the atmosphere. In the valley is growing an immense crop of grass, upwards of four feet high; the cabbage palm is still in the creek. We have seen a number of new shrubs and flowers. The course of the table land is north-north-west and south-south-east. The cliffs, from the camp in the valley, seem to be from two hundred and fifty to three hundred feet high. Beyond all doubt we are now on the Adelaide river. Light winds, variable. Latitude, 13 degrees 44 minutes 14 seconds.

Friday, 11th July, Adelaide River, North-west Side, Table Land. The horses being close at hand, I got an early start at 7.20, course north-west. In a mile I got greatly bothered by the boggy ground, and numbers of springs coming from the table land, which I am obliged to round. At two miles got clear of them, and proceeded over a great number of stony rises, very steep; they are composed of conglomerate quartz, underneath which is a course of slates, the direction of which is north-west, and lying very nearly perpendicular, and also some courses of ironstone, and a sharp rectangular hard grey flint stone. My horses being nearly all without shoes, it has lamed a great many of them, and, having struck the river again at fifteen miles, I camped. They have had a very hard day's journey. The country is nearly all burnt throughout, but those portions which have escaped the fire are well grassed. I should think this is a likely place to find gold in, from the quantity of quartz, its colour, and having so lately passed a large basaltic and granite country; the conglomerate quartz being bedded in iron, and the slate perpendicular, are good signs. The stony rises are covered with stringy-bark, gum, and other trees, but not so tall and thick as on the table land and close to it, except in the creek, where it is very large; the melaleuca is also large. Since leaving the table land we have nearly lost the beautiful palms; there are still a few at this camp, but they are not growing so high; the cabbage palm is still in the creek and valleys. Light winds from south-east. Country burning all round. Latitude, 13 degrees 38 minutes 24 seconds. This branch I have named the Mary, in honour of Miss Mary Chambers.

Saturday, 12th July, The Mary, Adelaide River. Started at 7.30; course, north-west. At one mile and a half came upon a running stream coming from the north-east; had great difficulty in getting the horses across, the banks being so boggy. One got fixed in it and was nearly drowned; in an hour succeeded in getting them all safe across. At six miles I ascended a high, tall, and stony hill; the view is not good, except to the westward. In that direction there is seemingly a high range in the far distance, appearing to run north and south; the highest point of the end of the range is west, to which the river seems to tend. My horse being so lame for the want of shoeing, I shall strike in for the river and follow it for another two miles, as it seems to run so much to the westward. I have resolved to use some of the horseshoes I have been saving to take me back over the stony country of South Australia. To enable McGorrerey to get them all shod on the front feet before Monday, I have camped. There is still a slaty range on each side of the river, with quartz hills close down to it; the timber the same as yesterday. The country has recently all been burned; but, judging from the small patches that have escaped, has been well grassed up to the pass of the hills. The valley and banks of the creeks are of beautiful alluvial soil. One new feature seen to-day is the growing of large clumps of bamboo on the banks of the river, from fifty to sixty feet in height and about six inches in diameter at the butt. I am now on one of the tributaries of the Adelaide River. There must have been a dreadful fire here a few days ago; it has destroyed everything before it, except the green trees, to the edge of the water. Slight winds, variable. Latitude, 13 degrees 35 minutes 58 seconds.

Sunday, 13th July, The Mary, Adelaide River. Shoeing horses. Wind blowing strong; variable from all points of the compass.

Monday, 14th July, The Mary, Adelaide River. One of the horses cannot be found this morning, and he has been for some time very ill and weak, and no appearance of getting better. It was my intention to have left him. We have been all round the tracks forward and backward over the feeding-ground and can see nothing of him. I am afraid he has gone off to some place and died; I shall therefore waste no more time in looking for him. If he is alive I may have a chance of recovering him on my return. Late start, in consequence of so long looking for him. As I have now got all the horses shod on the front feet, I shall proceed on a north-west course through the stony rises, which are still quartz and slate, splendidly grassed, with gums and other trees and bushes not too thick to get through with ease. Crossed six small creeks, one with holes with water in them; the third one, a large creek, which I crossed at nine miles, I have named William Creek, after the second son of John Chambers, Esquire, of Adelaide; all running at right angles to my course. Immediately after crossing this last creek the country changed to granite; the rises are composed of immense blocks of it, with occasionally some quartz. The country has been all burned. The valleys between the granite rises are broad and of first-rate soil, many of them are quite green, caused by springs oozing from the granite rock. We have passed a number of trees resembling the iron-bark, also some like new ones, and many shrubs which Mr. Kekwick has found. Wind, south-east. Latitude, 13 degrees 29 minutes 25 seconds.

Tuesday, 15th July, Billiatt Springs. I have named these springs in token of my approbation of Billiatt's thoughtful, generous, and unselfish conduct throughout the expedition. I started at 7.40 this morning, course north-west. Crossed granite and quartz rises, with broad valleys between, both splendidly grassed. At three miles crossed a small creek with water; at another mile the same creek again; one also to my line on the south-west side, and immediately went off to the south-west. At six miles the river came close to the line, and immediately went off to the west. Continued on my course through granite and quartz country, splendidly grassed, and timbered with stringy-bark and gums, pines, palms, nut-trees, and a wattle bush, which in some places was rather thick, but not at all difficult to get through. At ten miles again struck the river; it is now apparently running to the north. Changed to that course, but it soon left me. At three miles and a half on the north course struck another creek running from the range north-east; it has an abundance of water, and is rather boggy. King's horse fell with him in it, but did no further injury than giving him a wetting. A few of the other horses stumbled and rolled about in it for a short time, but we got them all across without accident. Changed to west of north; at half a mile reached a saddle between two hills, and ascended the one to the west, the river now running between ranges to the west; they seemed a good deal broken, with some high points to the north-west. There is a higher one, seemingly running north and south, with apparently a plain between about four miles broad, on which are four or five lines of dark trees; this leads me to suppose that the river is divided. The plain being very thickly timbered, I could not see distinctly which was the main channel. Descended, and proceeded on a north-west course. At one mile and a half struck the river, again running north; changed to that, and at two miles and a half camped. The country is now all burnt. I am obliged to stop where I can get feed for the horses. One of the channels comes close to the bank, east side, about six yards wide and two feet deep; bed sandy. The main channel must be in the middle of the plain. The hill I ascended to-day has been under the influence of fire; it is composed of quartz, and a hard dark-coloured stone; the quartz runs in veins throughout it, in places crystalline, and formed into spiral and many-sided figures; in places there is a crust of iron, as if it had been run between the stones, that is also crystalline. Wind, south-east. Latitude, 13 degrees 17 minutes 22 seconds.

Wednesday, 16th July, The Mary, Adelaide River. Started at 7.40, course north. The river runs off again to the north-west, and I have passed over an undulating country, all burnt, but the soil of the richest description. The rises are comprised of quartz and a hard white stone, with occasionally a little ironstone. At three miles crossed a creek with water holes. At five miles crossed another. At seven miles came close to a high hill—ascended it; at the foot it is composed of a hard slaty stone covered with a cake of iron; about the middle is quartz, and on the top conglomerated quartz. The view from south-west to north-west is extensive, but this not being the highest hill, the rest is hidden. To the west is a high hill, bluff at both ends, seemingly the last hill of the range; its course apparently north-west and south-east. At this bluff hill the range seems to cease, or drops into lower hills. A branch of the river lies between it and me, but there are still a number of stony hills before I can reach it. To the north-west and north there are high and stony hills. The river now seems to run to the west, on a bearing of 30 degrees north of west. From twenty to twenty-five miles distant is another range, at the foot of which there is a blue stripe, apparently water, which I suppose to be the main stream of the Adelaide. Descended, as the country is too rough and stony to continue either to the north or north-west. I changed to 3 degrees north of west, crossed some stony hills and broad valleys with splendid alluvial soil, the hills grassed to the top. On that course struck the branch of the river. Still very thick with the same kind of timber already mentioned. Most of the bamboos are dead. I suppose the fire has been the cause of it. I again find it running to the north; I turn to that course. At three miles struck a large creek coming from the east with large sheets of water; had to run it up half a mile before I could get across it. Crossed it all right, and passed through a beautiful valley of green grass. After that, found that I was again on the stony rise, where every blade of grass had been burned off, and not knowing how far this may continue, I have turned off again for the creek, to give the horses the benefit of the valley. The timber is the same as yesterday in some places; the stringy-bark is much larger. The banks of the river, when we first came upon it to-day, were high and stony. The range to the east seems to cease about here. We are now crossing low undulations. I have seen a number of kangaroos to-day; they do not seem to be as large as those in the south. The valleys are composed of conglomerated ironstone underneath the soil. A large number of new birds seen to-day, some of them with splendid plumage. Wind, south-east. Latitude, 13 degrees 7 minutes 21 seconds.

Thursday, 17th July, Tide Creek, Adelaide River. Started at eight o'clock, course north-west; passed over some stony hills, small creeks, and valleys well grassed. At three miles again met with the branch of the river, with bamboos and trees of the same description as before, a running stream, but not so rapid. At five miles, observing an open plain among the trees, and the river trending more to the westward, I changed my course to it, 15 degrees west of north; found it to be open plain, of rich alluvial soil in places; at times it seemed to be subject to inundation, I suppose the drainage from the range to the eastward, which is distant about four miles. I am pleased it has been burnt, but where it has not the grass is most abundant; where the water seems to remain it is rather coarse. The plains are studded with lines of green gum-trees, and the cabbage palms are numerous, which give them a very pretty park-like appearance. They continued for ten miles, when we made a small stony hill; we met with a large creek, with large holes of water in it, and supposing I had got upon the plain that ran to the sea-coast, and seeing those I had passed over so dry, camped; and having sent Thring to a rise to see where the river is, he returned, but can see nothing of it, but reports high hills to the north-west. I am glad of this, for it is not my intention to follow the river round if I can get water in other places, for it has already been well described south of this by Lieutenant Helpman when he came up in a boat, and I wish to see what the country is away from its banks. Wind south-east, with a few clouds from the north. For the last week the weather has been excellent, not too hot during the day, and cool and refreshing at night. The mosquitoes are very annoying, and the flies during the day are a perfect torment. This creek I have called Priscilla Creek. Latitude, 12 degrees 56 minutes 54 seconds.

Friday, 18th July, Priscilla Creek. Started at 8.15, course north-west. Passed over grassy plains and stony rise; when, at three miles, seeing the termination of a range in a bluff point, changed my course to 310 degrees. Proceeded, still crossing stony hills, consisting of ironstone, slate, and a hard white rock, which is broken into rectangular fragments; also over broad valleys, which are covered with grass that when green must have stood very high, but is now so dry that it breaks off before the horses. My horse being first, collects so much on his front legs that I have been obliged to stop, pull him back, and allow it to fall, so that he may step over it, go on, get another load, and do the same. At six miles and a half, after crossing a plain, crossed a deep bamboo creek; this I have named Ellen Creek. Proceeded over two other stony rises and valleys of the same description, and came upon extensive plains, well grassed, and of beautiful alluvial soil; crossing them towards the bluff point at fifteen miles, came upon the Adelaide between me and the bluff, which is about a mile further on; the river is about eighty yards wide, and so still that I could not see which way the current was. I suppose its being high tide was the cause of this. The banks are thickly lined with bamboo, very tall and stout, very steep, and twelve feet down to the water's edge; the water appeared to be of great depth, and entirely free from snags or fallen timber. The range on the opposite side of the river, for which I was directing my course, being the highest I have seen in this new country, I have named it after His Excellency the Governor-in-Chief of South Australia, Daly Range, and its highest peak to the north Mount Daly. Before reaching the river, at thirteen miles, we passed a high conspicuous tent hill, at right angle, north-east to our line; this I have named Mount Goyder, after the Surveyor-General of South Australia. Followed the river on a north course for about a mile, when I was stopped by a deep side creek of thick bamboo, with water; turned to the east, rounded the bamboo, but found myself in a boggy marsh, which I could not cross. This marsh is covered with fine grass, in black alluvial soil, in which is growing a new kind of lily, with a large broad heart-shaped leaf a foot or more across; the blossoms are six inches high, resemble a tulip in shape, and are of a deep brilliant rose colour; the seeds are contained in a vessel resembling the rose of a watering-pot, with the end of each egg-shaped seed showing from the holes, and the colour of this is a bright yellow. The marsh is studded with a great number of melaleuca-trees, tall and straight. As I could not cross, I had to round it, which took me a little more than an hour; when I got upon some low undulating rises, not far from Mount Goyder, composed of conglomerate ironstone and ironstone gravel, which seem to produce the springs which supply the marsh. Camped on the side of the marsh, to give the horses the benefit of the green grass, for some of them are still troubled with worms, and are very poor and miserable, and I have no medicine to give them, and there is not a blade of grass on the banks of the river—all has been burnt within the last four days. Native smoke in every direction. Wind south-east, with a few clouds. Latitude, 12 degrees 49 minutes 30 seconds.

Saturday, 19th July, Lily Marsh, Adelaide River. Started at 9.10, course 20 degrees east of north. At three miles crossed some stony rises and broad alluvial grassy valleys; at four miles met the river, had to go half a mile to the south-east to round it. Again changed to my first course; at seven miles and a half crossed a creek with water. The country to this is good, with occasionally a little ironstone and gravel, timber of stringy-bark, and a little low gum scrub. Having crossed this creek, we ascended a sandy table land with an open forest of stringy bark (good timber), palms, gums, other trees and bushes; it has been lately burnt, but the roots of the grass abound. This continued for about three miles. There is a small stony range of hills to the west, which at the end of the three miles dropped into a grassy plain of a beautiful black alluvial soil, covered with lines and groves of the cabbage palm trees, which give it a very picturesque appearance; its dip is towards the river; in two miles crossed it, and again ascended low table land of the very same description as the other. At fourteen miles struck another creek with water, and camped. The country gone over to-day, though not all of the very best description, has plains in it of the very finest kind—even the sandy table-land bears an abundant crop of grass. The trees are so thick that I can get no view of the surrounding country; the tall beautiful palm grows in this creek. Native smoke about, but we have not seen any natives. There are large masses of volcanic rock on the sides of this creek. At about a mile to the eastward is a large body of springs that supply water to this creek, which I have named Anna Creek. Camped at ten minutes to three o'clock. Wind variable. Latitude, 12 degrees 39 minutes 7 seconds.

Sunday, 20th July, Anna Creek. The mosquitoes at this camp have been most annoying; scarcely one of us has been able to close his eyes in sleep during the whole night: I never found them so bad anywhere—night and day they are at us. The grass in, and on the banks of, this creek is six feet high; to the westward there are long reaches of water, and the creek very thickly timbered with melaleuca, gum, stringy-bark, and palms. Wind, south-east.

Monday, 21st July, Anna Creek and Springs. Again passed a miserable night with the mosquitoes. Started at eight o'clock; course, north-north-west. At three miles came upon another extensive fresh-water marsh, too boggy to cross. There is rising ground to the north-west and north; the river seems to run between. I can see clumps of bamboos and trees, by which I suppose it runs at about a mile to the north-north-west. The ground for the last three miles is of a sandy nature, and light-brown colour, with ironstone gravel on the surface, volcanic rock occasionally cropping out. The borders of the marsh are of the richest description of black alluvial soil, and when the grass has sprung after it has been burnt, it has the appearance of a rich and very thick crop of green wheat. I am now compelled to alter my course to 30 degrees south of east, to get across a water creek coming into the marsh, running deep, broad and boggy, and so thick with trees, bushes, and strong vines interwoven throughout it, that it would take a day to cut a passage through. At three miles we crossed the stream, and proceeded again on the north-north-west course, but at a mile and a half were stopped by another creek of the same description. Changed to east, and at half a mile was able to cross it also, and again went on my original bearing. Continued on it for three miles, when we were again stopped by another running stream, but this one I was able to cross without going far out of my course. Proceeded on the north-north-west course, passing over elevated ground of the same description as the first three miles. At seventeen miles came upon a thick clump of trees, with beautiful palms growing amongst them; examined it and found it to have been a spring, but now dry. Proceeded on another mile, and was again stopped by what seemed to be a continuation of the large marsh; we now appeared to have got right into the middle of it. It was to be seen to the south-west, north-east, and south-east of us. Camped on a point of rising ground running into it. The timber on the rises between the creeks is stringy-bark, small gums, and in places a nasty scrub, very sharp, which tore a number of our saddle-bags: it is a very good thing the patches of it are not broad. The grass, where it has not been burned, is very thick and high up to my shoulder when on horseback. About a mile from here, to the west, I can see what appears to be the water of the river, running through clumps of trees and bamboos, beyond which, in the distance, are courses of low rising ground, in places broken also with clumps of trees; the course of the river seems to be north-north-west. On the east side of the marsh is also rising ground; the marsh in that direction seems to run five or six miles before it meets the rising ground, and appears after that to come round to the north. Nights cool. Latitude, 12 degrees 28 minutes 19 seconds. Wind, south-east.

Tuesday, 22nd July, Fresh-water Marsh. As the marsh seems to run so much to the east, and not knowing how much further I shall have to go to get across the numerous creeks that appear to come into it, I shall remain here to-day and endeavour to find a road through it to the river, and follow up the banks if I can. I have a deal of work to do to the plan, and our bags require mending. After collecting the horses Thring tried to cross the marsh to the river, and succeeded in reaching its banks, finding firm ground all the way; the breadth of the river here being about a hundred yards, very deep, and running with some velocity, the water quite fresh. He having returned with this information, I sent him, King, and Frew, mounted on the strongest horses, to follow the banks of the river till noon, to see if there is any obstruction to prevent my travelling by its banks. In two hours they returned with the sad tidings that the banks were broken down by watercourses, deep, broad, and boggy; this is a great disappointment, for it will take me a day or two longer than I expected in reaching the sea-coast, in consequence of having to go a long way round to clear the marsh and creeks. The edge of the marsh was still of the same rich character, and covered with luxuriant grass. The rise we are camped on is also the same, with ironstone gravel on the surface; this seems to have been a favourite camping-place for a large number of natives. There is a great quantity of fish bones, mussel, and turtle shells, at a little distance from the camp, close to where there was some water. There are three poles fixed in the ground, forming an equilateral triangle, on the top of which was a framework of the same figure, over which were placed bars of wood: its height from the ground eight feet. This has apparently been used by them for smoke-drying a dead blackfellow. We have seen no natives since leaving the Roper, although their smoke is still round about us. On and about the marsh are large flocks of geese, ibis, and numerous other aquatic birds; they are so wild that they will not allow us to come within shot of them. Mr. Kekwick has been successful in shooting a goose; it has a peculiar-shaped head, having a large horny lump on the top resembling a topknot, and only a very small web at the root of his toes. The river opposite this, about a yard from the bank, is nine feet deep. Wind variable. Night cool.

Wednesday, 23rd July, Fresh-water Marsh. Started at 7.40, course 22 degrees east of south, one mile, to round the marsh; thence one mile south-east; thence east for six miles, when we struck a large creek, deep and long reaches; thence three quarters of a mile south before we could cross it. This I have named Thring Creek, in token of my approbation of his conduct throughout the journey; thence east, one mile and a half; thence north for nine miles, when I again struck the large marsh. Thring Creek has been running nearly parallel with the north course until it empties itself into the marsh. The country gone over to-day, after leaving the side of the marsh, as well as the banks of the creek, and also some small plains, is of the same rich description of soil covered with grass; the other parts are slightly elevated, the soil light with a little sand on the surface of a brown colour; timber, mixture of stringy-bark and gums, with many others; also, a low thick scrub, which has lately been burnt in many places, the few patches that have escaped abounding in grass. I have come twelve miles to the eastward to try to round the marsh, but have not been able to do so; the plains that were seen from the river by those who came up it in boats is the marsh; it is covered with luxuriant grass, which gives it the appearance of extensive grassy plains. I have camped at where the Thring spreads itself over a portion of the marsh. There is rising ground to the north-west, on the opposite side, which I suppose to be a continuation of the elevated ground I passed before crossing the creek, and the same that I saw bearing north from the last camp. I suppose it runs in towards the river. Wind, south. Latitude, 13 degrees 22 minutes 30 seconds.

Thursday, 24th July, Thring Creek, Entering the Marsh. Started at 7.40, course north. I have taken this course in order to make the sea-coast, which I suppose to be distant about eight miles and a half, as soon as possible; by this I hope to avoid the marsh. I shall travel along the beach to the north of the Adelaide. I did not inform any of the party, except Thring and Auld, that I was so near to the sea, as I wished to give them a surprise on reaching it. Proceeded through a light soil, slightly elevated, with a little ironstone on the surface—the volcanic rock cropping out occasionally; also some flats of black alluvial soil. The timber much smaller and more like scrub, showing that we are nearing the sea. At eight miles and a half came upon a broad valley of black alluvial soil, covered with long grass; from this I can hear the wash of the sea. On the other side of the valley, which is rather more than a quarter of a mile wide, is growing a line of thick heavy bushes, very dense, showing that to be the boundary of the beach. Crossed the valley, and entered the scrub, which was a complete network of vines. Stopped the horses to clear a way, whilst I advanced a few yards on to the beach, and was gratified and delighted to behold the water of the Indian Ocean in Van Diemen Gulf, before the party with the horses knew anything of its proximity. Thring, who rode in advance of me, called out "The Sea!" which so took them all by surprise, and they were so astonished, that he had to repeat the call before they fully understood what was meant. Then they immediately gave three long and hearty cheers. The beach is covered with a soft blue mud. It being ebb tide, I could see some distance; found it would be impossible for me to take the horses along it; I therefore kept them where I had halted them, and allowed half the party to come on to the beach and gratify themselves by a sight of the sea, while the other half remained to watch the horses until their return. I dipped my feet, and washed my face and hands in the sea, as I promised the late Governor Sir Richard McDonnell I would do if I reached it. The mud has nearly covered all the shells; we got a few, however. I could see no sea-weed. There is a point of land some distance off, bearing 70 degrees. After all the party had had some time on the beach, at which they were much pleased and gratified, they collected a few shells; I returned to the valley, where I had my initials (J.M.D.S.) cut on a large tree, as I did not intend to put up my flag until I arrived at the mouth of the Adelaide. Proceeded, on a course of 302 degrees, along the valley; at one mile and a half, coming upon a small creek, with running water, and the valley being covered with beautiful green grass, I have camped to give the horses the benefit of it. Thus have I, through the instrumentality of Divine Providence, been led to accomplish the great object of the expedition, and take the whole party safely as witnesses to the fact, and through one of the finest countries man could wish to behold—good to the coast, and with a stream of running water within half a mile of the sea. From Newcastle Water to the sea-beach, the main body of the horses have been only one night without water, and then got it within the next day. If this country is settled, it will be one of the finest Colonies under the Crown, suitable for the growth of any and everything—what a splendid country for producing cotton! Judging from the number of the pathways from the water to the beach, across the valley, the natives must be very numerous; we have not seen any, although we have passed many of their recent tracks and encampments. The cabbage and fan palm-trees have been very plentiful during to-day's journey down to this valley. This creek I named Charles Creek, after the eldest son of John Chambers, Esquire: it is one by which some large bodies of springs discharge their surplus water into Van Diemen Gulf; its banks are of soft mud, and boggy. Wind, south. Latitude, 12 degrees 13 minutes 30 seconds.

Friday, 25th July, Charles Creek, Van Diemen Gulf. I have sent Thring to the south-west to see if he can get round the marsh. If it is firm ground I shall endeavour to make the mouth of the river by that way. After a long search he has returned and informs me that it is impracticable, being too boggy for the horses. As the great object of the expedition is now attained, and the mouth of the river already well known, I do not think it advisable to waste the strength of my horses in forcing them through, neither do I see what object I should gain by doing so; they have still a very long and fatiguing journey in recrossing the continent to Adelaide, and my health is so bad that I am unable to bear a long day's ride. I shall, therefore, cross this creek and see if I can get along by the sea-beach or close to it. Started and had great difficulty in getting the horses over, although we cut a large quantity of grass, putting it on the banks and on logs of wood which were put into it. We had a number bogged, and I was nearly losing one of my best horses, and was obliged to have him pulled out with ropes; after the loss of some time we succeeded in getting them all over safely. Proceeded on a west-north-west course over a firm ground of black alluvial soil. At two miles came upon an open part of the beach, went on to it, and again found the mud quite impassable for horses; in the last mile we have had some rather soft ground. Stopped the party, as this travelling is too much for the horses, and, taking Thring with me, rode two miles to see if the ground was any firmer in places; found it very soft where the salt water had covered it, in others not so bad. Judging from the number of shells banked up in different places, the sea must occasionally come over this. I saw at once that this would not do for the weak state in which my horses were, and I therefore returned to where I had left the party, resolving to recross the continent to the City of Adelaide. I now had an open place cleared, and selecting one of the tallest trees, stripped it of its lower branches, and on its highest branch fixed my flag, the Union Jack, with my name sewn in the centre of it. When this was completed, the party gave three cheers, and Mr. Kekwick then addressed me, congratulating me on having completed this great and important undertaking, to which I replied. Mr. Waterhouse also spoke a few words on the same subject, and concluded with three cheers for the Queen and three for the Prince of Wales. At one foot south from the foot of the tree is buried, about eight inches below the ground, an air-tight tin case, in which is a paper with the following notice:

"South Australian Great Northern Exploring Expedition.

"The exploring party, under the command of John McDouall Stuart, arrived at this spot on the 25th day of July, 1862, having crossed the entire Continent of Australia from the Southern to the Indian Ocean, passing through the centre. They left the City of Adelaide on the 26th day of October, 1861, and the most northern station of the Colony on 21st day of January, 1862. To commemorate this happy event, they have raised this flag bearing his name. All well. God save the Queen!"

[Here follow the signatures of myself and party.]

As this bay has not been named, I have taken this opportunity of naming it Chambers Bay, in honour of Miss Chambers, who kindly presented me with the flag which I have planted this day, and I hope this may be the first sign of the dawn of approaching civilization. Exactly this day nine months the party left North Adelaide. Before leaving, between the hours of eleven and twelve o'clock, they had lunch at Mr. Chambers' house; John Bentham Neals, Esquire, being present, proposed success to me, and wished I might plant the flag on the north-west coast. At the same hour of the day, nine months after, the flag was raised on the shores of Chambers Bay, Van Diemen Gulf. On the bark of the tree on which the flag is placed is cut—DIG ONE FOOT—S. We then bade farewell to the Indian Ocean, and returned to Charles Creek, where we had again great difficulty in getting the horses across, but it was at last accomplished without accident. We have passed numerous and recent tracks of natives to-day; they are still burning the country at some distance from the coast. Wind, south-east. Latitude, 12 degrees 14 minutes 50 seconds.

...

RETURN.

Saturday, 26th July, Charles Creek, Chambers Bay, Van Diemen Gulf. This day I commence my return, and feel perfectly satisfied in my own mind that I have done everything in my power to obtain as extensive a knowledge of the country as the strength of my party will allow me. I could have made the mouth of the river, but perhaps at the expense of losing many of the horses, thus increasing the difficulties of the return journey. Many of them are so poor and weak, from the effects of the worms, that they have not been able for some time to carry anything like a load, and I have been compelled to make the (symbol crescent over C) horses stand the brunt of the work of the expedition. As yet not one of them has failed; they have all done their work in excellent style. The sea has been reached, which was the great object of the expedition, and a practicable route found through a splendid country from Newcastle Water to it, abounding, for a great part of the way, in running streams well stocked with fish—and this has been accomplished at a season of the year during which we have not had one drop of rain. Started, following my tracks back. Passed my former camp on the Thring; went on and crossed it. Proceeded on my east course to the west, about one mile and a half, to some small green marshy plains of black alluvial soil, with a spring in the centre, covered with fine green grass. Camped. Wind, south. Latitude, 12 degrees 30 minutes 21 seconds.

Sunday, 27th July, Small Grassy Plains. Day rather warm; mosquitoes terrible; no sleep last night; never found them so bad before; not a breath of wind to drive them away.

Monday, 28th July, Small Grassy Plains. Started at 7.40, course 25 degrees west of south, for my camp of the eighteenth instant. At ten miles struck my tracks, thus avoiding the boggy creeks that flow into the large marsh. On this course passed five small black alluvial plains, covered with grass, three of them having springs with water on the surface. They lie between slightly elevated country of light-brown soil, having stringy-bark and gums, with occasionally a thin scrub abounding in grass. On the plains there is occasionally a little of the volcanic rock cropping out. Followed my former tracks to the camp on the Lily Marsh, and remained for the night. We all passed a miserable night with the mosquitoes. My hands, wrists, and neck, were all blistered over with their bites, and were most painful.

Tuesday, 29th July, Lily Marsh. At half-past seven o'clock proceeded on the track. Passed my camp of 17th instant, and arriving at the one of the 16th at four o'clock p.m., camped. Wind, south.

Wednesday, 30th July, Side Creek, Adelaide River. All were delighted with a comfortable night's rest—no mosquitoes. Proceeded to Billiatt Springs and camped. One of the horses, Jerry, has been ill for the last three weeks, and although he has not had anything to carry, it has been as much as we could do to get him into the camp. This afternoon he gave in altogether, and Mr. Kekwick was quite unable to get him a step further, and was compelled to leave him about three miles back, where there is some water and plenty of feed. Wind, south-east.

Thursday, 31st July, Billiatt Springs. Proceeded and passed our camps of 13th and 12th instant. Crossed the Mary branch of the Adelaide: went along the south side, expecting to avoid the boggy creek crossed on the 12th instant. When nearly opposite to it, camped. Found this part of the branch deep, broad, and boggy; but I think we will be able to cross in the morning by cutting down a number of cabbage palms, which are growing very thick here. Light winds from south-east.

Friday, 1st August, South Side of the Mary. Recrossed the Mary, which is very boggy on the banks. We were enabled to cross it safely by cutting a large quantity of long grass, laying it on the sides of the banks, with a few logs and pickets driven into the bed to prevent the current from carrying away the grass. In this we succeeded very well. After crossing I found we had still to encounter the other running and boggy creek of the 12th ultimo; but, by repeating the same operation, we were successful. Passed our camp of the 11th ultimo, and proceeded on towards the table land. On approaching it, where the springs come from underneath, found it very boggy; had some difficulty in getting the horses through it. Got them all through with the exception of Frew's horse, which stuck hard and fast in it, and we were obliged to pull him out, which was soon accomplished, and we got him safe on terra firma. Continued along the foot of the table land, and halted at our camp of the 10th ultimo. At about seven p.m. last night I heard something plunging in the river; sent down to see what it was; found two of the horses bogged, and unable to extricate themselves. Got ropes, and all the party to pull them out. After an hour's hard work succeeded. On coming near the table land the country is all on fire, causing a dense black smoke and heated atmosphere. Wind, south-east.

Saturday, 2nd August, North-west Side of Table Land. Proceeded up the creek to the gorge—where we came down from the top of the table land; ascended it, which they all did well except one horse, which refused to go up, and caused me to lose more than an hour with him; we had to take all the things off him and carry them to the top on our backs. We had to zigzag him backwards and forwards, and got him to the top after a deal of trouble. Crossing on the top we met with a large fire about two miles broad. The wind not being strong, nor the grass very long, we got through it well, but my weak eyes suffered much from the smoke coming from the burning logs, trees, and grass. The atmosphere very hot and almost overpowering before we got through it. One of the horses knocked up, but we were able to get him on to the running creek connected with Kekwick's large group of springs, where I am obliged to camp and try to recover him. This is the first one of the (symbol crescent over C) horses that has failed; but he has not had fair play, through the negligence of the man who had him. He has for some time been carrying a load of one hundred and forty pounds without my knowledge, far more than he was able to carry. He has been a good horse, and has done a deal of work. There are a number of native tracks both up and down our tracks. One of the natives seems to have a very large foot. Wind, south.

Sunday, 3rd August, Kekwick's Large Springs. Last evening, just as the sun was dipping, five natives made their appearance, armed with spears, and came marching boldly up to within eighty yards of the camp, where they were met by Mr. Kekwick and others of the party who had advanced to meet them. They were all young men, small, and very thin. Seeing so many approaching them they soon went off. They were all smeared over with burnt grass, charcoal, or some other substance of that description. This morning, shortly after sunrise, the same five again made their appearance. I went up to them to see what they wanted. Saw that they had painted their bodies with white stripes ready for war. As it is my intention to pass peaceably through the different tribes, I endeavoured to make friends with them by showing them we intended them no harm if they will leave us alone. One of them had a curious fish spear, which he seemed inclined to part with, and I sent Mr. Kekwick to get some fish-hooks to exchange with him, which he readily did; we then left them. They continuing a longer time than I wished, and gradually approaching nearer to our camp, thinking perhaps they really did not wish to part with the spear, I sent Mr. Kekwick back with it to them to see if that was what they wanted, and to take the fish-hooks from them. But when they saw what was intended, they gave back the spear and retained the hooks. They offered another with a stone head upon the same terms, which was accepted. Mr. Kekwick had a deal of trouble before he could get them to move off, when they were joined by another, and then went off by twos. In a short time they set fire to the grass all round us to try to burn us out. Two of them came again close to the camp under pretence of looking for game before the fire, at the same time setting fire to the grass closer to us. But Mr. Kekwick and one of the others, seeing their intention, ran up to them, who, on their approach, ran off, setting fire to the grass as they went along, which gave us a deal of trouble in putting out, as we wished to save as much feed for the horses as will do for them till to-morrow morning; we have managed that, if they do not come and set fire to it again. If they do I shall be compelled to use preventive means with them, for I can stand it no longer; they must be taught a lesson that we possess a little more power than they anticipate. I would have moved on, but some of my horses are so ill that they are unable to travel. If the natives we have seen to-day are a sample of those that inhabit this country, they are certainly the smallest and most miserable race of men that I have ever seen. In height about five feet, their arms and legs remarkably thin, they do not seem to want the inclination of doing mischief if they could get an opportunity, but they find we are rather too watchful to give them a chance. From their manner I have no doubt there were many more concealed, who intended attacking us under cover of the smoke—indeed if they see us unprepared they may yet do it before evening. At sundown they have not again made their appearance. Wind, south.

Monday, 4th August, Kekwick's Large springs. Proceeded to the Katherine and camped. The horse that knocked up on Saturday gave in again two miles before we arrived here, although the distance is only thirteen miles, and he had a rest all Sunday. I shall be compelled to leave him here; he only destroys other horses dragging him along, and as the season is so far advanced, I am doubtful of the water in some of the ponds, and therefore cannot stop with him. I have been so very unwell to-day, with symptoms of fever, that I could scarcely reach this place; but I hope I shall be better by to-morrow. Nights and mornings are now very cold, but the sun is very hot during the middle and afterpart of the day. Wind, south-east.

Tuesday, 5th August, The Katherine. Leaving the knocked-up horse behind, proceeded to the Fanny, and camped. It was as much as I could do to sit in the saddle this distance. Wind, south.

Wednesday, 6th August, The Fanny. Proceeded to the Waterhouse and camped. The natives have been along our track, and burned the grass to within three miles of our camp. On arriving here I was much disappointed on finding all the water gone, but, following back the north-west branch, I found enough for our use to-night and to-morrow morning. The country is all on fire to the south-east. Wind, variable. The journey has been rather rough and stony, and my weak horses feel it very much. I am afraid I shall be compelled to leave some more of them behind. I cannot now stay for them to recover, after seeing the rapidity with which this water has dried up. A long delay will cause my retreat to be cut off in the pond country. Wind, south-east. There is still permanent water up the north-west branch of this creek.

Thursday, 7th August, The Waterhouse. Started at half-past seven, and at two minutes past ten o'clock I arrived at the running stream (the Chambers) of the 4th ultimo and camped. Weak horses looking very bad. Country on fire round about us. A number of natives have been following on our former tracks. Wind, south.

Friday, 8th August, Running Stream, The Chambers. Crossed the hard sandstone range, and got upon the branch of the Chambers that I followed up, passing our camp of 3rd ultimo, with plenty of permanent water. Followed it down to our camp of the 2nd ultimo and remained there. Had to leave one of the done-up horses about two miles behind. Another horse gave in, and it was as much as Mr. Kekwick could do to get him thus far. The natives have burned all the grass throughout this day's journey. A little has escaped at this camp, and I am now compelled to give my horses a rest until Monday morning. I thought they would have been able to carry me across the Chambers before I gave them a rest, but, if I proceed further, I shall lose more of them. The weather is beginning to be again very hot in the middle of the day. Wind, south-east.

Saturday, 9th August, River Chambers. Resting horses. Day hot. Wind variable.

Sunday, 10th August, River Chambers. Resting horses. I have sent Thring to bring up the one that was left behind on Friday; in a short time he brought him up, looking a most deplorable picture; the other one that gave in the same day is quite as bad. I shall have to leave them behind; it is only destroying other horses to force them along. I must also reduce the weight the others are carrying, to enable them to get along. I have had all the saddle-bags overhauled, and shall leave everything we can possibly do without—even boots and clothes belonging to the party have not been spared; all were quite willing to sacrifice anything they had, with the exception of one who had a pair of new boots he had never put on. I told him to put them on, and leave the old ones, but he immediately told me that he had got a bad foot; I very soon cured him of that by telling him if that was the case he might leave the new ones. I have managed to leave about three hundredweight; many of the things I can ill spare, but I hope by doing this to be able in a short time to push on a little quicker. Light winds, variable.

Monday, 11th August, River Chambers. Two of the horses having strayed this morning, it was a quarter past nine before I could get a start. I had to proceed very slowly, in consequence of five of the horses being so ill that they were unable to walk quickly. Proceeded on my former tracks, cutting off the bends of the river. In some places it is very stony. Late in the afternoon managed to get all the horses to the first camp on this river. Light winds, south-east.

Tuesday, 12th August, River Chambers. Horses missing again this morning. Started at half-past eight. Proceeded to the south-east end of the reedy swamp, and at half-past three o'clock camped. An hour before halting, we surprised a number of native women and children who were preparing roots and other things for their repast. The moment they saw us they seized on their children, placed them on their shoulders, and ran off screaming at a great rate, leaving all their things behind them, amongst which we saw a piece of iron used as a tomahawk; it had a large round eye into which they had fixed a handle; the edge was about the usual tomahawk breadth; when hot it had been hammered together. It had apparently been a hinge of some large door or other large article; the natives had ground it down, and seemed to know the use of it. Left their articles undisturbed, and proceeded to the river Roper. My horses are still looking very bad. The cause must be the dry state of the grass; it is so parched up that when rubbed between the hands it becomes a fine powder, and they must derive very little nourishment from it. I can hear natives talking and screaming on the other side of the river, which at this place is a strong running stream about thirty yards wide and apparently deep. Wind, south-east, blowing strong.

Wednesday, 13th August, Roper River, Reedy Swamp. One of the horses missing again this morning; he is one that generally goes off and hides himself if he can find a place to do so. Searched all round, but could find nothing of him or his tracks. Thinking that he might be hidden amongst the thick bushes over the river, sent Frew to look through them on foot, and Mr. Kekwick to an open place up the river to see if he had got into it. Mr. Kekwick returned in a short time and reported that he saw him lying drowned in the middle of it. I am sorry for this: he was a good horse, in fair condition, was with me last year, and has always done his work well, although he has caused a deal of trouble and loss of time by so frequently concealing himself. I shall feel his loss very much, as so many of the other horses are so poor that they are able to carry but little of a load, and I am obliged to let four go without carrying anything; indeed it is as much as they can do to walk the day's journey, although the journeys are short. I shall be compelled to make them still shorter to try and get them round again. As we were saddling, one native man and two women made their appearance and came close to the camp. Mr. Kekwick and I went up to them; the man was middle-aged, stout and tall, the women were also tall, one especially. Their features were not so coarse as those we had seen before—a very great difference between this fellow and those I saw on the source of the Adelaide River. The man made signs that he would like to get a fishhook by bending his forefinger and placing it in his mouth, imitating the method of catching fish. I gave him one with which he was much pleased: I also gave a cotton handkerchief to each of the women; one of them no sooner got it than she held out the other hand and called out "more, more, more;" with that request I did not feel inclined to comply. They remained until we started. Proceeding about three quarters of a mile down the river to where I had crossed it before, I got all the horses over without difficulty. There is now no difference in the strength, depth, nor velocity of the stream since we were here; it is exactly in the same state as when we previously crossed it. After crossing it to the other side, I had to cross another deep although dry creek coming from the east; proceeded on a south-east course to avoid the deep boggy creek that comes into the river, but at two miles I was stopped by an immense number of springs, very boggy, and emitting a large quantity of water; they seem to come from the east, as far as I could see, in a wooded valley between two hills. I had to round them until I got upon the south-east course again. At seven miles came upon a large creek or chain of ponds, having long broad deep reaches of water; followed this, running nearly my course for seven miles in a straight line. Camped. My horses cannot do more. The country that I have travelled over to-day is of the very finest description, rich black alluvial soil, completely matted with grass, the water most excellent and abundant. The timber, gum and melaleuca, a few of the trees resembling the shea-oak also; a few of the fan palms growing among the springs, very tall, upwards of forty feet; the cabbage palm, and a number of other bushes. The general course to-day has been about east-south-east. Wind variable.

Thursday, 14th August, Springs and Chains of Ponds South of the Roper. Started at half-past seven, intending to follow a south-east course to make the Mussel Camp on the 23rd of June; but, meeting with another large creek with continuous water, deep, broad, and boggy, also a number of springs and water creeks, so boggy that I could not cross them, had to twist and turn about very frequently, and sometimes to go quite back again, before I could clear them—which brought me often close to the river again. About eleven o'clock, as I was approaching the east end of a low rocky range of hills, where I expected to get rid of all the boggy ground, I was again stopped by a broad, deep, and boggy sheet of water. A few minutes before coming to it, I was seized with a violent pain under the right shoulder-blade, which deprived me of breath and power of utterance: it darted through my body like lightning, causing the most excruciating pain that I have ever felt during my life. I had to halt the party, and was lifted from the saddle completely powerless. After dismounting, the pain became so violent, and the torture so excessive, that I thought my career in the world was coming quickly to a close. I was completely paralysed, and a cold perspiration was pouring in streams over my face and body. Recollecting I had got a mixture of laudanum and other strong aromatic tinctures, had it sought for and took a strong dose. After suffering an hour the extremes of torture, I began to feel the good effects of the medicine, and obtained a little relief from the pain ceasing for a few seconds; but still very bad. In a short time afterwards I was able to bear being lifted into the saddle; again my sufferings commenced, for every false step the horse made sent the pain through my body like a knife, and almost brought me to the ground. Being determined to reach the Mussel Camp to-night, and get quit of the Roper River, which has been so unfortunate to me in drowning two of my best horses, I kept my saddle until I reached it—which was not till near five o'clock. Such a day of torture I never experienced before. On reaching our tracks, about four miles from the Mussel Camp, another of the horses knocked up, and we could not get him a step further. I expected to have lost him long before this; he is one of those that failed on my last journey, and was sent back from Mount Margaret. Light winds from east.

Friday, 15th August, Mussel Camp. I have passed a miserable night, and feel but little better this morning, and as the horses require rest, I shall remain here to-day. Shortly after sunrise, three natives came close to the camp; Mr. Kekwick went up to them. Two were of the number of those who visited us the first time at the large reedy swamp. They were very quiet, and seemed very friendly; they had come to have a look at us, and satisfy their curiosity. I feel a little easier to-night. Light wind, variable.

Saturday, 16th August, Mussel Camp. Started at nine o'clock. Another of my horses very ill; I think that many of them must have eaten some poisonous plant on the Roper and its tributaries; I never saw horses fall away so rapidly before. The worst are those that have been in good condition throughout the journey, and the work they have been doing since I commenced my return journey any horses ought to have done with ease. I have never travelled more than eight hours a day, and frequently not more than six hours. In a day or two they fall away to perfect skeletons, are quite stupid, and hardly able to walk. I am glad that I am now quit of the Roper, and hope that I shall have no more of them taken ill. If I can only get the weak ones beyond Newcastle Water, where I expect to get some new grass for them (from the June and July rains), they would soon recover. My old horses are all looking well, although they have had to carry the heaviest loads throughout the journey. I should have been in a sad way without them—they are my mainstay. Arrived at the Rock Camp, River Strangways, at two o'clock without having to leave any more. I feel a little better to-day, but the motion of the horse has been very severe throughout the journey. The water at this camp is drying up very rapidly: it is reduced three feet in depth since we left, and I am very much afraid it will be all gone in Purdie Ponds—if such is the case, I shall lose all the weak horses. Wind in strong puffs, variable.

Sunday, 17th August, Rock Camp. Resting horses. Winds light and variable.

Monday, 18th August, Rock Camp. Three of the best horses are missing this morning—they are the three leading horses—while feeding; and I have never known them to be away from the others before. The three horse-keepers have returned at half-past ten, and can see nothing of them; the ground is so hard that their tracks leave but little impression, so that they might have passed them unseen. Mounted Thring and King on fresh horses to round the feeding-tracks again, and at half-past twelve they returned with them. They happened to come upon their tracks on a small piece of sandy ground on the opposite side of the creek; they traced them to a large permanent water lagoon, deep and broad, with water-lilies growing round it, and a number of ducks upon it; it is about three quarters of a mile west-south-west from this camp. Not seeing them there they followed their tracks for another mile, and there found them, at which I was very glad, for they are three of my very best horses, on which I am placing my dependence for carrying me back. I felt very uneasy at their being away, thinking that the natives might have cut them off during night. Saddled and proceeded to my first camp, north of the Rocky Gorge, but was disappointed to find all the water gone, which I did not expect. Proceeded a mile further, and found as much as will do for a drink for the horses to-night and to-morrow morning. Camped. Light winds, variable.

Tuesday, 19th August, First Camp North of Rocky Gorge. Started at eight o'clock, proceeding to the Rocky Gorge, and camped. This water has shrunk considerably since we left it, and I have now little hopes of there being any water in Purdie Ponds. If there is not I shall require to push through to Daly Waters. Light winds, south-east.

Wednesday, 20th August, Rocky Gorge, River Strangways. If there is no water in Purdie Ponds, I have six horses that will not be able to go through to Daly Waters; they must be two nights without it, and that they will not be able to stand. I have therefore determined to send Thring and King to Purdie Ponds to-morrow, to see if there is any water, and also to examine another place that I observed in coming through, where I think there may be water. If they find none at either of these places, I shall be compelled to leave the six weak horses at the camp, where there is and will be plenty of food and water for them. To attempt taking them through, and be compelled to leave them behind where there will be no chance of their getting a drop of water, would, I consider, be a great cruelty; here they are safe, and there is a chance of their being picked up by the next party. If Thring succeeds in getting water, I shall still endeavour to take them on. I am yet suffering very much from scurvy; my teeth and gums are so bad that it causes me excessive pain to eat anything, and what I do eat I am unable to masticate properly, which causes me to feel very ill indeed. Light winds, south-east.

Thursday, 21st August, Rocky Gorge, River Strangways. At 7.30 despatched Thring and King to see if there is any water in the Ponds. Resting horses, repairing saddle-bags, etc. Day hot, night and morning cool; wind, south-east. My sight has been very much impaired during the last month; after sundown, I am in total darkness. Even though the moon is full, and shining bright and clear to the others, to me it is darkness; I can see her dimly, but she gives me no more light than if she had been painted on a piece of canvas. I am now quite incapable of taking observations at night, and I am most thankful this did not happen before I was enabled to reach the ocean, as the most of my observations are taken at night. After the equinox the sun is too high to be measured by the sextant in the artificial horizon.

Friday, 22nd August, Rocky Gorge, River Strangways. Day exceedingly hot. Wind still from south-east, sometimes blowing in strong puffs. A little after two o'clock Thring and King returned with the good news that there is still water in

Purdie Ponds; there is as much as will do for us until Monday morning. I am very glad of it, for it will enable me to get the weak horses through to Newcastle Water. After that I hope they will soon recover, for I expect that rain has fallen to the southward of that, and trust I shall get some fresh feed for them, which they require very much. I still feel very unwell to-day.

Saturday, 23rd August, Rocky Gorge, River Strangways. Started at half-past seven, and at four o'clock arrived at the Ponds. The day has been extremely hot, but about noon some heavy clouds came up from the east and south-east, which made it a little cooler, and enabled me to get all the weak horses through; one of them showed symptoms of giving in before we reached the Ponds, but we got him in all right. I shall remain here until Monday morning, when I shall have again another long journey without water (thirty-five miles) to Daly Waters. At sundown the clouds all cleared away, without giving us any rain. Wind, south-east. This day's journey has completely knocked me up. At one time I thought I should never have been able to reach this water. I had no idea I was in such a weak state, and am very doubtful of my being able to stand the journey back to Adelaide; whatever may occur I must submit to the will of Divine Providence.

Sunday, 24th August, Purdie Ponds. Day hot. Wind light, from south-east. About noon a few clouds came up, but they all disappeared about sundown. Very little improvement in me to-day.

Monday, 25th August, Purdie Ponds. Started at seven o'clock on my former tracks towards Daly Waters. At seven miles south of the Blue-grass Swamp saw a heavy fog to the east, in the same place that I saw the black fog in coming up; it must be caused by a large body of water in that direction. The natives have been running our tracks, and have burnt the grass on both sides of it for some distance. There seem to be very few of them about this part of the country. At half-past four passed the large swamp that receives the surplus water of Daly Waters, with water still in it, but very much reduced. At a quarter past five o'clock arrived at Daly Waters; found them also very much reduced, but still an abundant supply. Got all the weak horses through, which is more than I expected. This long journey has again completely exhausted me, and I feel very ill. Wind, south-east, with a few clouds.

Tuesday, 26th August, Daly Waters. I feel a little better this morning, but still very weak and languid. I shall give the horses and myself a rest to-day, for I am quite unable to ride. Wind, south-east, with a few clouds from the same direction.

Wednesday, 27th August, Daly Waters. Last evening, about half-past seven, Thring observed a comet bearing about 20 degrees west of north, and about 15 degrees above the horizon; the tail is short and the nucleus large. I regret that I am unable to see it. I cannot now see a single star, everything at night is total darkness. I should like to take some observations of it, but I am quite debarred from doing so. Started at half-past seven and proceeded along the Daly Waters, in which we saw an abundant supply. On reaching McGorrerey Ponds, and finding plenty of water, camped. I feel a good deal better to-day, but the motion of travelling on horseback is still very severe. Although Daly Waters is much reduced, there is still enough to last six months longer, even should no rain fall. These ponds will also hold out about three months longer. Wind, strong from south-east, with a few clouds.

Thursday, 28th August, McGorrerey Ponds. Proceeded to King's Ponds and camped. Find that the natives have been running our tracks, and have burnt large patches of grass; at this camp they have burnt it round. The water here is nearly all dried up; a few days later and I should not have got a drop. There is enough to last me to-night and to-morrow morning. Strong wind from south-east. The natives have cut on one side of my initials, on a gum-tree by the water where we camp, a figure resembling (a stylised flying bird).

Friday, 29th August, King's Ponds. Started at quarter past seven; proceeded to Frew's Pond, but was disappointed to find it quite dry. Dug down two feet, but could find no water. Proceeded on a straight course for Newcastle Water. Crossed Sturt Plains, and after dark camped on them. I would have gone to Howell Ponds, but finding the others so nearly dry, I was doubtful of them. A little before sundown, after I had passed them some distance, I observed flocks of pigeons flying towards them, showing that there is water still there. It is too late for me to go there now, Newcastle Water being the nearest. Wind, south-east. I feel a little better than I did on the former long journey.

Saturday, 30th August, Sturt Plains. At dawn of day started, being still some eight miles from Newcastle Water. The horses look very wretched this morning, especially the weak ones. About half-past eight arrived there, and found an abundant supply of water, though much reduced. No rain seems to have fallen since we left this, upwards of four months ago. A short time before we arrived a number of natives were observed following at a distance behind the rear of the party. They followed us on to our old camp, when I sent Mr. Kekwick up to them to keep them amused until I had the horses unpacked and taken down to water. By giving them a handkerchief he obtained a stone tomahawk from them. They are a fine race of men, tall, stout, and muscular, but not very handsome in features. They were very quiet. By making signs they were made to understand they were not to come nearer to our camp than about one hundred and fifty yards. They remained until noon staring at us and our horses. Some who could not see us very well got into the gum-trees, and had a long look at us. They were seventeen in number; four of them were boys, one of them much lighter than the others, nearly a light yellow. At noon they all went off, after remaining for four hours. Once more have I returned, if I may so call it, into old country again, after an absence of four months and ten days, exploring a new and splendid country from this to the Indian ocean without receiving a single drop of rain, or without any hostilities from the natives. I have returned from the coast to this in one month and three days. The horses have been one night without water, but got it early next morning, between eight and nine o'clock, and they would not have been without it if I could have seen to have guided the party after sundown. After the rays of the sun have left the earth, all is total darkness to me, even if there is a moon; I was therefore compelled to camp until daylight. Had my horses been in anything like a fair condition to have done a day's journey, and my health permitting, I could have accomplished the journey from the coast to this in three weeks. Before sundown we were again visited by our black friends; this time two old men accompanied them, whom Mr. Kekwick recognised as among those who visited the Depot at Howell Ponds during my absence. They all came up this time painted in red and white, and after remaining a short time went quietly to their camp. Wind, south-east.

Sunday, 31st August, North Newcastle Water. The natives again visited us this morning, and after remaining some time went off quietly. Wind, south-east. Few clouds at sundown.

Monday, 1st September, North Newcastle Water. Whilst saddling the horses this morning the natives again came up, and were anxious to know if they might be permitted to visit the camp after we were gone; that of course I had no objection to. They have been very quiet and peaceable during our stay; but I suppose they observed that both night and day we were always prepared to resist any aggression on their part. Started at seven o'clock, and proceeded by the base of the Ashburton range to my former camp on the East Newcastle Water. Distance, twenty-five miles; course nearly south-east. Arrived at four o'clock and found the water much reduced, but still in great abundance. Not a drop of rain has fallen since we left. There are, apparently, two tribes of natives on this water, one inhabiting the north and the other the south; for, on those of the north visiting us, we could not recognise any of those we saw on the southern water. One of the natives was a very amusing little fellow, rather less than five feet high, having a very peculiar and comical countenance and antics that would have eclipsed Liston in his best days, and as supple in the movements of his joints as any clown on the stage. He imitated every movement we made, and burlesqued them to a very high degree, causing great laughter to his companions and us. He seems to be the buffoon of the tribe. The other natives delighted in making sport of him, by ridiculing the shortness of his stature and laughing at him behind his back. Wind, south-east.

Tuesday, 2nd September, East Newcastle Water. Proceeded to Lawson Creek, but found no water in the lower part. Went up into the gorge, and there found as much as will do; it also is nearly gone, but there are still a few feet of it. I had no idea that such a body of water could have evaporated so quickly, which now makes me very doubtful of the waters to the southward. Wind, south-east.

Wednesday, 3rd September, Lawson Creek. As I now do not expect to get water before I reach the Hunter or the Burke, a distance of upwards of forty miles, I shall give the horses one day's rest to enable them to do the journey. I expect to lose some of the weak ones; to delay longer is only making the risk the greater. This must be an uncommonly dry season; not a single drop of rain has fallen in this part of the country since we left it. Last year we had three days' rain about the middle of June, and I was in hopes there would be the same this year, but am very much disappointed. I shall lighten the horses as much as I can possibly do, by leaving the water-bags, which are nearly useless, blankets, rugs, and cloths, as well as any other articles that can be done without. Provisions I MUST carry. I sincerely hope the forthcoming equinox will give me some rain and enable me to return. I feel a little better, but very weak and feeble from the severe attack of scurvy. My mouth and gums are so sore that to eat any food gives me the greatest pain. I cannot chew it, and am obliged to swallow it as it is, which makes me very ill. I am the only one of the party that is at present troubled with it. Wind, east.

Thursday, 4th September, Lawson Creek. Started at 6.40, and proceeded to the Hawker, but found no water there; thence to Watson Creek, none there; thence to Powell, Gleeson, and a number of other creeks that had water in them last year, but there is not a drop. Continued on to the creek that I camped at coming up. Arrived at 6.45 p.m.; found that water also gone, although it was a large deep hole when we were here before. Camped. Weak horses nearly done up. About 8.30 p.m. sent Thring up the creek to see if he could find any water. In three hours he returned: he had followed it up into the rough rocky hills until he could get no further, without seeing a drop. Wind, east. A few clouds at sundown coming from the south and south-east.

Friday, 5th September, Branch Creek of Hunter. Had to watch the horses during the night to prevent them straying in search of water. Started at 5.40 a.m. for the Hunter; in an hour and three quarters found some water in its bed. Camped, and will give the horses the benefit of it to-day. Wind variable.

Saturday, 6th September, The Hunter. Proceeded to the Burke, and found an abundant supply of water in the large iron conglomerate water hole that I discovered last year; it is reduced about four feet, but is still deep, and will last yet a long time without rain. I should say it was permanent without a doubt. Camped. From here I shall require to send on in advance to see if there is water in the Tomkinson; if not, I shall require to rest the horses here for three or four days to enable them to do the journey to Attack Creek without it. If there is none in the Tomkinson, I do not expect to find any in the Morphett. Native smoke about. They have burnt a great portion of the grass about here. The day has been oppressively hot and close. Wind from the east-south-east, with heavy clouds from the south-east to the south-west at sundown.

Sunday, 7th September, The Burke. After sunrise the clouds all gone. At 6.30 despatched Thring and King to the Tomkinson to see if there is any water. The day again oppressively hot, with clouds from south and south-east. Wind variable.

Monday, 8th September, The Burke. The clouds continued to come up during the night, but after sunrise they cleared off; still no rain. Between one and two p.m. Thring and King returned with the disheartening tidings that there was no water in or about the Tomkinson. I shall give the horses two more days' rest, and push through to Attack Creek, where I am almost sure of there being water. The wind variable, sometimes north, east, and west. The clouds are broken up, and are nearly all gone, without leaving rain.

Tuesday, 9th September, The Burke. Resting horses, mending saddle-bags, etc. Wind, north and variable, with a few clouds from the west and south-west.

Wednesday, 10th September, The Burke. Thring on his return last Monday saw some water about four miles higher up this creek, nearly on our course for the Tomkinson; to that I shall go to-day, and make a start for Attack Creek to-morrow morning. Every mile now gained is of the utmost importance to me. Started early, to get there in the cool of the morning. In an hour and a half arrived at the water and camped. It is situated at the foot of some ironstone conglomerate rock, and will last a week or two longer. It has a number of small fish in it. The soil on its banks is light and a little sandy, with spinifex and grass mixed through it. Wind, north and north-west; the clouds have all disappeared. This morning I again feel very ill. I am very doubtful of my being able to reach the settled districts. Should anything happen to me, I keep everything ready for the worst. My plan is finished, and my journal brought up every night, so that no doubt whatever can be thrown upon what I have done. All the difficult country is now passed, and what remains is well known to those who have been out with me before; so that there is no danger of the party not finding their way back, should I be taken away. The only difficulty they will have to encounter is the scarcity of water, caused by the extreme dryness of the season.

Thursday, 11th September, The Upper Burke. Started at 6.40; crossed the Tomkinson and small grassy plains; ascended the north spur of the Whittington range. After sundown, it becoming quite dark to me, so that I could not see the horse's head before me, I was compelled to halt on the top of the range, four miles from my former camp on the Morphett. Day excessively hot; myself and horses have felt it very much. Wind variable, from the north and north-east.

Friday, 12th September, Top of Whittington Range. At break of day started over the range to my former camp, but found all the water gone. Proceeded down the Morphett, and at four miles found a little in the sandy bottom of what had once been a large hole. There is as much as will do for me until to-morrow by digging. All the clouds gone; not the slightest appearance of rain. The country on fire all round us. Wind, north-west and variable. Day exceedingly hot.

Saturday, 13th September, The Morphett. Started at 7.20, crossed the other spur of Whittington range, and at 11.20 arrived at Attack Creek. There is still an abundant supply of water, although much reduced—much lower than I have ever seen it. In about an hour and a half after camping, some native women came to the lower end of the hole where Billiatt was getting some water. The moment they saw him they went off at full speed. In a short time afterwards one man made his appearance and came marching up towards us. Sent Mr. Kekwick to meet him. As he approached him the black became stationary, and moving back a little, beckoned to some others to come up. Mr. Kekwick observed five or six others down at the lower end of the water hole, one of whom came up. I then sent Frew to Mr. Kekwick. They approached very cautiously, but as soon as they caught sight of Mr. Kekwick's gun, he could not get near them. On laying it down he got a little nearer; they shrank back when he attempted to touch them. Taking out a small strip of white calico which he had in his pocket, he tore it into two and held it out to them. They wished to possess it, but did not fancy coming too close to him for it. He made a sign that he wished to tie it round their wrists; they gradually approached nearer, holding out their arms at full length, and so frightened were they to come close, that he had to reach out his full length to tie them on; after which they gained a little more confidence, pointed towards the gun, imitated the report with their mouth, and held up three fingers, signifying that they recollected my first visit and number, which they do not seem to have forgotten, and seem to dread the appearance of a gun. The first one that came up had a very long spear, with a flat, sharp, and barbed point. They were two elderly stout men, one very much diseased and lame. They remained a long time looking at us. None of the others came up. In a little more than three hours they went off and we saw no more of them during the evening. Wind, south-west, with heavy clouds from the same direction and from the south.

Sunday, 14th September, Attack Creek. During the night the sky frequently became overcast with heavy clouds, which seemed to indicate rain, but none fell. About eight o'clock the wind changed to north-east, bringing up very heavy clouds, which led me to expect rain, but I was much disappointed, for at half-past twelve they all broke up and went off. This morning, at sunrise, I despatched Thring and Nash to see if there is water in Hayward, Phillip, Bishop, Tennant, or Goodiar Creeks. If there is none I shall require to rest the horses for three days, and then push on for the Bonney. It is a very long distance, and only the very best of them will be able to do it. I feel a little better this morning, but still very weak. The pains are increasing in my limbs, and my mouth is so bad I can eat nothing but a little boiled flour. How I am to get over such long pushes I do not know. I must trust entirely to Divine Providence. The natives have not visited us this morning. A little before four o'clock p.m. Nash returned. Thring had sent him back to report that there was water, by digging in the sand, at Hayward Creek, while he goes on to see if there is any other creek. Wind variable, with heavy clouds at sundown.

Monday, 15th September, Attack Creek. Started at 8.40. On crossing the creek, one of the weak horses, which had eaten some poison about the Roper, and which has been getting weaker every day, in attempting to get up the bank, which was not steep, fell and rolled back into the creek. There he had to be some time before he was able to get up. I saw that it was useless taking him any further, therefore left him where he will get plenty of feed and water. Proceeded to the Hayward, where I met Thring. There is some soft mud in Phillip Creek, but none in Bishop Creek. Camped, and cleared out a place for the horses to drink at. A number of natives have been camped on the opposite side of the creek, where they have left their spears, dishes, etc. Thring had arrived here some time before. About twenty of them coming closer to him than was safe, he mounted his horse and chased them to the hills, where they are now seated watching us. Some of them are approaching nearer. Mr. Kekwick could not get them to come near him until one of the old men who visited us at Attack Creek arrived and came up to him, which gave the others confidence. A number of them then came forward—tall, stout, well-made fellows, armed with long heavy spears, having bamboo at one end. One of them had also part of a large sea-shell, but it is so broken and ground down for a scoop that I cannot say of what description it is. The bamboo and the sea-shell show that this tribe has communicated with the sea-coast. They remained until sundown, and then did not seem inclined to go away, but prepared sleeping-places for the night—a proof that this is the only water near. There are upwards of thirty men, besides women and children. Wind, south-east. Clouds all gone.

Tuesday, 16th September, Hayward Creek. The natives showed themselves again at daybreak, but kept on the opposite bank of the creek, having a long look at us, and calling out something at the top of their voices which we could not understand. Watered our horses, saddled, and moved on amidst a succession of yells and screeches from old and young. Proceeded across Short ranges, and Phillip, and Bishop Creeks. Looked into every place I could think of, but could not find a drop. Moved on to Tennant Creek. Found that dry. Tried digging in the sand, without effect. Pushed on to the large rocky water hole in Goodiar Creek, where I made almost sure that I should find some. On arriving, was sadly disappointed to find that dry also. Proceeded across the McDouall range, and camped on a grassy plain between it and Mount Samuel. The natives followed us nearly to Tennant Creek, raising a line of smoke all the way. They kept about a mile to the east of us, on some rising ground that runs nearly parallel with my tracks. We have had to lighten a heavy cart-horse named Charley. When any hardship is to be undergone, he is always the first to show symptoms of giving in. He had only thirty pounds to carry to-day, and he looks ten times worse than those that are carrying one hundred and twenty. I shall require to let him go without anything to-morrow. We shall have to watch the horses during the night to prevent them from straying in search of water. Wind, south-east.

Wednesday, 17th September, McDouall Range. Started at daybreak for the Murchison range. About eleven o'clock the cart-horse gave in, and would not move a step further. I am obliged to leave him; he has been carrying nothing all the morning. Two others that have been very weak from eating some poisonous plant will, I fear, give in before the end of the day. A little after four o'clock I found I must leave them. At dark arrived at the Baker, which I found dry. Camped. This is another night the horses will be without water, and will require to be watched. A quantity of native smoke about. There must be permanent water about this range somewhere, but I have no time to look for it now. Tomorrow I must push on for the Bonney. If that fails me I shall be in a sad predicament, but I trust that the Almighty will still continue to show me the same great kindness that he has done throughout my different journeys. There is very little improvement in my health. I feel very much being in the saddle so long. Twelve hours is almost too much for my weak state, but I must endure it. Wind, south-east.

Thursday, 18th September, Murchison Range. Proceeded at daydawn to the Gilbert. Found it dry. Went on towards the Bonney; crossed the McLaren—no water. At two o'clock arrived at the Bonney, and am most thankful to Divine Providence that there is still a good supply of water that will last some time longer. My horses look very bad indeed. I expected to have lost more of them. They have got over this first difficulty very well. Towards the end of the journey my old horse took the lead. Day hot. Wind, south-east.

Friday, 19th September, The Bonney. From this camp Mount Fisher bears 119 degrees 30 minutes. I must remain here some time to get my horses round again. A large number of them are looking very ill this morning. Being so long without water and the dry state in which the grass is, has reduced them more than three months' hard work would have done. If the grass had any nourishment in it, two or three days would have done for them. Not a drop of rain seems to have fallen here for the last twelve months; everything is dry and parched up. This appears to be the driest part of the year. I am very doubtful of the water in the Stirling, the next place that I was depending upon. From the very reduced state in which this is, I have very little hope of there being any there. The day has been again oppressively hot. I trust we shall soon have rain. Wind variable. Native smoke about.

Saturday, 20th September, The Bonney. Resting horses. I feel very ill again; being so long in the saddle is very severe upon me. Day again very hot. Wind from the west, with a few clouds, which I trust will bring up rain.

Sunday, 21st September, The Bonney. Resting horses. Day very hot. Wind, west; clouds broken up.

Monday, 22nd September, The Bonney. This morning sent Thring up the creek to see if there is any larger water than this that can be depended on for some time to come. Very hot. Clouds all gone. Wind variable.

Tuesday, 23rd September, The Bonney. Recruiting horses, etc. About eleven o'clock Thring returned. He has been about twenty miles up the creek to where it became much narrower and was joined by a number of small ones coming from very rough and stony hills. Its general course is about east-south-east. At four miles from this he found a pool of water four feet deep, two hundred yards long, and thirty feet broad. There is a considerable quantity of water all the way up, but shallow, and none of the extent of the former one found. Should I be forced to retreat, that will be a safe place to fall back on until rain falls. Day again oppressively hot. Wind, east.

Wednesday, 24th September, The Bonney. Shortly after sunrise despatched Thring to see if there is any water in Thring Ponds, or any between them and this. I would have gone myself, but was quite unable to do so, being very little better. One of my good horses has met with an accident in feeding along the bank of the creek in places where it is very precipitous. A portion must have given way and thrown him into the creek, injuring him very much in the chest and other parts of the body. I am afraid he will not be able to travel with me, which will be a great loss, having so many weak ones already. Wind, south-east, with a few clouds.

Thursday, 25th September, The Bonney. Clouds all gone, no rain. Resting horses, etc. Day hot, morning and evening cool, with strong wind from east and south-east. I have been obliged to reduce the rations to five pounds of flour and one pound of dried meat per week for each man, which will leave me provisions at that rate until the end of January, in case I should be locked in with the dry state of the season. The flies at this place are a perfect torment. A little after three o'clock p.m. Thring returned. There was no water in the Barker, none in the Sutherland, and when he got to the ponds, found them quite dry also; he then returned two miles to where there was some good feed for the horse, and camped for the night without water, intending to return to this in the morning. In saddling he observed some crested pigeons fly past him to the south of east; he thought it would be as well to follow them some distance in that direction, as they might be going to water, as about that time in the morning is generally the time they fly towards it. After going a few miles he surprised fourteen natives at breakfast. As soon as they saw him they ran off at full speed. Observing some small wooden troughs with water in them, he collected it together and gave it to his horse. Examined the small creek for more, but could find none, and knowing the natives would not carry it very far, and that there must be some no great way off, went on a little further and found a fine pool of water with ducks on it, but shallow. He then returned. This will bring the Stirling within visiting distance. I shall remove the party down to the pool to-morrow. Strong wind, still from the south-east.

Friday, 26th September, The Bonney. In consequence of the horses separating during the night, I did not get a start before nine o'clock; followed my former tracks across Younghusband's range; thence on a bearing 25 degrees east of south; arrived at the pool of water at 5.15 p.m. Before reaching the water we crossed four red sand hills, with spinifex, running north-east and south-west, having broad valleys between, in which are growing melaleucas, gum-trees, and grass. After rain they retain water, but now are quite dry. This one that we are now camped at is much larger, having the same description of timber, with polyganum growing round about it; the water is shallow, and will not last long. There are a number of ducks, geese, and other water-fowl on it, but too shy to be approached. A quantity of native smoke about. I am very ill to-day; I am scarcely able to endure the motion of the horse thus far. The horse that injured himself so much knocked up about two miles from this water, but we were able to get him to it before sundown. I shall have to kill him and eat what is good of him; it is useless to attempt taking him on a long journey without water—he would never be able to do it; and, as we are now upon half rations of meat, I shall kill and eat him, so that he will not be lost altogether. Wind variable. Day exceedingly hot.

Saturday, 27th September, Pool of Water. Before attempting to see if there is water in the Stirling, I have sent Thring on course 20 degrees west of south, to see if there is any creek or water between two stony ranges of hills that lie east of Mount Morphett. At sundown he has not returned. Wind, west. Day very hot. After sundown we shot the black horse that was not able to travel; shall cut him up and dry him to-morrow; there are some parts very much injured by bruises he got in his tumble. He also showed evidence of having drunk too much water at the Bonney. Being so exhausted and knocked up on my arrival there, I was unable to go and see they did not drink too much, and had to leave it to others. In all my journeys (and my horses have been much longer time without water than this), this is the first horse that has injured himself in that way.

Sunday, 28th September, Pool of Water. About eight o'clock, Thring returned, being out all night without food or blankets; he had found a large gum creek in the place I had sent him to, with water in it, by sinking in its sandy bed. I shall move the party to it to-morrow morning. Wind variable, mostly from the north and north-east. Day very hot. Latitude, 20 degrees 47 minutes 59 seconds.

Monday, 29th September, Pool of Water. Started at seven o'clock, course 20 degrees west of south. For the first five miles we passed over a fine country, soil red, and in places a little sandy, with gums, grass, a little scrub, and in places a little spinifex. After this it became covered with spinifex until within five miles of the creek, where the mulga commenced, with plenty of grass, which continued to its banks, where we arrived after twenty-six miles, and had to dig six feet in the sand before we could get sufficient water for the horses; by ten o'clock p.m. however we got them all watered. I am inclined to believe this is a continuation of the Taylor and other creeks coming from Forster range more to the eastward. After my arrival here, I sent Thring up the creek to see if he could find any surface water. After dark he returned and informed me that he had followed it into the Crawford range, and that it came through the range; if such is the case, there is no doubt of it being the Taylor with the creeks from Forster range. There is no surface water, but apparently plenty by digging in the bed of the creek, judging from the number of native wells that he saw with water in them. At one of the wells he saw several natives, who ran off on his approach. Latitude, 21 degrees 9 minutes 30 seconds. Wind variable. Day oppressively hot.

Tuesday, 30th September, The Taylor. As soon as I could get the horses, I despatched Thring to the Stirling to see if there is water. I have sent King on with him, with a pack-horse carrying two bags of water for the horse that carries him to the Stirling. They are to follow this creek up, and, if it is the Taylor, they are to stop to-night at our last camp on it. Next morning King is to return to me, whilst Thring goes on to examine the Stirling. Still all hands engaged in sinking for water for the horses. Wind from the south-east, with heavy clouds from the north-west and south-west, showing every indication of rain, which I sincerely hope will fall before morning.

Wednesday, 1st October, The Taylor. About nine o'clock last night there were a few drops of rain, and almost immediately afterwards the clouds broke up and went off to the south-east, to our very great disappointment. This morning there are still a few light ones about, but very high, and no more appearance of rain. Wind still strong and blowing from the same quarter. We have now got enough water for the horses, and can water them all in about two hours. No natives have shown themselves since we have been here, although their smoke was quite close to us yesterday. In the afternoon Thring and King returned, having found a fine pool of water about fifteen miles up the creek, four feet deep, which will serve us for a short time. Sundown: still blowing strong from the south-east; clouds all gone.

Thursday, 2nd October, The Taylor. Started at five minutes to eight, course 3 degrees west of south; at five miles got through the gap in the range, then changed to 20 degrees west of south, and after ten miles on that course reached the water hole. The journey to-day has been over first-rate travelling-ground, avoiding crossing the range at Mount Morphett. The country in many places along the creek has large grassy plains with mulga, gum-trees, and scrub, not too thick to get easily through. Native smoke under the hills to the east. Strong cool wind blowing all day from the south-east. A little before sundown three natives came within three hundred yards of the camp, setting fire to the grass as they came along. We could not get them to come any nearer. Latitude 21 degrees 22 minutes 12 seconds.

Friday, 3rd October, Surface Water, The Taylor. Shortly after sunrise despatched Thring and King in search of water higher up the creek. I feel so weak and ill that I am now scarcely able to move about the camp. This morning Frew, in searching for some of the horses, came upon the three natives we saw last night; the moment they saw him off they went at full speed, and he saw no more of them. They must have been sneaking about and watching our camp during the night. Wind still blowing strong south-east.

Saturday, 4th October, Surface Water, The Taylor. It still continues to blow very strong from the same quarter. A little before two p.m. King returned. They had followed up this creek for a considerable distance beyond where the Taylor joined it, and as it came more from the south-east than I had expected, and approached near to Forster range, Thring changed his course to the Stirling, according to my instructions. A little before sundown they arrived at my former camp on the Stirling; found the water hole quite dry; dug down, but could find no moisture. They had not seen a drop of water during the whole day. In the morning King returned to me, giving Thring's horse the water that he had carried with him to enable him to search the Stirling down and round about the adjoining country. Still blowing strong from the same direction. No clouds visible.

Sunday, 5th October, Surface Water, The Taylor. Still blowing strong and cool from the same quarter. About half-past one o'clock Thring returned; he could find no surface water, neither any to be had by digging. He then crossed over to the foot of the Hanson, where he saw some native smoke; on his arriving at it he surprised a native busily engaged in sinking for water, about six feet deep, in the bed of the creek, who, as soon as he saw him, jumped out of the well and ran off as fast as he could. He then tried to see what quantity of water was in the bottom of the well, but having nothing but a quart pot to clear it out with, he was unable to form a correct opinion, but from all appearances he thinks there will be sufficient for our use for some time, only it will require an immense deal of labour and time to remove the great body of sand to enable the horses to get down to it. To-morrow I shall send Thring with McGorrerey and Nash, with four horses and sufficient provisions for a fortnight. On their arrival at the native well on the Hanson they will be able easily to get water enough for their four horses that night. McGorrerey and Nash will then clear out the well and see what quantity there is in it, while Thring will proceed up the Hanson to see if there is water in the springs that I discovered on my first journey through the centre. If they are dry he will proceed with the examination of the Hanson to above where we crossed it; he will then return to the diggers; by that time they will be able to judge if there is sufficient water for the whole party. If there is sufficient he will leave them to dig, and come on to me; if not, and there is no more water higher up, he will bring them on with him, and I shall require to try a course more to the south-east. In the afternoon the three natives again made their appearance, bawling out as they came near, but retreated as Mr. Kekwick went towards them to see what they wanted. Wind still south-east.

Monday, 6th October, Surface Water, The Taylor. Shortly after sunrise despatched Thring with McGorrerey and Nash to the Hanson. Day very hot. I am still very ill—no improvement whatever. Wind strong from the south-east.

Tuesday, 7th October, The Taylor. What a miserable life mine is now! I get no rest night nor day from this terrible gnawing pain; the nights are too long, and the days are too long, and I am so weak that I am hardly able to move about the camp. I am truly wretched. When will this cease? Wind, south-east.

Wednesday, 8th October, The Taylor. Wind still blowing from the south-east; no appearance of rain.

Thursday, 9th October, The Taylor. Last night, about sundown, a native woman and youngster came to the waterhole, rushed down, had a drink, and were running off again, when I cooed and made signs of friendship; in a few seconds the woman gained confidence, and, not seeing any of us approach, went down to the hole again, and fetched up a large troughful of water. Mr. Kekwick tried to induce her to stop, in order to gain some information from her, but it was of no use; the faster he walked the faster she did the same, chatting all the time, pointing to the south; so he left her to walk at her leisure. They do not seem to be at all frightened of us; but we cannot get any of them to come near, although we have tried every time they have come. The day again oppressively hot. I still feel very ill. Wind from south-east. Nothing particular has occurred during the day. This is dreadful work to be detained here so long. I am afraid soon I shall not be able to sit in the saddle, and then what must I do? I feel myself getting weaker and weaker every day. I hope the Almighty will have compassion on me, and soon send me some relief. He is the only one that can do it—my only friend.

Friday, 10th October, The Taylor. Last night, a little before sundown, until after dark, we were amused by a farce enacted by the natives, apparently to keep us quiet and render us powerless, while they approached the water hole and got what water they required. They commenced at some distance off, raising a heavy black smoke, (by setting fire to the spinifex), and calling out most lustily at the top of their voices. As the sun got lower I had the party prepared for an attack; on they came, the fire rolling before them. We could now occasionally see them; one was an old man with a very powerful voice, who seemed to be speaking some incantations, with the most dreadful howl I ever heard in my life, resembling a man suffering the extremes of torture; he was assisted in his horrid yell by some women. As the evening got darker and they were within one hundred and fifty yards of us, and nearly opposite our camp, the scene was very pretty—in fact grand. In the foreground was our camp equipment with the party armed, ready to repel an attack. On the opposite side of the creek was a long line of flames, some mounting high in the air, others kept at a low flickering light. In the midst of the flames the natives appeared to be moving about, performing all sorts of antics; behind them came the old man with his women. At every high flame he seemed to be performing some mysterious spell, still yelling in the former horrid tone, turning and twisting his body and legs and arms into all sorts of shapes. They appeared like so many demons, dancing, sporting, and enjoying themselves in the midst of flames. At last they and their fire reached the water hole after continuing this horrid noise for nearly two hours without intermission; as soon as they came in sight of the water, those in front rushed down into it, satisfied themselves, filled their troughs and bags, except the old man, who kept up his howl until he was stopped by a drink of water. This seemed to satisfy them, for they went off from us about three quarters of a mile and camped, I suppose thinking they had done great things in keeping us so quiet. Shortly after this something started the horses which made them all rush together. I kept the party under arms till nine o'clock p.m. and then, everything appearing to be quiet, I sent them all to bed except the one on guard. The natives were quiet during the night. This morning the blacks watched us collecting the horses and watering them; they then very quietly slipped down to the water, filled their troughs, etc., and in about half an hour went off and left us in possession of the water. They must certainly think we are very much to be frightened by fire and a great noise, or they would never have come in the way they did last night; they would have been rather surprised had they attacked us, to find that we could both speak and injure by fire. I am better pleased that they went away quietly; it is far from my wish to injure one of them if they will let me pass peaceably through. About two o'clock p.m. Thring returned; he had examined up the Hanson, but could not find a drop of water, either on the surface or by digging. On his return to where he had left the two men to dig, he found there would not be enough water for the whole party, as it came in so slowly; it is on the top of hard burnt sandstone; he therefore came on to inform me of the result, leaving the two men still there. They had been visited by the natives, who appeared to be inclined to be rather unfriendly at first, but on showing them they were welcome to use the water as well as the party, they became friendly, and came over night and morning to fill their troughs and bags. They pointed to the south-south-east, and made signs, by digging with a scoop, that there was water in that direction, but how far he could not make out. This is a sad disappointment to me. I dare not move the party on to where they are digging, there is too little water. To-morrow morning I must send Thring and King on to Anna Reservoir to see if there is any there; if that is dry I shall be locked in until rain falls, and that may not be before the equinox, in March, a very dismal prospect to look forward to. I shall start Thring and King to-morrow morning; they will reach where the diggers are to-morrow night, and will rest their horses there on Sunday. On Monday morning start for Anna Reservoir—King, with a pack-horse carrying water, will go on one day with Thring. The water to be given to Thring's horse night and morning. Thring will proceed to the Reservoir. King will return to the diggers with the empty bags, have them filled, and next morning start with fresh horses and the water to meet Thring on his return in case the Reservoir is dry; this is the only way that I see it can be done. I now begin to feel the want of my health dreadfully. Although Thring is a good bushman and does his best, poor fellow, yet he wants experience and maturer judgment; he has had hard work of it lately, but he is always ready to start again at any moment that I wish. Wind, south-east. A few light clouds about.

Saturday, 11th October, The Taylor. The natives camped last night at their former place; they seem to have given up all their buffoonery. I suppose they see it has no effect upon us. Shortly after sunrise despatched Thring and King. The day again oppressively hot, with a few light clouds from the south. Wind, south-east.

Sunday, 12th October, The Taylor. The natives again encamped in their former place last night. They came in late and started early this morning. They always seem to go off to the westward. Day again oppressively hot. Wind, south-east.

Monday, 13th October, The Taylor. Can see nothing of the natives this morning; they must have gone off during the early part of last night. We tried to get near to them yesterday afternoon by making friendly signs, etc., but the moment we approached them they ran off, and everything we can think of will not induce them to come near us or allow us to get near them; they are the most timid race I have ever met with, which I think is a very bad feature—such are often very treacherous. I should have a much higher opinion of them if they would come boldly forward and see if we were friends or foes. Wind from the north; heavy clouds from south and south-west.

Tuesday, 14th October, The Taylor. During the night there was a deal of lightning in the south and south-west; clouds about, but high and much broken. About two o'clock p.m. they collected together and gave a very promising appearance of a heavy fall of rain; they seemed to be coming up all round, but the heaviest from the south and south-west. At four o'clock p.m. it began to lighten and thunder, accompanied by a shower which did not last above a few minutes. Sundown: still the same promising dark, heavy, gloomy appearance. Wind, south-east.

Wednesday, 15th October, The Taylor. During the night we had a terrific storm of lightning and thunder, which continued throughout the night and morning at intervals, but little rain has fallen, it has merely damped the surface of the ground. At twelve o'clock to-day it has nearly cleared all away, leaving only a few light clouds, which is another very great disappointment. At sundown it again became overcast. Wind variable.

Thursday, 16th October, The Taylor. Still cloudy during the night and morning, but no rain has fallen; the heavy clouds pass south of us to the eastward. I am now nearly helpless; my legs are unable to support the weight of my body, and, when I do walk a little way, I am obliged to have the assistance of one of the party, and the pains caused by walking are most excruciating. I get little sleep night or day. I must endure my sufferings with patience, and submit to the will of the Almighty, who, I trust, will soon send me some relief. Wind variable.

Friday, 17th October, The Taylor. Still heavy clouds during the night and day, but no rain will fall. Still very ill. About three o'clock p.m. Thring returned; he has been to Anna Reservoir and found plenty of water, and a number of natives camped at it, who ran off the moment they saw him; he watered his horse and recrossed the range, not thinking it prudent to camp where there were so many of them. He has met with the same description of weather that we have had up here, thunder and lightning with a heavy, cloudy sky, but nothing but a light shower or two of rain. I shall move the party on to the Hanson to-morrow, and, if I am able to ride, shall push on to-morrow. Wind variable; sky still overcast.

Saturday, 18th October, The Taylor. Started at twenty minutes to eight for the Hanson; sky still overcast with heavy clouds. We had two light showers during the journey. I am now so helpless that I have to be lifted into the saddle. I endured the pain of riding for the first seventeen miles far better than I expected; after that it became almost unbearable, and camped at twenty-four miles, having found as much water in the rocks of the Stirling as will do for the horses to-night and to-morrow morning, left from a shower of rain, for which I am very thankful. I could not have gone on more than three miles. I was then enduring the greatest pain and agony that it is possible for a man to suffer. On being lifted from the horse, all power was gone out of my legs, and when I attempted to put the weight of my body on them the pain was most excruciating. Still heavy clouds about, indicating rain. Wind, south-east.

Sunday, 19th October, The Stirling. I had a few hours' sound sleep last night, which I find has done me a deal of good. During the early part of the night two heavy showers of rain fell, and left plenty of water for the horses; got them up, and saddled and proceeded to the Hanson. At eight miles arrived there, finding the party all well; they had not been troubled with the natives except by their coming down to the water during the night time, and bringing into the hole a quantity of sand with them. I had to be taken from horseback nearly in the same state as yesterday. Wind, south-east.

Monday, 20th October, The Hanson. Started early; passed the Centre; crossed the upper part of the Hanson, and at five miles beyond it camped. Distance, thirty-five miles. Not a drop of rain seems to have fallen for a long time. During the whole day's journey this has been a terrible day of agony for me; nine hours and a half in the saddle. I had to be taken from my horse in the same helpless state as before. My feet and legs are now very much swollen; round the ankles they are quite black, and the pain is dreadful. I still continue to take the bicarb of potash, but it has little or no effect. Wind variable.

Tuesday, 21st October, South of the Centre. About sunrise started for Anna Reservoir, and at 5.30 p.m. arrived there, completely exhausted. Wind, variable. Heavy clouds from the south-east.

Wednesday, 22nd October, Anna Reservoir. Last night I was so completely overcome by fatigue and exhaustion that I had no sleep during the whole of the night, which makes me feel very ill indeed this morning. I shall be obliged to remain here to-day and to-morrow, to see if that will recruit my strength and enable me to perform the long journeys to the McDonnell range. About twelve o'clock heavy thundery weather to the west and south.

Saturday, 23rd October, Anna Reservoir. I shall rest to-day and have what shoes there are left put on the horses. I, with William Auld, will proceed to-morrow about ten miles in advance, to divide the long journey into two, for I have not strength to do it in one day. Wind variable.

Friday, 24th October, Anna Reservoir. Started early, taking with me Thring, King, and Auld, with one pack-horse to carry my tent, water, etc. Proceeded through the thick mulga scrub, and at ten miles camped, which I find is quite as much as I am able to do. Had my tent put up, and myself carried into it. Sent Thring and King back with the horses to the Reservoir, keeping Auld with me. The party will start from the Reservoir early to-morrow morning, pick me up, and proceed to Mount Harris. Wind, east.

Saturday, 25th October, Mulga Scrub South of Anna Reservoir. A few minutes before ten o'clock a.m. the party arrived all right. I was soon ready and lifted up into the saddle, and started at 10.10. During the day it has been excessively hot. At 5.45 p.m. arrived at Mount Harris, being nearly eight hours in the saddle, which is far more than I am able to endure in my terribly weak state. It is between my shoulder-blades and the small of my back that I am so much affected while riding. When the pain from them becomes unbearable I endeavour to get on as far as I can by supporting my weight upon my arms until they give way. I arrived here in a state of utmost exhaustion; so much so that I was quite unable to eat a single mouthful of anything. After we had the horses unpacked, a few natives made their appearance on the side of the mount, calling out something and pointing to the north-east. Sent Thring and King to see if they could make anything of them, but they soon ran down the other side of the mount, and, when seen again, were marching off in the direction they had pointed out. They had taken good care before leaving to use nearly all the water in the crevices of the granite rocks; they left about a quart. Finding it quite impossible to remain so long in the saddle as I have done to-day, I got Mr. Kekwick and some of the others to construct a stretcher during the night, which I hope will enable me to do a long journey to-morrow. Wind, south-east.

Sunday, 26th October, Mount Harris. Had the stretcher placed between two horses. Had great difficulty before we could get two that would allow it to be passed between them. At last succeeded in getting two that we thought would do very well, as they seemed to go very quietly with it. I shall continue on horseback until I find that I have got enough of it. Started a little after sunrise. I found I could continue two hours and a half in the saddle without fatiguing myself too much. Having done this, I sent to the rear of the party for the stretcher, when, to my great disappointment and vexation, I found that a short time before something had annoyed one of the horses, which set to and kicked it all to pieces, which is a great misfortune. I continued in the saddle, and proceeded until I was exhausted, which happened at the end of fifteen miles, when I was compelled to stop. Keeping Auld with me, and some water, I sent on the party and all the horses to Mount Hay. If they find water they are to camp and return for me to-morrow; if not, they are to push on to the Hamilton Spring; if that is gone, they will have to cross the range to Brinkley Bluff. I find myself getting weaker and weaker every day. I am very ill indeed. Wind, south-east.

Monday, 27th October, Hills North of Mount Hay. About 11.30 a.m. King and Nash returned for me. Thring had found water in one of the gullies, but the approach to it was very rough and stony indeed. Thring had gone to see if there was any water in the clay-pans that I had camped at on my journey up, and if there is, will take the party over there, and will send one of the men to meet me and inform me of it. The distance from here to the water is ten miles. Had the horses saddled; mounted, and proceeded towards it. At the end of two hours the motion of the horse became so dreadful to me, and the pain I was suffering from was such as no language can describe; but I still continued in the saddle, and, within a mile and a half of the water, met Frew, whom Thring had sent to say that he had found plenty of water in the clay-pans, with green grass, and that the party had moved on to it. Distance from where we were then to the clay-pans, six miles further. I could no more sit in the saddle that distance than I could fly; I am now already completely exhausted, and have still a mile and a half to ride before I can reach the other water. To that I must go, and see what a night's rest will do in the morning. While taking a drink of water, I was seized with a violent fit of vomiting blood and mucus, which lasted about five minutes, and nearly killed me. Sent Frew on to the party. Went on the best way I could with the other three to the water. Arrived there feeling worse than I have ever done before. I have told King and Nash to remain with me in case of my dying during the night, as it would be lonely for one young man to be here by himself. Wind, south-east.

Tuesday, 28th October, Mount Hay. Started in the cool of the morning, and in two hours reached where the party were camped, so much exhausted and so completely done up that I could not speak a word—the power of speech has completely left me. I was lifted from the saddle and placed under the shade of a mulga bush. In about ten minutes I recovered my speech. I find that I can no longer sit on horseback; gave orders for some of the party to make a sort of reclining seat, to be carried between two horses, one before the other; also gave orders that a horse was to be shot at sundown, as we are getting rather short of meat, and I hope the change of beef tea made from fresh meat will give me some increase of strength, for I am now reduced to a perfect skeleton, a mere shadow. At sundown had the horse shot; fresh meat to the party is now a great treat. I am denied participating in that pleasure, from the dreadful state in which my mouth still is. I can chew nothing, and all that I have been living on is a little beef tea, and a little boiled flour, which I am obliged to swallow. To-night I feel very ill, and very, very low indeed. Wind, south-east, with a few clouds.

Wednesday, 29th October, Clay-pans East of Mount Hay. This morning I feel a little relieved in comparison with my exhausted state of yesterday. I had a very troubled night's rest. All hands cutting up the horse, and hanging up the meat to dry. Thring and Nash out for two long poles to fix the chair in, which they succeeded in finding. At twelve o'clock had all the meat of the horse cut up and hung up to dry. Day oppressively hot. Wind, south-east. Clouds.

Thursday, 30th October, Clay-pans East of Mount Hay. I think I am a little better this morning, but still very weak and helpless. Find that the chair will not answer the purpose, and must have a stretcher instead. Wind, south-east.

Friday, 31st October, Clay-pans East of Mount Hay. I felt a little improvement this morning, which I hope will continue; and I think I have reached the turn of this terrible disease. On Tuesday night I certainly was in the grasp of death; a cold clammy perspiration, with a tremulous motion, kept creeping slowly over my body during the night, and everything near me had the smell of decaying mortality in the last stage of decomposition and of the grave. I sincerely thank the Almighty Giver of all Good, that He, in His infinite goodness and mercy, gave me strength and courage to overcome the grim and hoary-headed king of terrors, and has kindly permitted me yet to live a little longer in this world. Auld, who was in attendance upon me on that night, informed me that my breath smelt the same as the atmosphere of a room in which a dead body had been kept for some days. What a sad difference there is from what I am now and what I was when the party left North Adelaide! My right hand nearly useless to me by the accident from the horse; total blindness after sunset—although the moon shines bright to others, to me it is total darkness—and nearly blind during the day; my limbs so weak and painful that I am obliged to be carried about; my body reduced to that of a living skeleton, and my strength that of infantine weakness —a sad, sad wreck of former days. Wind variable.

Saturday, 1st November, Clay-pans East of Mount Hay. Although in such a weak state, I shall try if I can ride in the stretcher as far as Hamilton Springs. Started early; found the stretcher to answer very well. On arriving at the springs, saw that there was not sufficient water for the horses, and, as I had stood this part of the journey so well, made up my mind to cross the range to Brinkley Bluff. Proceeded, and arrived there about five o'clock p.m. I have stood the long journey far better than I expected, but feel very tired and worn out. Wind variable. Cloudy.

Sunday, 2nd November, Brinkley Bluff, The Hugh. Got a few hours' good sleep during the night, and feel a good deal better this morning. Day still cloudy. Wind variable.

Monday, 3rd November, Brinkley Bluff, The Hugh. Started at 7.30 a.m. for Owen Springs. Saw where one of the horses died that I was compelled to leave behind on coming up. As there is only the hair of his mane and tail to be seen, and not a single bone, I am inclined to think that he has been killed, carried off, and eaten by the natives. I expect the other one has shared the same fate. At 2.20 p.m. arrived at the springs. Plenty of water. I have stood the journey very well, but am very tired. Wind, south-east.

Tuesday, 4th November, Owen Springs, The Hugh. Started at 7.20 a.m., passing through the gorge of the Waterhouse range. At 1.20 arrived at the springs under the conglomerate rock, a mile and a half north-east of the gorge in James range. I feel the shaking of the stretcher very much, and am again very tired, but am glad to find that I am getting a little stronger. Wind, south-east. The clouds are all gone.

Wednesday, 5th November, Spring, Conglomerate Rock, The Hugh. Started at 7.25 a.m. Passed through the gorge of James range and proceeded to the side creek in which water was obtained on coming up. Found some still there. Camped. Sent four of the party to clear out the hole; in the meantime sent Thring up the side creek to see if there is any surface water left from the showers of rain that have fallen here some short time ago. Since leaving the McDonnell range we have had plenty of green grass, showing that rain has fallen some time back; it has made no impression upon the large creek, which is quite dry. In a short time Thring returned; he has seen as much as will do for forty horses to-night, which is a good thing. Sent him up with them, and watered the remainder at this hole, into which the water comes very slowly, in consequence of the main creek having none in its bed below the sand. I again feel tired from the shaking of the horses and the stretcher. The swelling of my gums and the black blisters, which have been so very painful for such a long time back, are slowly giving way before some vegetable food which I have been able to get since coming into the green, grassy country; I hope it will soon cure me. My teeth are still loose, but it is a great thing to get a little relief from a great mouthful of swollen, blistered, and most painful gums. When my mouth was closed I had scarcely room for my tongue; the blisters are now much reduced. Wind, south-east.

Thursday, 6th November, The Hugh. Started at 7.20 towards the Finke; at five p.m. met with some water in a clay-pan, and camped. I am a little stronger to-day, and feel that I am gradually improving. Wind, south-east. Night and morning cool.

Friday, 7th November, North of the Finke. Proceeded to Pascoe Springs in the Finke; found plenty of water and camped. Day oppressively hot. Wind, south-east.

Saturday, 8th November, Pascoe Springs, The Finke. Proceeded to Sullivan Creek and found sufficient water to do for us until Monday morning, and this being a place for feed for the horses, I shall remain here until that time. I feel very tired and sore after this rough week's work, and am glad of a day's rest. I feel a gradual improvement in my health and strength, which I hope will continue to increase. Wind variable, mostly from south-east.

Sunday, 9th November, Sullivan Creek. During the night had a few drops of rain; heavy clouds to the west, north-west, north, north-east, and east. Wind blowing strong and variable. Sundown: the sky overcast with heavy clouds.

Monday, 10th November, Sullivan Creek. Some of the horses missing this morning. Did not get a start till nine o'clock a.m. Day oppressively hot. Crossed the Finke three times, and arrived at Polly Springs, where there is plenty of water. Camped. Wind, south-east.

Tuesday, 11th November, Polly Springs, The Finke. Proceeded to Marchant

Springs. Camped. The water is low and rather boggy. Dug a place about eighteen inches deep in the firm ground, and the water came boiling up. I am happy to find that I am gaining a little strength again. I was able to walk two or three steps by leaning upon two of the party, but the pain was very severe. Wind, south-east; a few clouds about.

Wednesday, 12th November, Marchant Springs, The Finke. As I am not certain of water at the next two camps, I will rest the horses as well as myself here to-day, for we both require it very much; it will enable them to stand a long push if required. A number of showers of rain seem to have fallen here this month. Wind, south-east.

Thursday, 13th November, Marchant Springs, The Finke. Started at 7.40. Proceeded towards the Goyder, and at nine miles found myself in as dry a country as ever; not a drop of rain seems to have fallen here for upwards of twelve months. On arriving at the Goyder found a little moisture at the bottom of the sand in the rocks—not enough for the horses. Pushed on towards the Coglin, and at dark camped in the mulga scrub without water. Day most oppressively hot. Light wind from south-east.

Friday, 14th November, Mulga Scrub. Started at six o'clock a.m. Examined the different creeks in which I found water on my journey to the north but there was not a drop. At twelve miles reached the Coglin—none there. Country all in the same dry state. Proceeded on to the Lindsay, where I am sure of water. At four o'clock arrived there and found plenty. Camped. Thanks be to God, I am once more within the boundary of South Australia! I little expected it about a fortnight ago. If the summer rain has fallen to the south of this, there will be little difficulty in my getting down. I am again suffering very much from exhaustion, caused by a severe attack of dysentery, which has thrown me back a good deal in the strength I was collecting so quickly, but I hope it will not continue long. Wind, south-east.

Saturday, 15th November, The Lindsay. At day-break I have sent Thring to the Stevenson to see if there is water there, either on the surface or by digging in the sand; if there is I shall move the party over there to-day, and on Monday morning start for the Hamilton (I expect no water between); and if not, I shall remain here till that time and push for the Hamilton. About ten o'clock a.m. he returned and reported no water, only a little moisture on the top of the clay beneath the sand. Day very hot. I still continue to be very unwell. Wind, south-east.

Sunday, 16th November, The Lindsay. Day oppressively hot. Light winds, south-east.

Monday, 17th November, The Lindsay. Started soon after sunrise, crossed the Stevenson and the Ross; both quite dry. Proceeded across Bagot range to the gum water-hole; that is also dry. Found a little rain water in one of the small creeks, but not enough for all the horses. The day being excessively hot, the journey very rough and stony, and many of them lame from want of shoes, also it being near sundown, and there being a little green grass about, I have camped. Wind variable.

Tuesday, 18th November, The Gums, Bagot Range. Started at 5.40 a.m. to the large waterhole in the Hamilton; in about a mile found some rain water, which I allowed the horses to drink. At 10 a.m. arrived at the large water-hole, and found it very low indeed; a great number of dead fish all round it. This must certainly be a very unprecedentedly dry season indeed; this water-hole does not seem to have received any water for the last two years. The water being old and stagnant, I am afraid will make us ill; we have all already been suffering much from stagnant waters we have been compelled to use. I, however, must give the horses a day's rest to enable them to make the next and last push, nearly a hundred miles, to the first springs. From the dryness of the season, I scarcely expect to find water before I reach them, which will be a severe trial for the horses, the weather being so extremely hot. I am still suffering very much from the effect of the stagnant waters; they have sent me back again nearly to my former state of weakness, and have assisted in checking my recovery from the scurvy, which is now again gaining ground upon me since I lost the vegetable food. The country being now so dry, there having been no late rain, there is not a blade of grass to be seen. Hot wind from the north. This is the first and only hot wind I have felt during the whole journey from Mount Margaret to the sea-coast, and back to this place. In the afternoon the sky became overcast with heavy clouds. At sundown the wind changed to west, and blew very strong till eleven o'clock p.m.; we then had a few drops of rain, but not enough to moisten the surface of the ground; after this it became calm, the clouds broken, and there was no more of it.

Wednesday, 19th November, The Hamilton. This morning still cloudy, but excessively close and hot. I am glad that I resolved to remain here to-day, for the poor horses would have felt it very much travelling over the high and heavy sand hills that we have to go over in the first day's journey. In the afternoon the sky again became overcast with heavy clouds, and there was a great deal of thunder and lightning to the west and north, and again, at the same time as last night, we were favoured with a few drops of rain; the result the same as it was then. Wind variable and squally.

Thursday, 20th November, The Hamilton. This morning the clouds have cleared away, but there is a nice cool strong breeze from the south-east and east—a fine thing for the horses crossing the heavy sand hills. Started at six o'clock a.m. Got over them very well, and reached the mulga plain. About twelve the wind ceased, and it became very hot. In the afternoon one of the horses (Trussell) began to show symptoms of being very ill. One of the party was riding him at the time. I had him changed immediately and allowed him to run loose, but he seemed to have lost all spirit and soon dropped behind. I then had him led and driven for upwards of two miles until I reached the Frew or Upper Neale. The dreadfully dry state of the country since leaving the sand hills—it being completely parched up—leaving me no hope of getting water until I reached the gap in Hanson range or the Freeling Springs, and it being quite impossible for us to drag him on there, I was compelled to abandon him, as it would only knock up the other horses to drive him on. Proceeded through a still parched-up country to the large dry lagoon, and at dark camped without water. Wind, south-east.

Friday, 21st November, Large Dry Lagoon. Started at break of day through some low sand hills, with valleys and clay-pans, all dry. At a little more than six miles after starting, I was rather surprised to find recent tracks of horses that had been feeding on and about our tracks. Thinking it might be a party out looking for us, as I have now been some time longer than I anticipated at starting, I sent Thring to examine and see how many horses there were. In about half an hour he returned, and said that he could only make out two, and those I immediately concluded were two of the horses that had given in near this place on my journey to the north. Proceeded on to the camp where I had buried the two hundred pounds of sugar, frequently meeting their tracks, apparently in search of water. Arrived at the camp, but there is not a drop there, and no appearance of the two horses, but only their tracks in the bed of the creek, following it down to the eastward, where there must be permanent water that has supplied them during the past year. A thunder-shower must have brought them out to visit the spot where they were first left. I should have liked very much to have regained them, but the dry state of the country and the want of water will not allow me to look for them. Found that the things buried had been disturbed, and most of them carried away by the natives—the others all destroyed—the sugar all gone, except about five pounds, which was left in the hole and covered up. Proceeded, crossing side branches of the Neale, but not a drop of water in any of them—everything dried up. Went on towards the gap in Hanson range. At about eight miles before reaching it, Frew's horse (Holland) knocked up with him; he could not get him on a step further, and had to leave him. On reaching the Lindsay, this horse had been allowed by Frew to drink too much water, and had not recovered from the effects of it. At dark arrived at the gap, and found plenty of water, for which I am very thankful, for there are many of the horses that would not have stood another day's journey without it. Day exceedingly hot. Wind, south-east.

Saturday, 22nd November, Gap in Hanson Range. Resting horses, etc. Sent Frew in search of his horse shortly after sunrise. About half-past two he returned, and reports that he cannot be found; that he had searched round about the creeks and gullies where he had been left, but could find nothing of him, and the country was too stony to track him. Day again very hot.

Sunday, 23rd November, Gap in Hanson Range. Started at six o'clock a.m., intending to get to Freeling Springs, but one of the horses that had eaten poison about the Roper country, and has never recovered from it, but was always very poor, and of no use whatever, knocked up, and would not move a step further; being only six miles from where we started, we left him and proceeded on our journey. About this time the wind changed to the north, and it came on to blow a fierce hot wind, and by the middle of the day it was almost unbearable. Two more of the horses knocked up, and being nearly opposite the McEllister Springs, I turned to them and camped. These springs required to be dug out before we could get water enough for all the horses. After opening two of them, we found them to yield a sufficient supply. Still continuing to blow a terrific hot wind from the north. A little before sundown it changed, and came on to blow from the south, and blew the hot wind back again. For three hours it was as hot as when coming from the north.

Monday, 24th November McEllister Springs. Proceeded to the Freeling Springs and camped. This journey was as much as the horses are now able to do. The stagnant and spring waters have weakened them so much that I shall be compelled to rest them some time at Mr. Jarvis's, Levi's station, before they will be able to perform the remainder of the journey to Adelaide, that is, if I can get them that length.

Tuesday, 25th November, Freeling Springs. Found one of the chestnut horses that was left here. The other one seems to have been taken on to Mr. Jarvis's. Started shortly after sunrise. Proceeded to the Milne Springs and camped. The day again extremely hot. Wind still from the south-east. Twenty miles a day is now as much as my horses can accomplish.

Wednesday, 26th November, Milne Springs. Proceeded to Mr. Jarvis's station, Mount Margaret, which I expected to reach without losing any more horses, but I am disappointed, for I had to leave four behind knocked up, which I shall be able to recover to-morrow or the next day. Mr. Jarvis being from home, we were received by his men with a hearty welcome, and were shown every kindness and attention that was in their power. Day again very hot. Wind, south-east.

Thursday, 27th November, Mount Margaret Station. Resting horses. Sent out and had the one that knocked up about two miles from here brought in. I am still very ill, but am able to walk a few yards without assistance. I hope a few days will benefit me much. Day very hot. Wind, south-east. Clouds.

Friday, 28th November, Mount Margaret Station. Resting horses. Still cloudy. Promising rain. Sent out and had the other three knocked-up horses brought in all right. Yesterday got in the other chestnut horse left at the Freeling Springs, and brought down here by Woodforde. Clouds breaking up. No rain. Wind, south-east.

Saturday, 29th November, Mount Margaret Station. Resting horses, etc. I find the scurvy is fast gaining upon me, although I have had fresh meat for the last few days. I must therefore push on as fast as possible down the country, in order to get some vegetables. I shall start to-morrow evening, and travel during the night to the William Spring to avoid the great heat of the day, taking with me the stretcher (for I am not yet able to ride), three men, and the strongest horses, leaving the rest here for another week to recover with remainder of the party in command of Mr. Kekwick, who, as soon as the horses are sufficiently strong, will conduct the party to Adelaide. Clouds all gone. Wind, south-east.

Sunday, 30th November, Near Mount Margaret Station. Started at five p.m. for the William Spring with fourteen horses, leaving the weak and done-up ones at Mount Margaret for another week to recover. I have also brought on with me Auld, King, and Billiatt. The others I have left with Mr. Kekwick, to whom I have given command of the party, and who will conduct them to Adelaide by easy stages, as soon as the horses are able to travel. I travelled during the night, and arrived at the spring a little before six a.m. Camped, unsaddled the horses, and turned them in amongst the young reeds to feed, which they seemed very eager for.

Monday, 1st December, William Spring. During the day the horse that I was compelled to leave here on my northward journey came towards the others, but appeared very shy. I left him alone till nearly sundown, when I sent King to see if he had joined them, and to see if the others were all right. At dark he returned, and reported them to be all right, and that the other had joined them. He tried to catch him, but that he would not allow, so he left him with the others during the night. The day has been very close and oppressive, with heavy clouds and distant thunder. I am glad I performed this long journey during the night. Wind, south-east. Clouds all gone.

Tuesday, 2nd December, William Spring. Got all the horses into camp, and attempted to catch the stranger, but could not without roping him; I therefore drove him along with the others to the Beresford Springs, and then he allowed himself to be caught and hobbled. The journey has quieted him. It is the longest journey he has had for nearly twelve months. I arrived about four o'clock p.m., and there being plenty of young reeds, camped. The day has been again very hot, but occasionally strong breezes from the south-east and east.

Wednesday, 3rd December, Beresford Springs. Proceeded to Mount Hamilton Station, where I received a very kind reception from Mr. Brown, and was treated with the greatest possible kindness. Toward evening I again felt very ill. Day very hot. Wind, south-east.

Thursday, 4th December, Mount Hamilton Station. I have been very ill during the night, but started for Chambers Creek. Arrived there about mid-day, where I again experienced a like hospitable reception and great kindness from Mr. Lee. Wind variable. Day extremely hot.

Friday, 5th December, Chambers Creek. I shall require to rest my horses here to-day. I was in great hopes that when I reached this place I should have been again able to have ridden on horseback, but the waters of the spring country through which I have just passed have reduced me nearly to my former state of weakness, and I shall still be compelled to continue in the ambulance a little longer. I feel a little better this morning—I suppose in consequence of drinking fresh water. Hot wind from the north. Towards evening a heavy thunderstorm coming from the westward.

Saturday, 6th December, Chambers Creek. Started at eight o'clock with the ambulance towards Termination Hill. After crossing numerous sand hills, we frequently found rain water. Towards sundown arrived at the south side of Porter Hill. Found rain water, and camped, one of the horses being nearly knocked up. I shall be compelled to take to the saddle to-morrow, for the ambulance horses will not be able to carry me further. I must send them back to the creek, there to rest till the others come down. Cloudy. Wind variable.

Sunday, 7th December, Porter Hill. Mounted and started at six a.m. I find that I can endure the motion of the horse better than I expected; but about mid-day began to feel it very much. Towards four o'clock found some rain water about ten miles from Termination Hill, for which I am very thankful, for I could not have continued the journey any further. Camped. Wind variable.

Monday, 8th December, Termination Hill. During the night had a heavy thunderstorm and shower from the south-east. Started at six a.m. and arrived at Mr. Glen's Station at sundown, quite done up; received a hearty welcome. Encountered a heavy storm of thunder and lightning a few miles from the station. Wind, south-east.

Tuesday, 9th December, Mr. Glen's Station. Proceeded to Mount Stuart Station, where I had the pleasure of meeting Mr. John Chambers, who received me with great kindness. There has been some heavy rain here lately. Wind, south-east. Day hot.

Wednesday, 10th December, Mount Stuart Station. Accompanied by Mr. Chambers, proceeded to Moolooloo, and arrived there in the afternoon completely tired and exhausted from riding in the saddle. Day hot. Wind, east.

In conclusion, I beg to say, that I believe this country (i.e., from the Roper to the Adelaide and thence to the shores of the Gulf), to be well adapted for the settlement of an European population, the climate being in every respect suitable, and the surrounding country of excellent quality and of great extent. Timber, stringy-bark, iron-bark, gum, etc., with bamboo fifty to sixty feet high on the banks of the river, is abundant, and at convenient distances. The country is intersected by numerous springs and watercourses in every direction. In my journey across I was not fortunate in meeting with thunder showers or heavy rains; but, with the exception of two nights, I was never without a sufficient supply of water. This will show the permanency of the different waters, and I see no difficulty in taking over a herd of horses at any time; and I may say that one of our party, Mr. Thring, is prepared to do so. My party have conducted themselves throughout this long and trying journey to my entire satisfaction; and I may particularly mention Messrs. Kekwick and Thring, who had been with me on my former expedition. During my severe illness every attention and sympathy were shown to me by every one in the party, and I herewith beg to record to them my sincere thanks.

I may here mention that the accident which occurred to me at the starting of the Expedition from Adelaide has rendered my right hand almost useless for life.

The Journal concludes with the following letter:

To the Honourable H.B.T. Strangways, Commissioner of Crown Lands and Immigration.

Adelaide, December 18, 1862.

Sir,

For the information of His Excellency the Governor-in-Chief, I have the honour to report to you my return to Adelaide, after an absence of twelve months and thirteen days; and I herewith beg to hand you my chart and journals of the Expedition from which I have just returned.

To you, Sir, and the Government, my especial thanks are due for the liberal manner in which the supplies were voted, and for the kind and ready assistance I at all times experienced. Also to George Hamilton, Esquire, Chief Inspector of

Police, for the efficient manner in which my party was fitted out. The original promoters of my various expeditions, Messrs. James Chambers and William Finke, have always shown the most lively interest in my success, to which they cheerfully contributed. How much I regret the unexpected decease of the first-named gentleman I need here hardly state, for he was indeed heart and soul in the result, and no one would have felt so proud of my success as my much-lamented and best friend James Chambers. To Mr. John Chambers I am also under many obligations for assistance in many instances, and I hereby tender him my best thanks.

I have the honour, etc.,

J.M. STUART.

APPENDIX

[FROM THE PROCEEDINGS OF THE ZOOLOGICAL SOCIETY OF LONDON. JUNE 9, 1863

ON A COLLECTION OF BIRDS FROM CENTRAL AUSTRALIA.
BY JOHN GOULD, F.R.S., ETC.

The Board of Governors of the South Australian Institute having liberally forwarded for my inspection a selection from the ornithological collection made by Mr. Frederick G. Waterhouse during Mr. Stuart's late Exploratory Expedition into Central Australia, I have thought the matter of sufficient interest to bring these birds under the notice of the Society, the more so as it will enable me to make known through our Proceedings a new and very beautiful species of Parrakeet pertaining to the genus Polyteles, of which only two have been hitherto known. Every ornithologist must be acquainted with the elegant P. melanurus and P. barrabandi, and I feel assured that the acquisition of an additional species of this lovely form will be hailed with pleasure. The specific appellation I would propose for this novelty is alexandrae, in honour of that Princess who, we may reasonably hope, is destined at some future time to be the Queen of these realms and their dependencies, of which Australia is by no means the most inconspicuous.

Polyteles alexandrae, sp. nov.

Forehead delicate light blue; lower part of the cheeks, chin, and throat rose-pink; head, nape, mantle, back, and scapularies olive-green; lower part of the back and rump blue, of a somewhat deeper tint than that of the crown; shoulders and wing-coverts pale yellowish green; spurious wing bluish green; external webs of the principal primaries dull blue, narrowly edged with greenish yellow; the remaining primaries olive-green, edged with greenish yellow; under wing-coverts verditer-green; breast and abdomen olive-grey, tinged with vinous; thighs rosy red; upper tail-coverts olive, tinged with blue; two centre tail-feathers bluish olive-green; the two next on each side olive-green on their outer webs and dark brown on the inner ones; the remaining tail-feathers tricoloured, the central portion being black, the outer olive-grey, and the inner deep rosy red; under tail-coverts olive; bill coral red; feet nearly brown.

Total length 14 inches; bill 1/2; wing 7; tail 9; tarsi 7/8.

Habitat. Howell Ponds, Central Australia, 16 degrees 54 minutes 7 seconds South latitude.

Remark. This is in every respect a typical Polyteles, having the delicate bill and elegantly striped tail characteristic of that form. It is of the same size as P. barrabandi, but differs from that species in having the crown blue and the lower part of the cheeks rose-pink instead of yellow.

The following is a list of the other species of birds comprised in the collection:

Trichoglossus rubritorquis. Rare.

Aprosmictus erythropterus.

Platycercus brownii. Rare.
Struthidea cinerea.
Climacteris melanura.
Pomatorhinus rubecula. Rare.
Cincloramphus cruralis.
Artamus leucopygialis.
Artamus cinereus. Rare.
Colluricincla brunnea.
Petroica bicolor.
Pardalotus rubricatus. Extremely rare: the second specimen seen.
Graucalus melanops.
Tropidorhynchus argenteiceps.
Geopelia cuneata.
Geopelia humeralis.
Erythrogonys cinctus.

...

[FROM THE PROCEEDINGS OF THE ZOOLOGICAL SOCIETY OF LONDON, NOVEMBER 10, 1863.]

DESCRIPTIONS OF NEW SPECIES OF FRESHWATER SHELLS COLLECTED BY MR. F.G. WATERHOUSE, DURING J. McDOUALL STUART'S OVERLAND JOURNEY FROM ADELAIDE TO THE NORTH-WEST COAST OF AUSTRALIA. BY ARTHUR ADAMS, F.L.S., AND G. FRENCH ANGAS, CORRESPONDING MEMBER OF THE ZOOLOGICAL SOCIETY.

1. Vivipara waterhousii, Adams & Angas.

V. testa turbinata, globoso-conica, late umbilicata, spira elatiuscula, epidermide tenui fusco-viridi obtecta; anfractibus convexis, ad suturas subplanatis, faciis tribus vel quatuor angustis olivaceo-viridibus transversis ornatis; anfractu ultimo inflato, lineis duabus impressis ad peripheriam instructo; apertura ovata, postice subangulata; labio simplici; labro acuto.

Long. 2 inches, lat. 1 inch 8 lines.

Habitat. Newcastle Waters, Arnhem's Land (Coll. Angas):

This fine species most nearly resembles Vivipara ussuriensis, Gerst.; but the last whorl is more inflated, and the surface of the shell is not malleated or lirate. It is the largest species yet discovered on the Australian continent. We have great pleasure in dedicating it to F.G. Waterhouse, Esquire, who, under great difficulties during the expedition, succeeded in making many valuable additions to science.

2. Vivipara kingi, Adams & Angas.

V. testa turbinata, globoso-conica, umbilicata, spira mediocri erosa nodulosa, epidermide tenui pallide fusco-viridi obtecta, ad apicem purpurascente; anfractibus convexis, lineolis transversis et longitudinalibus elevatis decussatis, anfractu ultimo ad basin sulcis impressis spiralibus instructo; apertura ovata, antice subeffusa; labio vix reflexo.

Long. 1 inch, lat. 8 lines.

Habitat. King's Ponds, Arnhem's Land (Coll. Angas).

This is a neat, finely-decussated, concolorous species, with the upper whorls nodulous from erosion, as in Vivipara praerosa, Gerst. It is named after Mr. Stephen King, one of the gentlemen who accompanied the expedition.

3. Melania (Melasma) onca, Adams & Angas.

M. testa fusiformi-turrita; spira elata, conica; epidermide pallide olivaceo induta, rufo-fusca, pulcherrime maculata, maculis saepe in lineis undulatis longitudinalibus dispositis; anfractibus planis, longitudinaliter plicatis, plicis aequalibus regularibus subdistantibus, ad suturas nodulosis; apertura oblongo-ovata, antice effusa; labio subincrassato; labro simplici, acuto.

Long. 1 inch, lat. 4 lines.

Habitat. Tributary of Adelaide River, Arnhem's Land (Coll. Angas).

A species remarkable both for the elegance of its form and the beauty of its painting. The whorls are plicate, with a necklace-like series of nodules at the sutures; and the shell is covered with dark red-brown spots, suggestive of its specific name.

4. Amphipeplea vinosa, Adams & Angas.

A. testa ovata; spira mediocri, tenui, semipellucida, vinosa; anfractu ultimo magno, ventricoso, postice ad suturas gibboso; apertura ovata; labio callo tenui mediocri obtecto, columella spiraliter tortuosa; labro convexo, margine acuto.

Long. 9 lines, lat. 5 lines.

Habitat. Tributary of Adelaide River, Arnhem's Land (Coll. Angas).

This species may readily be distinguished on account of its peculiar vinous colour. The whorls are posteriorly gibbose or tumid at the sutures, and the callus is less spreading than in others of the genus.

5. Amphipeplea phillipsi, Adams & Angas.

A. testa ovata; spira elata, acuta, tenui, cornea; anfractu ultimo magno, non ventricoso, transversim creberrime striato; apertura oblongo-ovali; labio callo tenui expanso obtecto; labro simplici, acuto.

Long. 9 lines, lat. 4 lines.

Habitat. Arnhem's Land (Coll. Angas).

A neat, horn-coloured, finely transversely striated species, with an acute elevated spire. We have named it after Mr. T. Phillips, who has assiduously collected many new Australian shells.

6. Physa newcombi, Adams & Angas.

P. testa ovata, umbilicata; spira mediocri, acuta, ad apicem integra, cornea, viridescente aut pallide fulva; anfractibus quinque, convexis, saepe plus minusve transversim subliratis; apertura ovata; labio reflexo, umbilicum partim tegente; labro vix incrassato, peristomate nigrescente.

Long. 10 lines, lat. 7 lines.

Habitat. Ponds at Mount Margaret (Coll. Angas.)

We have much pleasure in naming this noble Physa after Dr. Newcomb, the distinguished American conchologist, who has contributed so much, by his researches in the Sandwich Islands, to our knowledge of the genus Helicter or Achatinella. The species is widely umbilicated, and the peristome is usually dark-coloured.

7. Physa ferruginea, Adams & Angas.

P. testa ovata, rimata, ferruginea; spira mediocri, apice eroso; anfractibus tribus, convexis, simplicibus, transversim crebre crenato-striatis; apertura ovata, intus purpurascente; labio tenui, late reflexo; labro acuto.

Long. 5 lines, lat. 4 lines.

Habitat. Arnhem's Land, North-west Australia (Coll. Angas.)

This is a small ferruginous species, with the whorls finely transversely striated.

8. Physa badia, Adams & Angas.

P. testa elongato-ovata, imperforata, solida, badia; spira elata, apice obtuso eroso; anfractibus quinque, convexiusculis, longitudinaliter strigillatis; apertura elongato-ovata; labio albo, excavato, lirula antica subspirali instructo; labro arcuato, in medio producto, intus fusco tincto.

Long. 1 inch, lat. 6 lines.

Habitat. Tributaries of Adelaide River, Arnhem's Land (Coll. Angas.)

A fine, solid, brown species, generally more or less eroded, and with a peculiarly strongly plicate columella.

9. Physa olivacea, Adams & Angas.

P. testa elongato-ovata, imperforata, solidiuscula, olivacea; spira elata, attenuata, apice eroso; anfractibus quinque, convexiusculis; apertura ovato-acuta; labio incrassato, flexuoso; labro acuto, margine arcuato.

Long. 6 lines, lat. 3 lines.

Habitat. Arnhem's Land (Coll. Angas.)

A neat, olive-coloured species, somewhat resembling in form the British Aplexa hypnorum, but without the polished exterior of the latter.

10. Physa concinna, Adams & Angas.

P. testa ovata, imperforata, solidiuscula, cornea; spira brevi, acuta, apice interdum papilloso; anfractibus quinque, convexiusculis; transversim striatis; apertura acuto-ovata; labio incrassato, spiraliter valde tortuoso; labro intus incrassato et fusco tincto, margine acuto, arcuato.

Long. 6 lines, lat. 3 lines.

Habitat. Arnhem's Land (Coll. Angas.)

A pale horn-coloured, somewhat solid species, with a moderately elevated spire, acute (not eroded) at the apex, and with the terminal whorls sometimes papillary.

11. Physa (Ameria) reevii, Adams & Angas.

P. testa ovali, postice abrupte truncata, imperforata, cornea; spira plana, tenui; anfractibus quatuor, planis, ultimo permagno, postice acute angulato, transversim obsolete striato; apertura oblongo-truncata; labio antice valde tortuoso; labro postice angulato.

Long. 6 lines, lat. 4 lines.

Habitat. Arnhem's Land (Coll. Angas.)

We have much pleasure in dedicating this singular species to Mr. Lovell Reeve, who has evinced much interest in the shells of this group. The last whorl is acutely angulate posteriorly, and the spire is tabulated, giving to the shell a peculiar truncate appearance.

12. Physa (Ameria) bonus-henricus, Adams & Angas.

P. testa ovata, rimata, tenui, cornea; spira vix elata, plana; anfractibus tribus, planis, postice angulatis, ultimo magno, inflato, ventricoso, postice subangulato, longitudinaliter plus minusve plicato; apertura ovata; labio tenui, subtortuoso; labro simplici, margine arcuato.

Long. 4 lines, lat. 2 1/2 lines.

Habitat. Arnhem's Land (Coll. Angas.)

This is a small inflated species, with a short truncate spire. We have dedicated it to the founder of the section Ameria, a gentleman well known for his deep researches in conchology.

13. Unio (Alasmodon) stuarti, Adams & Angas.

U. testa transversim elongato-ovata, tenui, compressa, epidermide olivaceo-fusca induta, postice corrugato-plicata, latere antico breviore rotundato, postico longiore oblique subtruncato, margine ventrali regulariter arcuato; umbonibus parvis, erosis, dentibus cardinalibus elongatis valde divergentibus, postico bifido, antico prominulo; intus iridescente.

Alt. 1 1/2 inch, lat. 3 inches 2 lines.

Habitat. Lagoon, Mount Margaret, Central Australia (Coll. Angas.)

This species, which we have named after Mr. J. McD. Stuart, the leader of the expedition, is the only Naiad, besides Alasmodon angasana of Lea, yet discovered in the regions traversed by the explorers.

...

Description of a new Helix from the interior of Australia, by Dr. L. Pfeiffer.

Helix perinflata, Pfr.

T. umbilicata, globosa, solida, striis incrementi rugosis et lineis impressis antrorsum descendentibus decussata, isabellino-albida; spira convexo-conoidea, apice obtusa; anfr. 4 1/2, ultimus magnus, ventrosus, subtus, perinflatus, striis spiralibus obsolete sculptus, antice deflexus; apertura diagonalis, lunari-rotundata; perist. breviter expansum margine columellari supra umbilicum angustum fornicatim dilatato.

Diam. mag. 23 1/2, min. 20, alt. 20 mill. (Coll. Angas.)

Habitat. McDonnell Range, Central Australia. Waterhouse, on Stuart's expedition.

ENUMERATION OF THE PLANTS COLLECTED DURING MR. J. McDOUALL STUART'S EXPEDITIONS ACROSS THE AUSTRALIAN CONTINENT IN 1860, 1861, AND 1862. BY FERDINAND MULLER, M.D., Ph.D., F.R.S.

Dilleniaceae.

Pachynema macrum, F.M. Purdie Ponds. Waterhouse. Hibbertia glaberrima, F.M. Fragmenta, Phyt. Austr. iii. 1. Brinkley Bluff, McDonnell Range. J.M. Stuart.

Nymphaeaceae.

Nymphaea gigantea. Hook. Botanical Magazine 4647. Strangways River.

Nelumbium speciosum, W. Sp. Pl. ii. 1258. Arnhem's Land.

Capparideae.

Capparis nummularia, Cand. Prodr. i. 246. Central Australia.

Capparis lasiantha, R. Br in Cand. Prodr. i. 247. Near Central Mount Stuart.

Busbeckea umbonata (Capparis umbonata, Lindl. in Mitch. Trop. Austr. 275). Near Newcastle Waters and Attack Creek. Flowers similar to those of B. Mitchellii.

Droseraceae.

Drosera indica, Linn. Sp. Pl. 403. On the Bonney and Finke Rivers and Attack Creek, also in Central Australia.

Violaceae.

Ionidium enneaspermum, Vent. Malmais. page 27. Burke Creek. An allied species with a blue labellum occurs in the collection gathered at Purdie Ponds.

Frankeniaceae.

Frankenia laevis, Linn. Sp. 473 var. Finke River.

Zygophylleae.

Zygophyllum apiculatum, F.M. in Linnaea, 1852, page 373. Stevenson River.

Tribulus terrestris, Linn. Sp. 554. Mount Morphett. A large flowering variety with petals 1 inch long. At Marchant Springs, Burke River, and Attack Creek.

Malvaceae.

Hibiscus brachysiphonius, F.M. Fragm. Phyt. Austr. i. 67. Near the Strangways Range.

Hibiscus pentaphyllus, F.M. Fragm. Phyt. Austr. ii. 13. Newcastle Waters and Daly Waters.

Hibiscus radiatus, Cav. Diss. iii. 150, t. 54, fig. 2. Purdie Ponds, Newcastle Waters. Attack Creek.

Hibiscus sturtii, Hook. in Mitch. Trop. Austr. page 363. North of McDonnell Range.

Hibiscus solanifolius, F.M. Fragm. ii. 116. Mount Denison.

Hibiscus panduriformis, Burm. Fl. Ind. page 151, t. 47, f. 2. Burke River.

Gossypium Australe, F.M. Fragm. i. 46. Newcastle Waters, Waterhouse. Between Mount Woodcock and the Davenport Ranges.

Gossypium Sturtii, F.M. Fragm. iii. 6. as far north as the Stevenson River.

Abutilon tubulosum, All. Cunn. in Mitch. Trop. Austr. 390. Burke River.

Abutilon leucopetalum, F.M. Fragm. iii. 12. Daly Waters.

Sida corrugata, Lindl. in Mitch. Three Exped. ii. 12. Var. filipoda. Attack Creek. J.M. Stuart.

Sida cryphiopetala, F.M. Fragm. ii. 4. Brinkley Bluff, McDonnell Range. J.M. Stuart.

Tiliaceae.

Corchorus sidoides, F.M. Fragm. iii. 9. McDonnell Range. J.M. Stuart.

Triumfetta plumigera, F.M. Fragm. i. 69. Purdie Ponds. F. Waterhouse.

Buettneriaceae.

Kerandrenia nephrosperma, Benth. in Proceedings of the Linnean Society; Seringea nephrosperma, F.M. in Hook. Kew Miscell. 1857, 15. Towards Arnhem's Land.

Kerandrenia Hookeri, Walp. Annal. Bot. Syst. ii. 164. Near the Roper River.

Rulingia loxophylla, F.M. Fragm. i. 68. Towards Arnhem's Land.

Melhania incana, Heyne in Wall. List. 1200. Burke River and Purdie Ponds.

Sterculiaceae.

Brachychiton ramiflorum, R. Br. in Horsf. Plant. Savan. rarior. 234. From Burke Creek onward to Arnhem's Land.

Cochlospermeae.

Cochlospermum Gregorii, F.M. Fragm. Phyt. Austr. i. 71. Strangways River.

Cochlospermum heteronemum, F.M. in Hook. Kew Miscell. 1857, 15. Strangways River.

Meliaceae.

Owenia acidula, F.M. in Hook. Kew Miscell. ix. 304. Central Mount Stuart.

Sapindaceae.

Thouinia variifolia, Fragm. Phyt. Austr. i. 45. Crawford Range.

Diplopeltis Stuartii, F.M. Fragm. iii. 12. Between Mount Morphett and the Bonney River. J.M. Stuart.

Distichostemon phyllopterus, F.M. in Hook. Kew Miscell. ix. 306. Purdie Ponds. Var. serrulatus; leaves tender, lanceolate, acute, serrulated; stamens about 44. Burke River.

Dodonaea lanceolata, F.M. Fragm. Phyt. Austr. i. 73. Purdie Ponds, Waterhouse. Mount Woodcock. Stuart.

Dodonaea platyptera, F.M. Fragm. i. 73. Strangways River.

Dodonaea physocarpa, F.M. Fragm. i. 74. Daly Waters.

Dodonaea microzyga, F.M. Somewhat viscid, almost glabrous; leaves with 1 to 2 pairs of small obovate-cuneate leaflets; in front rounded, or truncate, or retuse, or sometimes 3-toothed, flat at the margin; rachis dilated; fruit-bearing pedicels solitary; capsules 3 to 4-celled; valves cymbeo-semiorbicular, all around broadly winged; the wing rounded-blunt on both extremities; dissepiments persistent with the columella. On the River Neale. J.M. Stuart.

A shrub with spreading and rigid branches. Most leaves about 1/2 an inch long; leaflets 1 to 2 inches long; flowers unknown; capsule with the wings added about 1/2 an inch long, shining, reddish; valves ceding from the septa; ripe seeds unknown.

The fruit of this species is almost like that of Dodonaea viscosa.

Mollugineae.

Mollugo trigastrotheca, F.M. Plants indigenous to Victoria, i. 201. Arnhem's Land.

Caryophylleae.

Polycarpoea corymbosa, Lam. Mount Samuel. J.M. Stuart.

Portulaceae.

Portulaca oleracea, Linn. Sp. Pl. 638. Common in the interior and in North Australia.

Calandrinia Balonnensis, Lindl. in Mitch. Trop. Austr. page 148. River Finke.

Phytolacceae.

Codonocarpus cotinifolius, F.M. Plants of Victoria, i. 200. From 300 to 800 miles north of Adelaide, F. Waterhouse; Central Mount Stuart, J.M. Stuart.

Gyrostemon ramulosus, Desf. in Memoir. du Museum, vi. 17 River Finke. J.M. Stuart.

Didymotheca pleiococca, F.M. Plants indigenous to Victoria, i. 198. Between the River Bonney and Mount Morphett. J.M. Stuart.

Leguminosae.

Acacia retivenea, F.M. Fragm. iii. 128. Short Range.

Acacia dictyophleba, F.M. Fragm. iii. 128. Mount Humphries.

Acacia aneura, F.M. in Linnaea, xxvi. 627. Mulga. Over the whole of Central Australia. F. Waterhouse.

Acacia tumida, F.M. in Proceedings of the Linnean Society iii. 144. Attack Creek.

Acacia impressa, F.M. in Proceedings of the Linnean Society iii. 133. Short Range.

Acacia lycopodifolia, A. Cunn. in Hook. Icon. ii. t. 172. Towards Arnhem's Land.

Acacia umbellata, A. Cunn. in Hook. London Journal of Botany i. 378. Robinson River. Stuart.

Acacia holosericea, A. Cunn. in Don. Gen. Syst. ii. 407. Near Newcastle Waters.

Pithecolobium moniliferum, Benth. in Hook. Journal of Botany iii. 211. Arnhem's Land.

Neptunia spicata, F.M. Fragm. Phyt. Austr. iii. 151. Arnhem's Land.

Erythrophloeum Laboucherii, Laboucheria chlorostachya, F.M. in Proceedings of the Linnean Society iii. 159. Newcastle Waters, Stuart; Strangways River, Waterhouse.

Cassia venusta, F.M. Fragm. Phyt. Austr. i. 165. Newcastle Waters and Mount Freeling. J.M. Stuart.

Cassia notabilis, F.M. Fragm. ii. 28. Between the River Bonney and Mount Morphett.

Cassia Absus, Linn. Spec. Plant. 537. Arnhem's Land.

Cassia oligoclada, F.M. Fragm. iii. 49. Attack Creek.

Cassia desolata, F.M. in Linnaea, 1852. Central Australia.

Cassia eremophila, A. Cunn. in Sturt's Centr. Austr. Append. ii. 77. Central Australia.

Petalogyne labicheoides, F.M. in Hook. Kew Miscell. 1856. From latitude 30 degrees South to latitude 17 degrees 58 minutes South. J.M. Stuart. Petalogyne cassioides forms merely a variety of this species.

Erythrina biloba, F.M. in Hook. Kew Miscell. 1857, page 21. Common to most creeks, from latitude 22 degrees to 19 degrees South. Wood soft, corky. J.M. Stuart. Stuart's Bean-tree is a species of Erythrina.

Bauhinia Leichartdtii, F.M. in Transact. Phil. Inst. Vict. iii. 50. Hayward Creek. J.M. Stuart.

Gastrolobium grandiflorum, F.M. Fragm. Phyt. Austr. ii. 17. Whittington Range, J.M. Stuart; Purdie Ponds, where it attains a height of 8 feet, Waterhouse.

Gompholobium polyzygum, F.M. Fragm. ii. 29. Between Mount Morphett and the Bonney River.

Jacksonia odontoclada, F.M. Between Newcastle Water and Attack Creek. J.M. Stuart.

Isotropis atropurpurea, F.M. Fragm. Phyt. Austr. ii. 16. Attack Creek, and between Mount Morphett and the Bonney River. J.M. Stuart.

Leptosema Chambersii, F.M. Essay on the Plants of the Burdekin Expedition page 8. Near Davenport Range, and between the Rivers Finke and Stevenson.

Crotalaria medicaginea, Lamb. Dict. ii. 201. Newcastle Waters. J.M. Stuart.

Crotolaria dissitiflora, Benth. in Mitch. Trop. Austr. 386. Newcastle Waters and McDonnell Range. Stuart.

Crotalaria Mitchellii, Benth. l. c. 120. Central Australia.

Crotalaria Cunninghami, R. Br. in Sturt's Central Austr. Append. 71. Burke Creek, Waterhouse; Mount Humphries, Stuart.

Indigofera hirsuta, L. Sp. Pl. 1862. Arnhem's Land.

Indigofera viscosa, Lam. Encyl. Menth. iii. 247. Brinkley Bluff. Stuart.

Indigofera oxycarpa, F.M. Fragm. Phyt. Austr. iii. 103. Burke Creek. Waterhouse.

Indigofera brevidens, Benth. in Mitch. Trop. Austr. 385. Central Australia.

Indigofera lasiantha, F.M. Report on Gregory's Plants from Cooper Creek, page 6. Denison Range. J.M. Stuart.

Swainsona phacoides, Benth. in Mitch. Trop. Austr. 363. River Neale. Stuart.

Swainsona campylantha, F.M. Report on Gregory's Plants from Cooper Creek. Bagot Range. J.M. Stuart.

Psoralea patens, Lindl. in Mitch. Three Exped. ii. 8. Attack Creek. Var. cinerea. Mount Kingston.

Psoralea balsamica, F.M. in Proceed. Phil. Inst. Vict. iii. 55. Attack Creek and McDonnell Range. J.M. Stuart.

Psoralea leucantha, F.M. l. c. iii. 54. Attack Creek.

Clianthus Dampierii, All. Cunn. in Transact. Horticult. Soc. ii. Ser. Vol. i. 522. Near Mount Humphries.

Onagreae.

Jussioea suffruticosa, Linn. Sp. Pl. 555. Attack Creek and Strangways River.

Rhamnaceae.

Alphitonia excelsa, Reiss. in Endl. Gen. Plant. page 1098. Daly Waters.

Euphorbiaceae.

Euphorbia hypericifolia, Linn. Sp. Plant. Attack Creek.

Flueggea leucopyris, W. Sp. Plant. McDouall Range and Roper River.

Petalostigma quadriloculare, F.M. in Hook. Kew Miscell. ix. 17. Near Mount Blyth.

Combretaceae.

Macropteranthes Kekwickii, F.M. Fragm. iii. 151. Newcastle Waters, near Ashburton Range.

Terminalia circumalata, F.M. Fragm. Phytogr. Austr. ii. 91. Attack Creek.

Terminalia bursarina, F.M. Fragm. Phytogr. Austr. ii. 149. Newcastle Waters.

Rhizophoreae.

Carallia integerrima, Cand. Prodr. iii. 33. Roper River. Waterhouse.

Cucurbitaceae.

Cucumis jucunda, F.M. in Transact. Phil. Inst. Vict. iii. 45. Central Australia.

Melastomaceae.

Osbeckia Australiana, Naudin in Annal. des Scien. Naturell. Ser. iii. xiv. 59. Arnhem's Land.

Melastoma Novae Hollandiae, Nand. l. c. xiii. 290. Adelaide River.

Myrtaceae.

Carega arborea, Roxb. Coromand. iii. t. 218. Billiatt Springs. Waterhouse.

Melaleuca leucadendron. L. Mant. 105. Attack Creek. Roper River.

Melaleuca dissitiflora, F.M. Fragm. iii. 153. Between the Bonney River and Mount Morphett.

Eucalyptus setosa, Schauer in Walp. Report, ii. 926. Sandy Scrub near the River Bonney.

Calycothrix microphylla, All. Cunn. in Botanical Magazine 3323. Sources of the River Roper.

Boeckea polystemonea, F.M. Fragm. Phyt. Austr. ii. 124. Brinkley Bluff, McDonnell Range.

Umbelliferae.

Didiscus glaucifolius, F.M. in Linnaea, 1852, page 395. Var. cyanopetalus. Finke River. J.M. Stuart. The colour of the petals varies likewise blue and white in Didiscus coeruleus and in one species of Dimetopia.

Rubiaceae.

Canthium oleifolium, Hook. in Mitch. Trop. Austr. 397. Var. latifolium. Central Australia, in Mulga Scrub. J.M. Stuart.

Compositae.

Calotis Waterhousii. F.M. Purdie Ponds. Waterhouse.

Eurybia Ferresii, F.M. Fragm. Phyt. Austr. iii. 18. t. xviii. Brinkley Bluff. J.M. Stuart.

Pluchea ligulata, F.M. Enumeration of Plants of Babbage's Expedition page 12. Strangways River. Waterhouse.

Monenteles globifer, Cand. Prodr. v. 455. McDonnell Range, Stuart. Attack Creek, Waterhouse.

Helichrysum Davenportii, F.M. Fragm. Phyt. Austr. iii. 32. (Sect. Acroclinium.) On the River Neale.

Helichrysum Cassianum, Gaudichaud Voyage Freycenet. page 466, t. 87. (Sect. Pteropogon.) River Finke. J.M. Stuart. The capitula are rather smaller than those figured by Gaudichaud; but in Mr. Oldfield's collection from the Murchison River we observe analogous specimens, with intermediate gradations. The involucre-scales are sometimes delicately rose-coloured.

Senecio Gregorii, F.M. Report on Gregory's Plants from Cooper Creek, page 7. Finke River. J.M. Stuart.

Goodeniaceae.

Goodenia grandiflora, Sims, Botanical Magazine 890. Mount Freeling. Stuart.

Goodenia hirsuta, F.M. Fragm. iii. 35. Central Australia.

Goodenia heterochila, F.M. Fragm. iii. 142. Newcastle Water.

Goodenia Vilmoriniae, F.M. Fragm. Phyt. Austr. iii. 19. Between the River Bonney and Mount Morphett. Stuart.

Goodenia Ramelii, F.M. Fragm. iii. 20. Attack Creek. Stuart.

Vellega connata, F.M. Transactions of the Phil. Soc. i. 18. Between the River Bonney and Mount Morphett. Stuart.

Scaevola microcarpa, Cavan. Icon. vi. 6, t. 509. Towards Central Australia.

Lobeliaceae.

Isotoma petroea, F.M. in Linnaea, 1852, page 420. James Range and Hugh River.

Asclepiadeae.

Leichardtia Australis, R. Br. in Sturt's Central Australia. ii. Append. page 81. Daly Water.

Apocyneae.

Carissa lanceolata, R. Br. Prodr. 468. Strangways River.

Acanthaceae.

Dipterancanthus Australasicus, F.M. Report on Gregory's Plants from Cooper Creek, page 8. Near Anna Reservoir.

Rostellularia procumbens, Nees in Wall. Plant. Asiat. rarior. iii. 101. Purdie Ponds.

Solaneae.

Solanum pulchellum, F.M. Transact. Phil. Soc. Vict. i. 18. Purdie Ponds.

Soluanum chenopodinum, F.M. Fragm. ii. 165. On Stuart Creek, and between Mount Blyth and Mount Fisher. Stuart.

Scrophularineae.

Buchnera linearis, R. Br. Prodr. 437. King's Ponds.

Vandellia plantaginea, F.M. in Trans. Vict. Inst. iii. 62. Arnhem's Land.

Morgania floribunda, Benth. in Mitch. Trop. Austr. Var. glandulosa. Central Australia.

Rhamphicarpa adenophora, F.M. Near Attack Creek.

Bignoniaceae.

Spathodea heterophylla, R. Br. Prodr. 470. King's Chain of Ponds.

Tecoma Australis, R. Br. Prodr. 471. Var. angustifolia. McDonnell Range, and distributed over a wide range of latitude in the interior, according to Mr. Stuart. Tecoma Oxleyi, Tecoma floribunda, and Tecoma diversifolia are mere varieties of Tecoma Australis.

Asperifoliae.

Halgania solanacea, F.M. in Hook. Kew Miscell. 1857. page 21. Between Bonney River and Mount Morphett.

Halgania strigosa, Schlecht. Linnaea, xx. 640. Brinkley Bluff.

Trichodesma Zeilanicum, R. Br. Prodr. 496. Newcastle Water.

Labiatae.

Prostanthera striatiflora, F.M. in Linn. 1852, page 376. Mount Morphett.

Convolvulaceae.

Evolvulus linifolius, Linn. Sp. Pl. 392. Brinkley Bluff.

Ipomoea reptans, Poir. Encycl. Suppl. iii. 460. A white-flowering variety. Purdie Ponds.

Ipomoea pannosa, R. Br. Prodr. 487. Newcastle Water, Attack Creek, and Strangways River.

Jasminiae.

Jasminum calcarium, F.M. Fragm. i. 212. Common to most creeks of the interior. Stuart. The lobes of the calyx are narrower than in the specimens from the Murchison River; the lobes of the corolla likewise narrower, and occasionally augmented to nine. The leaflets sometimes ovate. Transient forms are sent from Champion Bay by Mr. Walcott.

Myoporinae.

Avicennia officinalis, L. Sp. Pl. page 110. Var. angustifolia. Daly Water.

Eremophila Goodwinii, F.M. Report on Babb. Plants, page 17. Mount Freeling, Attack Creek, and Mount Samuel. Stuart. Var. angustifolia; leaves linear; calyx and pedicel glabrous; corolla outside glabrous or scantily hairy. Marchant Springs.

Eremophila Macdonellii, F.M. Report on Babb. Plants, page 18. Var. glabra. Valley of the Elizabeth River.

Eremophila Latrobei, F.M. in Papers of Royal Society of Tasmania 1858. Arnhem's Land, and near Anna Reservoir. J.M. Stuart.

Eremophila Brownii, F.M. in Papers of Royal Society of Tasmania 1858. McDonnell Range. Stuart.

Eremophila Willsii, F.M. Fragm. Phyt. Austr. ii. 21, t. xx. River Finke. Stuart.

Eremophila Sturtii, R. Br. in Sturt's Central Austr. App. page 85. Daly Water.

Eremophila longifolia, F.M. in Papers of Royal Society of Tasmania 1858. Strangways Range, Stuart; Billiatt Springs, Waterhouse.

Eremophila maculata, F.M. in Papers of Royal Society of Tasmania 1858. Sandy scrub country from the south through Central Australia to Attack Creek. Waterhouse.

Verbenaceae.

Clerodendron cardiophyllum, F.M. Fragm. iii. 144. Mulga Scrub, Stuart; Daly Water, Waterhouse.

Newcastlia spodiotricha, F.M. Fragm. Phyt. Austr. iii. 21. Between the Victoria River and the Gulf of Carpenteria, from 17 to 19 degrees South latitude.

Lentibulariae.

Utricularia fulva, F.M. in Trans. Phil. Inst. iii. 63 Strangways River.

Laurineae.

Gyrocarpus sphenopterus, R. Br. Prodr. page 405. Short Range.

Thymeleae.

Pimelea sanguinea. F.M. Fragm. Phyt. Austr. i. 84. Purdie Ponds.

Proteaceae.

Grevillea mimosoides, R. Br. Prodr. page 380. Roper River.

Grevillea agrifolia, All. Cunn. in R. Br. Suppl. page 24. McDonnell Range, Short Range. Var. lancifolia. Central Australia.

Grevillea Sturtii, R. Br. in Sturt's Centr. Austr. Append. page 24. Central Mount Stuart. Var. pinnatisecta; segments usually five. Scrub near Forster Range. J.M. Stuart.

Grevillea lineata, R. Br. in Sturt's Centr. Austr. Append. page 24. Scrub near Forster Range.

Grevillea chrysodendron, R. Br. 379. Billiatt Springs. Waterhouse.

Grevillea refracta, R. Br. Prodr. page 380. Newcastle Water, Billiatt Springs, and Short Range.

Grevillea dimidiata, F.M. Fragm. Phyt. Austr. iii. 146. Roper River. Waterhouse.

Hakea arborescens, R. Br. Prodr. page 386. Arnhem's Land.

Hakea lorea, R. Br. Suppl. page 25. Central Australia. Bark corky.

Amaranthaceae.

Alternanthera denticulata, R. Br. Prodr. 417. Burke River.

Alternanthera nana, R. Br. Prodr. 417. Burke River.

Gomphrena humilis, R. Br. Prodr. 416. Attack Creek. The upper pair of leaves stand either next to the flower-heads or remote from them. The same species has been found by Dr. Muller on the Dawson River, and by Mr. Fitzalan at Port Denison.

Gomphrena canescens, R. Br. Prodr. 416. Attack Creek. J.M. Stuart. (Victoria River and Sturt Creek, F. Muller; Sweer's Island, Henne; Nickol Bay, Walcot.) Capsula usually beautifully pink, sometimes purple or white. Peduncles occasionally more than 6 inches long; the staminodia sometimes excel the anthers in length.

Ptilotus corymbosus, R. Br. Prodr. 415. Var. spicatus. Attack Creek.

Trichinium gracile, R. Br. 415. Tropical Australia.

Trichinium nobile, Lindl. in Mitch. Three Exped. ii. 22. Short Range.

Trichinium brachytrichum, F.M. Fragm. iii. 157. Central Australia. J.M. Stuart.

Urticeae.

Ficus Stuartii, F.M. McDonnell Range; Brinkley Bluff. Several other undescribed species of fig-trees occur in the collection, but cannot be satisfactorily characterised from the material extant.

Cycadeae.

A cycadeous plant, seemingly distinct from the seven Australian species, occurs on McDonnell Range, and is mentioned as a palm in the Journal of the explorers. Only leaves being now submitted for examination, it remains for future researches to throw light on this plant.

Amaryllideae.

Calostemma luteum, Sims, in Botanical Magazine 2101. Mount Margaret. Stuart. The edge of the corona is sometimes rather undulated than toothed.

Crinum angustifolium, R. Br. 297. From latitude 22 to 32 degrees South. J.M. Stuart.

Orchideae.

Cymbidium canaliculatum, R. Br. Prodr. 331. Strangways River.

Commelyneae.

Commelyna ensifolia, R. Br. Prodr. 269. McDonnell Range, and near Mount Freeling. J.M. Stuart.

Commelyna agrostophylla, F.M. Arnhem's Land.

Liliaceae.

Bulbine semibarbata, Haw. Revis. 33. Thring River. Stuart.

Gramineae.

Eriachne obtusa, R. Br. Prodr. 184. Short Range.

Ectrosia leporina, R. Br. Prodr. 186. Purdie Ponds.

Perotis rara, R. Br. Prodr. 172. Purdie Ponds, Waterhouse; Short Range, Stuart.

Andropogon bombycinus, R. Br. Prodr. 202. Central Australia, McDonnell Range.

Chloris ventricosa, R. Br. Prodr. 186. Arnhem's Land.

Lappago racemosa, W. Sp. l. 484. Attack Creek.

Panicum decompositum, R. Br. Prodr. 191. Stevenson River.

Oryza sativa, L. Sp. Pl. Newcastle Water. J.M. Stuart.

Pappophorum commune, F.M. Enumeration of Greg. Plants from Cooper Creek, page 10. Central Australia.

Cyperaceae.

Hypaelyptum microcephalum, R. Br. Prodr. 221. Attack Creek.

Filices.

Marsilia quadrifolia, L. Sp. Pl. Var. hirsuta. Nardoo. Through Central and North Australia, on localities subject to inundation.

Lygodium semibipinnatum, R. Br. Prodr. 162. Roper River.

Blechnum Orientale, L. Sp. Pl. 1535. River Adelaide. This fern was not previously recorded as existing in Australia.

Cheilanthes tenuifolia, Swartz Filic. 129. River Roper, Mount Freeling.

LaVergne, TN USA
27 March 2011

221773LV00002B/140/A